Introduction
to
Dynamics of Rotor-Bearing Systems

DyRoBeS© (**Dy**namics of **Ro**tor-**Be**aring **S**ystems)
is a trademark of Eigen Technologies, Inc.

Introduction
to
Dynamics of Rotor-Bearing Systems

Wen Jeng Chen

Eigen Technologies, Inc.
Davidson, NC 28036

www.DyRoBeS.com

and

Edgar J. Gunter, Ph.D.

RODYN Vibration Analysis, Inc.
Charlottesville, VA 22903

www.RODYN.com

DyRoBeS© developed by Wen Jeng Chen, Ph.D., is a trademark of Eigen Technologies, Inc.

For additional information on DyRoBeS, contact
RODYN Vibration Analysis, Inc.
Charlottesville, VA 22903
www. RODYN.com
Tel: 434-296-3175

Limits of Liability and Disclaimer of Warranty
Th e authors and publisher have used best eff orts in preparing this book, the program, and data on the electronic media accompanying this book. Th ese eff orts include the development, research, and verifi cation of the theories and programs. But, due to the complex nature of this type of software, the author and publisher make no expressed or implied warranty of any kind with regard to these programs nor the supplemental documentation in this manual, including but not limited to, their accuracy, eff ectiveness, or fi tness for a particular purpose. In no event shall the author, publisher, or program distributors be liable for errors contained herein or for any incidental or consequential damages in connection with, or arising out of the furnishing, performance, or use of any of these materials. Th e information provided by these programs is based upon mathematical assumptions that may or may not hold true in a particular case. Th erefore, the user assumes all of the risks in acting on or interpreting any of the program results.

Order this book online at www.trafford.com
or email orders@trafford.com

Most Trafford titles are also available at major online book retailers.

Print information available on the last page.

ISBN: 978-1-4120-5190-3 (sc)
ISBN: 978-1-4269-9050-2 (e)

Trafford rev. 11/26/2018

 www.trafford.com

North America & international
toll-free: 1 888 232 4444 (USA & Canada)
fax: 812 355 4082

Preface

DyRoBeS© (**Dy**namics of **Ro**tor-**Be**aring **S**ystems) is a powerful and sophisticated computer software tool based on Finite Element Analysis for complete rotor dynamics analysis and comprehensive bearing performance calculations. Since the first release in 1991, it has become the most widely used software for rotor dynamics analysis and bearing design. This program has been rigorously validated and extensively tested by many academic researchers and industrial engineers. *DyRoBeS* is now commonly used as a teaching tool by many universities, equipment manufacturers, and consulting companies due to its integrated modeling and simulation tools, user-friendly features, excellent graphic capabilities, and robust numerical algorithms.

Over the years, I have frequently been asked about the theory and the use of *DyRoBeS* from many users on various applications. This book is written as an introduction to rotor-bearing dynamics for practicing engineers and students who are involved in rotor dynamics and bearing design. The goal of this book is to provide a step-by-step approach to the understanding of fundamentals of rotor-bearing dynamics by using *DyRoBeS*. Therefore, the emphasis of this book is on the basic principals, phenomena, modeling, and interpretation of the results. Numerous examples, from a single-degree-of-freedom system to complicated industrial rotating machinery, are employed throughout this book to illustrate these fundamental dynamic behaviors. The concepts in the text are reinforced by parametric studies and numerous illustrative examples and figures.

The book begins with a brief discussion of the mathematical modeling of physical dynamic systems and an overview of the basic vibration concepts in Chapter 1. The generalized coordinates and Lagrange's equation are introduced. Lagrange's equation and the finite element formulation are the basis of the derivation of equations of motion for the rotating structures. A single degree-of-freedom (SDOF) system is employed to discuss the fundamentals of the free and forced vibrations. The concept of assumed mode is introduced to approximate a continuous system by a discrete SDOF generalized parameter model. Numerous examples with many illustrative figures and parametric studies are extensively utilized to emphasize the concepts in the text.

The coordinate systems and the kinematics of the rotor motion are presented in Chapter 2. The emphasis is on the lateral vibration. The coordinate transformation between the fixed reference frame and the rotating reference frame is included. The overall steady state rotor lateral motion is reviewed. The simplest rotor steady harmonic motion, elliptical motion, is discussed in detail. The definition of forward and backward whirls is explained and illustrated by many figures and related equations. The chapter ends with numerous examples to demonstrate the rotor motion. The directions of whirling and spinning are easily identified and explained by visualization of the rotor motion.

A simple two-degrees-of-freedom (2DOF) rotor system, the Laval-Jeffcott rotor model, is utilized in Chapter 3 to demonstrate many important phenomena in rotor dynamics. The fundamentals of rotor bearing dynamics are discussed with this simple rotor model. Although many assumptions made in this simple model are not practical and do not correspond with reality, they can simplify the solution and allow for parametric studies be performed. This allows us to understand the effects of each parameter on rotor dynamics behaviors. This simple 2DOF model also provides many valuable physical insights into more practical and complicated systems. The destabilizing effects of the cross-coupled stiffness terms that produce the non-conservative circulatory force are discussed. These circulatory forces, which destabilize the rotor system, are commonly caused by fluid film bearings, liquid seals, and other fluid interactions. The conservative gyroscopic effects on the natural frequencies or whirl speeds are presented. These two unique features, cross-coupled stiffness and gyroscopic effects, make the rotor dynamics study so fascinating and different from the general non-rotating structural dynamics. Many phenomena in rotor dynamics are illustrated and explained from an energy point of view. The transient motion during startup and shutdown is presented here by using a simple rotor model. The effects of acceleration on the rotor motion are discussed, and the results are compared with the steady state response. The effects of flexible support are also discussed using the equivalent impedance.

Chapter 4 discusses the rotating disk equations and rigid rotor dynamics. The governing equations for a rotating rigid or flexible disk are derived. A rigid disk can be modeled as a 4-degrees-of-freedom (4DOF) system. This equation is extended to the simulation of a rigid rotor dynamics. The dynamic behavior of a rigid rotor on flexible bearings can be simulated with an equivalent 4DOF system. A rigid disk on a flexible massless rotor with rigid bearings is also presented. The dynamics of a centrally and off-centrally mounted disk on a flexible rotor is discussed in great details. The effects of gyroscopic moments introduced previously are reinforced in this chapter. The mathematical model for a flexible disk is presented. If the attachment of the disk to the shaft is not stiff, or the natural frequency of the first diametral mode of a disk is close to the operating speed range of the rotor system, the disk is considered to be flexible. Additional two degrees-of-freedom is required for the flexible disk, which has a total of six degrees-of-freedom (6DOF). The offset disk is also discussed in this chapter. The offset disk can be either rigid or flexible, depending on the attachment method and disk flexibility.

Chapter 5 covers the finite element formulation for a rotating shaft element. The concept and the use of shape functions are introduced to convert a continuous system with infinite DOF into a discrete finite DOF model. Cylindrical element, tapered element, and user's supplied element are considered in this chapter. Guyan reduction, commonly used in the finite element method, is employed for the condensation of sub-elements. It is also utilized to condense out the additional shear deformation coordinates in the formulation of tapered element, resulting in the conventional eight degrees-of-freedom (8DOF) elements for the consistence with the cylindrical element. The importance of the mode shapes is discussed. The mathematical model for the coupling, which connects the rotors, is discussed. Again, the chapter ends with several examples to illustrate the effects and refresh the concepts discussed in the text. The effects of rotatory inertia, shear deformation, and axial loads are discussed. The effect of the number of finite elements is also addressed. The classification of rigid rotor and flexible rotor is illustrated by using a uniform beam supported by elastic springs. When the spring stiffness is small compared to the shaft bending stiffness, the first two modes can be characterized as rigid rotor modes where the potential energy is mainly in the bearings, and the shaft possesses much less potential energy. The third mode can be characterized as a flexible rotor mode where the shaft possesses most of the potential energy, and the bearings have little contribution. When the spring stiffness is very high compared to the shaft bending stiffness, the bearing stiffness has little effect on the natural frequencies and it becomes the pinned-pinned boundary conditions. The first two modes now are characterized as flexible rotor modes. There are no rigid rotor modes for bearings with very high stiffness.

Chapter 6 deals with various types of bearings, dampers, seals, and other interconnection components. All the reaction forces from these components are nonlinear in nature. The concept of linearization around the static equilibrium is discussed. The advantage of linearization is to decouple the rotor equations and the lubrication equations, which allows for rapid linear analysis to be performed in the preliminary design stage. The fundamentals of hydrodynamic journal bearings are presented. The finite element formulation for the solutions of Reynolds equation and the derivation of the perturbed pressure equations are extensively discussed with many illustrative figures and examples. The importance of the bearing static and dynamic characteristics is stressed. Various types of bearings and bearing-like components are covered in this chapter. General design guidelines on the fluid film bearings are provided. Linear and nonlinear analyses on a single journal bearing are extensively discussed. This example provides a basis for understanding the difference between the linear and nonlinear analyses. The proper use of these analyses is discussed. The solution techniques for the incompressible and compressible Reynolds equation are discussed.

Chapter 7 summarizes the lateral vibration study with several practical examples. The governing equations of motion for the entire system are presented. Various solution techniques and interpretations of the results are discussed. It provides an overview of the lateral vibration presented in previous chapters. The important concepts and fundamentals of rotor-bearing dynamics discussed previously are summarized and illustrated through these examples. Both linear and nonlinear analyses are utilized to demonstrate the effects discussed in the text.

Chapter 8 is devoted to the important subject of torsional vibration. In contrast to lateral vibration, which has been carefully analyzed by the manufacturers and intensively monitored by the equipment users, the torsional vibration for the complete geared train has very often been ignored until catastrophic failures occur. The steady state and transient motion of geared train systems are discussed in this chapter. Examples are also provided to demonstrate the analysis discussed in the text.

Finally, a brief description of the balancing method, influence coefficient method, is presented in Chapter 9. The influence coefficient method has been widely used by practicing engineers in dealing with many field vibration problems. Therefore, the detailed calculation procedure is also presented, in case the computational tools are not available and hand calculation has to be performed.

Acknowledgement

Writing this book would not have been possible without the encouragement of Dr. Edgar J. Gunter, who wrote one of the earliest rotordynamics books, entitled *Dynamic Stability of Rotor-Bearing Systems*, sponsored by NASA in 1966. Dr. Gunter has been my mentor throughout my career. **DyRoBeS** has been constantly improved under his guidance and assistance.

I would like to express my sincere appreciation to Dr. Harold D. Nelson, who brought me into this exciting field while I was pursuing my graduate studies. Without his stimulating instruction, guidance, and support, I would not have had the confidence to accomplish my educational goals.

The authors also wish to thank Toni Miller for her efforts as we worked on the manuscripts these past few years.

Finally, I wish to express my gratitude to Mr. Malcolm E. Leader of Applied Machinery Dynamics Company, Mr. Jim Kay of General Electric Company, Dr. Leo Lu of Northrop Grumman Corporation, Professor Gordon Kirk of Virginia Tech, and the numerous **DyRoBeS** users throughout various industries, government agencies, and universities for their valuable suggestions and constructive criticism. Without their comments, **DyRoBeS** would not be what it is today.

W. J. Chen

Table of Contents

Introduction 1

1.1 Basic Concepts

All physical dynamic systems are continuous in nature. Mathematical models of dynamic continuous systems result in partial differential equations described by variables depending on time and space over a given domain and boundary conditions. Analytical solutions are not available or simply not feasible in most practical applications. The difficulties in obtaining closed-form solutions to problems of continuous distributed-parameter systems can be traced to the inherent difficulty of solving partial differential equations and in satisfying boundary conditions. These difficulties can be avoided by eliminating the spatial dependence from the problems through discretization in space. Discretization implies certain approximations, and is the first step utilized in the finite element method to convert a continuous model into a discrete model. As a result of this discretization, a physical model governed by partial differential equations, is transformed into a mathematical model governed by a set of simultaneous ordinary differential equations described by variables, which are function of time only. Due to the rapid advance in the speed and memory of digital computers and numerical algorithms, the finite element method has become the principal technique for solving complex structural and fluid problems. The finite element method is employed throughout this book in rotordynamics and bearing analysis.

The minimum number of independent variables required to describe the motion of a discrete system is known as the number of *degrees of freedom* (DOF) of the system. Thus, discretization converts a continuous model with infinite DOF into a finite DOF system. These variables, also known as coordinates used to describe the behavior of the system, can be the physical quantities, such as displacements and slopes in lateral rotordynamics, but they can also be other quantities, such as pressure and pressure gradients in hydrodynamic bearing analysis. The n independent coordinates, are necessary to describe the motion of a discrete system of n degrees of freedom. These independent coordinates are also known as *generalized coordinates*. Sometimes the physical coordinates used to describe the system are not independent, and some of them are related by constraints such as the gear ratio in torsional vibrations. The number of independent generalized coordinates (*NDOF*) needed to describe the system is equal to the number of coordinates (n) minus the number of independent constraint equations (m). That is $NDOF = n\text{-}m$. In rotordynamics, the constraints are relatively simple and the independent coordinates can always be identified. The excess coordinates, due to the constraints, can always be eliminated during the assembly process. The coordinates used in this book, in general, are independent unless specified in the text.

The equations of motion for simple systems are commonly derived by Newton's law, which utilizes the vector relationships for the forces and accelerations. It offers many physical insights into the system behaviors. As the system becomes more complicated, the direct application of Newton's law becomes increasingly difficult. The Lagrangian approach utilizes the scalar quantities of energy and work. It offers considerable advantages in the derivation of equations of motion for complicated systems.

Consider an n degrees of freedom system with n independent generalized coordinates q_i. The Lagrange's equation can be written in the form:

$$\frac{d}{dt}\left(\frac{\partial T}{\partial \dot{q}_i}\right) - \frac{\partial T}{\partial q_i} + \frac{\partial V}{\partial q_i} + \frac{\partial \mathfrak{I}}{\partial \dot{q}_i} = Q_i \quad (i = 1,2,...n) \tag{1.1-1}$$

where T is the total kinetic energy of the system which is a positive definite expression and can be expressed in terms of the generalized coordinates and their first time derivatives (generalized velocities):

$$T = T(q_1, q_{2,}...,q_n, \dot{q}_1, \dot{q}_{2,}...,\dot{q}_n) \tag{1.1-2}$$

The potential energy V is a function of position only and can be expressed in terms of generalized coordinates alone:

$$V = V(q_1, q_{2,}...,q_n) \tag{1.1-3}$$

The Rayleigh's dissipation function $\mathfrak{I}$ consists of viscous damping forces, which depend on the generalized velocities, and are derived from the quadratic function:

$$\mathfrak{I} = \mathfrak{I}(\dot{q}_1, \dot{q}_{2,}...,\dot{q}_n) = \frac{1}{2}\sum_{i=1}^{n}\sum_{j=1}^{n}c_{ij}\dot{q}_i\dot{q}_j \tag{1.1-4}$$

The generalized forces Q_i are derived from the virtual work. The virtual work of non-conservative forces acting on the system has the form:

$$\delta W = \sum_{i=1}^{n}Q_i\delta q_i \tag{1.1-5}$$

Where Q_i should not include the forces provided by the potential energy and dissipation function. In general, the Lagrange's equations constitute a set of n non-homogeneous nonlinear second order ordinary differential equations. In rotordynamics, the equations of motion derived from the Lagrange's equations for a constant rotational speed can be written in the matrix form:

$$M\ddot{q} + (G + D)\dot{q} + (K_s + K_c)q = Q \qquad (1.1\text{-}6)$$

or expressed in a more general form for vibration problems:

$$M\ddot{q} + C\dot{q} + Kq = Q \qquad (1.1\text{-}7)$$

where M is known as the mass/inertia matrix and is a positive definite real symmetric matrix. C is the general damping matrix that consists of a skew-symmetric gyroscopic matrix G, derived from kinetic energy due to gyroscopic moments, and a symmetric damping matrix D, derived from Rayleigh's dissipation function. K is the general stiffness matrix which consists of a symmetric stiffness matrix K_s, derived from potential energy and a skew-symmetric circulatory matrix K_c, contributed mainly from the fluid film bearings. Including all the fluid interacting forces between the structures, such as bearings, seals, impeller-diffuser interaction, etc., the damping C and stiffness K matrices in equation (1.1-7) are general real matrices.

1.2 Single-Degree-of-Freedom Systems

The simplest model of a vibrating system is composed of a single mass element that is connected to a rigid support through a spring and damping as sketched in Figure 1.2-1. The mathematical model of this single (one) degree of freedom system is described by a single second-order ordinary differential equation.

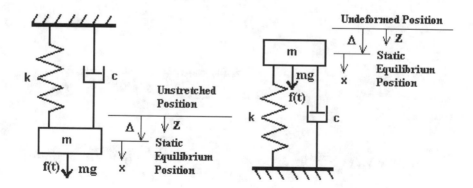

Figure 1.2-1 Single-Degree-of-Freedom system

Assume the spring and dashpot are linear with a spring constant k and a viscous damping coefficient c. The external force applied to the mass element is $f(t)$ and the gravitational force acting on the mass element is mg. Let z be the displacement of the mass element measured from the unstretched (undeformed) position. In rotordynamics, it can be the geometric center of the bearing. The equation of motion of the mass element can be derived from Newton's law:

$$m\ddot{z} + c\dot{z} + kz = mg + f(t) \qquad (1.2\text{-}1)$$

The above equation can be simplified by describing the mass motion from the static equilibrium position. Let Δ be the static displacement of the mass element due to gravity, it is also the deformation of the spring. The displacement of the mass element measured from the static equilibrium position is x ($z=\Delta+x$). At static equilibrium, the spring force is equal to the gravitational force acting on the mass element:

$$k\Delta = mg \tag{1.2-2}$$

Since Δ is a constant, the equation of motion of the mass element described from the static equilibrium position can be written as:

$$m\ddot{x} + c\dot{x} + kx = f(t) \tag{1.2-3}$$

Thus, the weight force mg can be eliminated if the motion is described from the static equilibrium position. In practices, this single-degree-of-freedom model cannot describe the dynamics of the real machinery, since it only has the first fundamental frequency. However, it provides an insight into many important dynamic characteristics of a vibrating system.

A simply supported elastic beam, with a uniform cross-section loaded at midspan with a concentrated mass M_c as shown in Figure 1.2-2, is employed to demonstrate the procedure of approximation of a continuous system by a SDOF generalized-parameter model. To obtain a single-degree-of-freedom model for this continuous system, the concepts of assumed-modes or shape function are introduced.

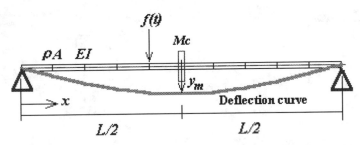

Figure 1.2-2 Simply supported beam with a concentrated mass

If the mass of the beam is small in comparison with the lumped mass (M_c) at midspan, it can be assumed with reasonable accuracy, that the deflection curve of the vibrating beam has the same shape as the static deflection curve, due to a concentrated force applied at midspan. The deflection curve of the vibrating beam at the axial position of x may be approximated by:

$$Y(x,t) = \Phi(x)\, y_m(t) = \left(\frac{3xL^2 - 4x^3}{L^3} \right) y_m(t) \qquad (x \le L/2) \tag{1.2-4}$$

where $y_m(t)$ is the maximum displacement of the concentrated mass at midspan. This assumption transforms a continuous system to be a single DOF model with $y_m(t)$ being the *generalized coordinate*. The static deflection curve, due to a concentrated load at midspan, is employed to approximate the deflection of the continuous beam. The maximum displacement at the concentrated mass, also at center of the beam, during vibration is chosen to be the generalized coordinate for this SDOF.

The kinetic energy and potential (strain) energy of the system are:

$$T = \frac{1}{2} \int_0^{L/2} 2\rho A \, (\dot{Y})^2 \, dx + \frac{1}{2} M_c \dot{y}_m^2 = \frac{1}{2}\left(\frac{17}{35} \rho AL + M_c \right) \dot{y}_m^2 \tag{1.2-5}$$

$$V = \frac{1}{2} \int_0^{L/2} 2EI \, (Y'')^2 \, dx = \frac{1}{2}\left(\frac{48EI}{L^3} \right) y_m^2 \tag{1.2-6}$$

The "dot" denotes the derivative with respect to time and the "prime" represents the derivative with respect to space (x in this case). If the beam is subject to the gravitational load and an external force $f(t)$ at x_i, the virtual work is:

$$\delta W = \int_0^{L/2} 2\rho Ag \, \delta Y(x,t) \, dx + M_c g \, \delta Y(L/2,t) + f(t) \delta Y(x_i,t)$$

$$= \left(\frac{5}{8} \rho ALg + M_c g + f(t) \Phi(x_i) \right) \delta y_m = Q \delta y_m \tag{1.2-7}$$

Therefore, the governing equation of motion derived from Lagrange's equation for this SDOF is:

$$\left(\frac{17}{35} \rho AL + M_c \right) \ddot{y}_m + \left(\frac{48EI}{L^3} \right) y_m = Q(t) \tag{1.2-8}$$

where Q is the *generalized force* derived from the virtual work. It should be noted that the assumed mode is an admissible function that satisfies the geometric boundary conditions and possesses derivatives of an order larger or equal to that appearing in the strain energy expression. It is very important that the assumed mode (admissible function) be in close approximation to the shape of the vibration mode under consideration. When the displacement at the middle of the beam is solved through the equation of motion (1.2-8), the displacement at any location can be obtained from the approximation equation (1.2-4).

Let us consider a torsional system sketched in Figure 1.2-3 as a second example. The rigid disk, with a mass polar moment of inertia I_p, is connected to the rigid ground by an elastic shaft with a circular cross section; a length of L, a radius of r, and a shear modulus of G. The torsional stiffness of the shaft is given by the torque and angle relationship from the strength of materials:

$$k_T = \frac{GJ}{L} = \frac{G\pi r^4}{2L} \tag{1.2-9}$$

For circular cross section, $J = \dfrac{\pi r^4}{2}$

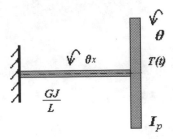

Figure 1.2-3 Simple torsional system

If the mass (inertia) of the shaft is negligible in comparison to that of the rigid disk, the equation of motion at the disk can be determined from the strength of materials and Newton's law.

$$I_p \ddot{\theta} + \left(\frac{GJ}{L}\right)\theta = T(t) \tag{1.2-10}$$

where $\theta(t)$ is the angular displacement at the disk and $T(t)$ is the external torque acting on the disk. If the mass of the shaft is not negligible, assumed-mode method is again utilized to convert this continuous system to a single-degree-of-freedom model. The rotation (angle of twist) along the shaft is assumed to be linear distribution along its length:

$$\theta_x(x,t) = \frac{x}{L}\theta(t) \tag{1.2-11}$$

This approximation satisfies the two boundary conditions:

$$\theta_x(0,t) = 0 \tag{1.2-12a}$$
$$\theta_x(L,t) = \theta(t) \tag{1.2-12b}$$

The angular displacement at the disk $\theta(t)$ is selected to be the *generalized coordinate*. The kinetic energy and potential (strain) energy of the system are given:

$$T = \frac{1}{2}\int_0^L \rho\left(\frac{\pi r^4}{2}\right)\left(\dot{\theta}_x\right)^2 dx + \frac{1}{2}I_p\dot{\theta}^2 = \frac{1}{2}\left(\frac{\rho AL}{6}r^2 + I_p\right)\dot{\theta}^2 \tag{1.2-13}$$

$$V = \frac{1}{2}\int_0^L G\left(\frac{\pi r^4}{2}\right)\left(\theta_x'\right)^2 dx = \frac{1}{2}\left(\frac{G\pi r^4}{2L}\right)\theta^2 \tag{1.2-14}$$

The governing equation of motion is then:

$$\left(\frac{\rho AL}{6}r^2 + I_p\right)\ddot{\theta} + \left(\frac{G\pi r^4}{2L}\right)\theta = T(t) \tag{1.2-15}$$

It shows that one-third (1/3) of the polar moment of inertia of the shaft ($\frac{\rho AL}{2}r^2$) contributes to the effective inertia in the final equation of motion.

1.3 Free and Forced Vibrations

As previously discussed, the governing equation of motion for a single DOF system is a second-order ordinary differential equation (ODE). Let us consider a general linear system with constant coefficients. The equation of motion is:

$$m\ddot{x}(t) + c\dot{x}(t) + kx(t) = f(t) \tag{1.3-1}$$

with two initial conditions at $t = 0$

$$x(0) = x_0$$
$$\dot{x}(0) = \dot{x}_0 \tag{1.3-2}$$

The complete solution of this second-order ODE is the sum of the homogeneous solution and particular solution. The homogeneous solution, obtained by eliminating the excitation in the right hand side of equation ($f(t) = 0$), is called *free vibration*, or *natural motion* due to initial conditions. The particular solution with ($f(t) \neq 0$) is called *forced vibration*, or *steady state response* due to the excitation irrespective of the homogeneous solution and initial conditions. Therefore, the total response is the sum of the free vibration and the forced response.

$$x(t) = x_h(t) + x_p(t) \tag{1.3-3}$$

Individual response and total motion will be discussed in details in the following sessions.

1.4 Free Vibration

Let us first examine the homogeneous equation, that is, the free vibration due to initial conditions. It is convenient to divide the homogeneous equation by m and rewrite it as:

$$\ddot{x} + 2\xi\omega_n\dot{x} + \omega_n^2 x = 0 \tag{1.4-1}$$

where

$$\omega_n = \sqrt{\frac{k}{m}} \tag{1.4-2}$$

and

$$\xi = \frac{c}{c_{cr}} = \frac{c}{2\sqrt{km}} = \frac{c}{2m\omega_n} \tag{1.4-3}$$

ω_n is called the *undamped natural frequency* (radians per second). ξ is a non-dimensional parameter, called *damping ratio* or *damping factor*, and c_{cr} is called the *critical damping coefficient*. The homogeneous solution is also known as the *transient solution*. From the ordinary differential equation, the solution has the general form:

$$x_h = A_1 e^{\lambda_1 t} + A_2 e^{\lambda_2 t} \tag{1.4-4}$$

where λ_1, λ_2 are the roots of the characteristic equation

$$\lambda^2 + 2\xi\omega_n\lambda + \omega_n^2 = 0 \tag{1.4-5}$$

and they are:

$$\lambda_{1,2} = -\xi\omega_n \pm \omega_n\sqrt{\xi^2 - 1} \tag{1.4-6}$$

Depending upon the value of the damping ratio, ξ, the roots λ_1 and λ_2 must both be real numbers or both be imaginary numbers, or a complex conjugate pair. The magnitude of the damping ratio can also be used to determine if the motion is oscillatory or non-oscillatory. For the underdamped case ($\xi < 1$), the motion is oscillatory with a decaying ($1 > \xi > 0$) or growing ($\xi < 0$) amplitude. For the overdamped case ($\xi > 1$), the motion is non-oscillatory with an exponentially decaying function of time. For the critically damped case ($\xi = 1$) which separates the underdamped and overdamped cases, the motion is non-oscillatory and the amplitude decays faster than in any other cases. The three cases are studied in detail. For vibration analysis, the underdamped case is the most important case to study and it will be examined first.

Case 1: Underdamped $(\xi < 1)$ Oscillatory Motion

λ is a complex conjugate pair and can be written in a simple form:

$$\lambda_{1,2} = -\xi\omega_n \pm j\omega_d \tag{1.4-7}$$

where ω_d is called the *damped natural frequency* given by

$$\omega_d = \omega_n \sqrt{1 - \xi^2} \tag{1.4-8}$$

The transient solution can be written in the following two forms by using the Euler's formula to convert the complex exponentials to trigonometric functions.

$$x_h(t) = e^{-\xi \omega_n t} \left(A_1 \cos \omega_d t + A_2 \sin \omega_d t \right) \tag{1.4-9a}$$

or

$$x_h(t) = e^{-\xi \omega_n t} \left(A_1 \cos(\omega_d t - \phi) \right) \tag{1.4-9b}$$

The two arbitrary constants A_1 and A_2 are determined through two initial conditions at $t=0$ (or at some other specific time). Note that these constants are evaluated when the complete solution is known. The homogeneous solution is simply an oscillatory motion with a circular frequency of ω_d and is an exponentially decaying $(0 < \xi < 1)$ or growing $(\xi < 0)$ function of time. For the undamped case when $\xi = 0$, a pair of imaginary roots occurs, the homogeneous solution is a simple harmonic motion with a circular frequency of ω_n.

In the study of rotordynamics, the system stability is a very important dynamic characteristic. The value of ξ has an effect on the damped natural frequency. Its importance, however, is mainly in the system stability. As previously shown, when $(\xi > 0)$, the motion is exponentially decaying with time, and the system is said to be stable. When $(\xi < 0)$, the motion is exponentially increasing with time, and the system is said to be unstable in the linear theory. Eventually the motion will be constrained by using the non-linear theory.

Another quantity commonly used in the study of system stability is the *logarithmic decrement*. The logarithmic decrement is a measure of the rate of decay or growth of free oscillations, and is defined as the natural logarithm of the ratio of any two successive amplitudes as illustrated in Figure 1.4-1:

$$\delta = \ln\left(\frac{x_i}{x_{i+1}} \right) = \xi \omega_n \tau_d \tag{1.4-10}$$

where the *damped natural period* is given as

$$\tau_d = \frac{2\pi}{\omega_d} = \frac{2\pi}{\omega_n \sqrt{1 - \xi^2}} \tag{1.4-11}$$

The logarithmic decrement can also be written as:

$$\delta = \frac{2\pi \xi}{\sqrt{1 - \xi^2}} \tag{1.4-12}$$

The logarithmic decrement can be obtained by measurement and once it is known, the damping ratio can be obtained from the following equation:

$$\xi = \frac{\delta}{\sqrt{4\pi^2 + \delta^2}}$$

(1.4-13)

Figure 1.4-1 shows the free motion for a SDOF system with the following parameters:

$$m = 0.1, \ c = 0.4, \ k = 40, \ x_0 = 1, \ \dot{x}_0 = 20$$
$$\omega_n = 20.0 \ rad/\sec, \quad \xi = 0.1$$
$$\omega_d = 19.9 \ rad/\sec, \quad \delta = 0.631$$

The above values can be verified from the figure where

$$\delta = \ln\left(\frac{x_1}{x_2}\right) = \ln\left(\frac{x_2}{x_3}\right) = \ln\left(\frac{x_i}{x_{i+1}}\right) = 0.631, \quad \xi = 0.1$$

$$\tau_d = t_2 - t_1 = t_3 - t_2 = t_{i+1} - t_i = 0.3157, \quad \omega_d = 19.9 \ rad/\sec$$

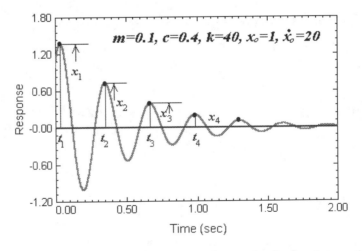

Figure 1.4-1 Logarithmic decrement for underdamped case ($\xi < 1$)

Case 2: Critically-Damped $(\xi = 1)$ Non-oscillatory Motion

The two roots are identical in this case and the solution is:

$$x_h(t) = (A_1 + A_2 t)e^{-\omega_n t}$$

(1.4-14)

The solution is the product of a linear function of time and a decaying exponential.

Case 3: Overdamped $(\xi > 1)$ Non-oscillatory Motion

There are two distinct, negative real roots. One root increases and the other decreases as the damping ratio increases. The homogeneous solution is:

$$x_h(t) = A_1 e^{\left(-\xi+\sqrt{\xi^2-1}\right)\omega_n t} + A_2 e^{\left(-\xi-\sqrt{\xi^2-1}\right)\omega_n t} \qquad (1.4\text{-}15)$$

The motion is an exponentially decaying function of time.

Figure 1.4-2 shows the two non-oscillatory motions for the critically damped and overdamped cases.

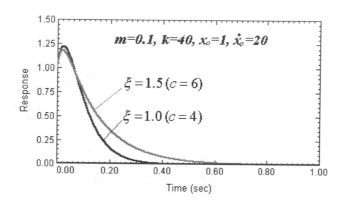

Figure 1.4-2 Critically damped and overdamped cases

For a typical structural system, it usually falls into the underdamped system, with the damping ratio of 0.5% to 5%. However, for the rotating machinery, due to the complexity of the working fluid interaction and supporting mechanism, all three cases are possible.

The arbitrary constants in the homogeneous solutions are evaluated when the complete solution is known. However, when $f(t)=0$, these constants can be determined immediately with given initial conditions.

For the underdamped case without excitation, the solution has the following forms:

$$x(t) = e^{-\xi\omega_n t}\left[x_0 \cos\omega_d t + \left(\frac{\dot{x}_0 + \xi\omega_n x_0}{\omega_d}\right)\sin\omega_d t \right] \qquad (1.4\text{-}16\text{a})$$

or

$$x(t) = e^{-\xi\omega_n t}\left(X\cos(\omega_d t - \phi)\right) \qquad (1.4\text{-}16\text{b})$$

where

$$X^2 = x_0^2 + \left(\frac{\dot{x}_0 + \xi\omega_n x_0}{\omega_d}\right)^2 \qquad (1.4\text{-}17)$$

and

$$\tan\phi = \frac{\dot{x}_0 + \xi\omega_n x_0}{\omega_d x_0}$$ (1.4-18)

For the critically damped case without excitation, the solution takes the form:

$$x(t) = e^{-\omega_n t}\left[x_0 + (\dot{x}_0 + \omega_n x_0)t\right]$$ (1.4-19)

For the overdamped case without excitation, the solution can be written in the form:

$$x(t) = A_1 e^{\left(-\xi+\sqrt{\xi^2-1}\right)\omega_n t} + A_2 e^{\left(-\xi-\sqrt{\xi^2-1}\right)\omega_n t}$$ (1.4-20)

where

$$A_1 = \frac{\dot{x}_0 + \left(\xi + \sqrt{\xi^2-1}\right)\omega_n x_0}{2\omega_n\sqrt{\xi^2-1}}$$ (1.4-21)

and

$$A_2 = \frac{-\dot{x}_0 - \left(\xi - \sqrt{\xi^2-1}\right)\omega_n x_0}{2\omega_n\sqrt{\xi^2-1}}$$ (1.4-22)

Example 1.1: SDOF- Free Vibration

Consider the following equation of motion for a single degree-of-freedom linear system without excitation force:

$$m\ddot{x} + c\dot{x} + kx = 0$$

with specified initial conditions (displacement and velocity) at $t=0$:

$$x(0) = x_0 = 1$$
$$\dot{x}(0) = \dot{x}_0 = 0$$

The mass and stiffness constants are chosen to be:

$$m = 0.1$$
$$k = 40$$

The undamped natural frequency is:

$$\omega_n = \sqrt{\frac{k}{m}} = 20 \, rad/\sec$$

Damping effect will be studied in this example; therefore, the viscous damping coefficient is expressed in terms of damping factor for easy data input: $c = 2\xi\omega_n m$.

Four cases are studied and they are:

ξ	C	Comment
1.5	6	Overdamped system – non-oscillatory motion
1.0	4	Critically damped system – non-oscillatory motion
0.2	0.8	Underdamped system – oscillatory motion
0.0	0	Undamped system – oscillatory motion

The responses for various damping level are shown in Figure 1.4-3:

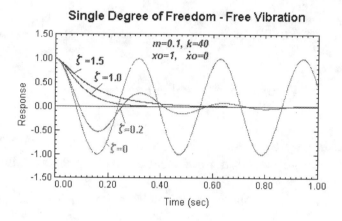

Figure 1.4-3 Damping effect on free vibration

For the undamped system ($\xi=0$), once the mass element is in motion with initial conditions, the natural motion will continue indefinitely in theory. For systems with positive damping, the natural motion will eventually die out with time. Since the underdamped system ($\xi <1$) is the most common in practice, the response for various damping level is further studied in this case as shown in Figure 1.4-4.

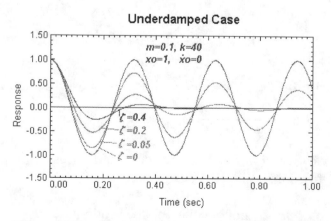

Figure 1.4-4 Damping effect – underdamped case

For the oscillatory motion as shown in Figure 1.4-4, although damping has an effect on the oscillatory frequency and period, the most pronounced effect of damping is on the response amplitude and the rate of decaying. In the dynamics of rotating machinery, the unstable system (in the linear sense) is also commonly seen in many applications. The responses for various negative damping are shown in Figure 1.4-5:

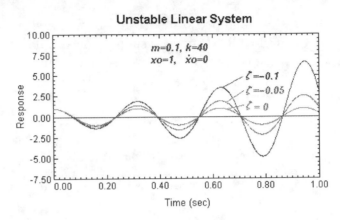

Figure 1.4-5 Unstable linear systems

It shows that for the unstable systems with negative damping, the natural motion increases exponentially with time, once it is set into motion. In reality, the motion cannot grow indefinitely. For large vibration, non-linear theory applies and the response will be restrained by *limit cycles*.

The natural motions for an undamped system with various initial conditions are shown in Figure 1.4-6.

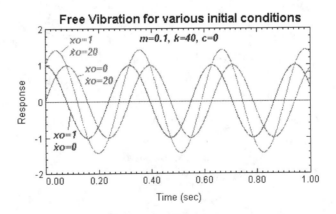

Figure 1.4-6 Free Vibration with various initial conditions

Although **DyRoBeS** has been developed mainly for complex rotor bearing systems, it can also be used for analysis of a single-degree-of-freedom system. The above results are obtained by using **DyRoBeS-Rotor**. The procedures and necessary inputs used to model this single-degree-of-freedom system are described below:

1. Start **DyRoBeS-Rotor** by double clicking the program icon
2. Select *Model* from the Main menu

3. Select *Data Editor* from the *Model* menu
4. Select (0) - Consistent Units and input descriptions under *Units/Description* tab
5. Create a dummy material under *Material* tab and a dummy shaft element under *Shaft Elements* tab
6. Input mass (m=0.1) under *Disks* tab and input disk length and diameter for graphic presentation
7. Apply boundary conditions to eliminate the unwanted degrees-of-freedom under *Constraints* tab, only 1 DOF is retained
8. Input stiffness (k=40) and various damping (c) under *Bearings* tab. Various damping factor can be obtained by changing C_{xx}
9. Save data file (Example-1-1.rot)
10. Create a data ASCII file (Example-1-1.ics) containing the initial conditions by using Notepad or any text editor and save it in the same directory as the rotor file The data format is: Station no, x, y, $xdot$, $ydot$.
    For example:
    1   1   0   -20   0   indicates at station 1, x=1, $xdot$=-20
11. Select *Analysis* from the Main menu
12. Select *Lateral Vibration* from the *Analysis* menu
13. Enter necessary analysis parameters
14. Run the analysis and use Post-Processor to view the results

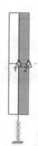

Figure 1.4-7 SDOF Model

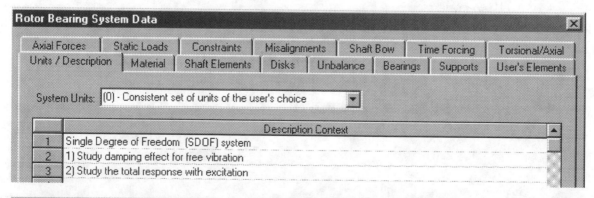

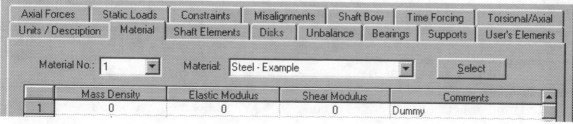

Rotor Bearing System Data [×]

| Axial Forces | Static Loads | Constraints | Misalignments | Shaft Bow | Time Forcing | Torsional/Axial |
| Units / Description | Material | Shaft Elements | Disks | Unbalance | Bearings | Supports | User's Elements |

Shaft: 1 of 1   Starting Station #: 1   [Add Shaft] [Del Shaft] [Previous] [Next]

Speed Ratio: 1   Axial Distance: 0   Y Distance: 0

Comment: Dummy shaft element

	Ele	Sub	Mat	Length	Mass ID	Mass OD	Stiff ID	Stiff OD	Comments	
1	1	1	1	1.000000	0.000000	10.000000	0.000000	0.000000		

| Axial Forces | Static Loads | Constraints | Misalignments | Shaft Bow | Time Forcing | Torsional/Axial |
| Units / Description | Material | Shaft Elements | Disks | Unbalance | Bearings | Supports | User's Elements |

	Type	Stn	Mass	Dia. Inertia	Polar Inertia	Skew x	Skew y	Length	ID	OD	Disk Density	
1	Rigid	1	0.1	0	0	0	0	2	0	10	0	

| Units / Description | Material | Shaft Elements | Disks | Unbalance | Bearings | Supports | User's Elements |
| Axial Forces | Static Loads | Constraints | Misalignments | Shaft Bow | Time Forcing | Torsional/Axial |

	Stn	x	y	Theta x	Theta y	Shear	Moment	Comments	
1	1	0	Fixed	Fixed	Fixed	0	0	SDOF	
2	2	Fixed	Fixed	Fixed	Fixed	0	0		
3									

Rotor Bearing System Data [×]

| Axial Forces | Static Loads | Constraints | Misalignments | Shaft Bow | Time Forcing | Torsional/Axial |
| Units / Description | Material | Shaft Elements | Disks | Unbalance | Bearings | Supports | User's Elements |

Bearing: 1 of 1   [Add Brg] [Del Brg] [Previous] [Next]

Station I: 1   J: 0   Angle: 0

Type: 0-Linear Constant Bearing

Comment: Change Cxx to obtain the damping factor. c=0.8 indicates zeta=0.2

Change Cxx for various damping levels

Translational Bearing Properties

| Kxx: 40 | Kxy: 0 | Cxx: 0.8 | Cxy: 0 |
| Kyx: 0 | Kyy: 0 | Cyx: 0 | Cyy: 0 |

Rotational Bearing Properties

| Kaa: 0 | Kab: 0 | Caa: 0 | Cab: 0 |
| Kba: 0 | Kbb: 0 | Cba: 0 | Cbb: 0 |

To simulate the system motion, select *Time Transient Analysis* and input the necessary data in the transient analysis as shown below. For free vibration, uncheck all the excitation effects and select initial conditions (Yes). Readers are encouraged to make changes in the input data, run the analysis, and examine the results by using Postprocessor functions. Since the closed form solution for the free vibration is presented before, readers are encouraged to verify the numerical results obtained from **DyRoBeS-Rotor** with the analytical solutions.

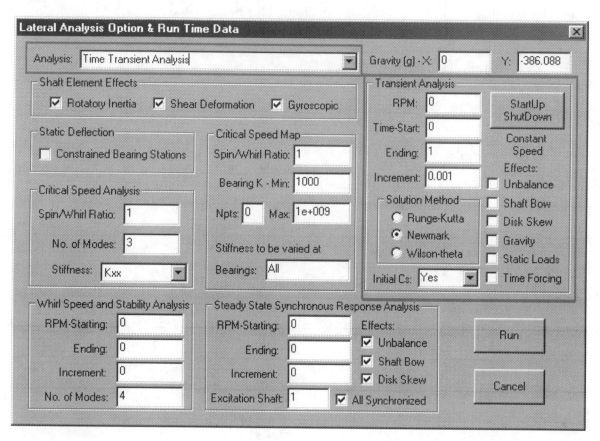

1.5 Forced Vibration

There are many types of excitations that are encountered often in the rotating machinery. Let us examine the steady state unbalance response first, since the mass imbalance is the most common source of excitation in the rotating machinery. The unbalance excitation is a harmonic excitation, which has an excitation frequency coincident with the rotor rotational speed (spin speed). The unbalance excitation amplitude varies with the square of the speed. The unbalance excitation is sometimes referred to as *synchronous excitation*, because the excitation frequency is synchronized with the rotor rotational speed. In general, consider a single-degree-of-freedom system subject to a harmonic excitation, the differential equation of motion is:

$$m\ddot{x} + c\dot{x} + kx = F_0 \cos \omega_{exc} t \qquad (1.5\text{-}1)$$

where ω_{exc} is the excitation frequency and F_0 is the excitation amplitude. For unbalance excitation, ω_{exc} will be the rotor spinning speed ($\omega_{exc} = \Omega$) and $F_0 = me\Omega^2$, with m being the mass and e being the mass eccentricity. The total response of a linear system consists of the superposition of a free vibration (natural motion) and a forced vibration. For stable systems, the forced response is also called *steady state response*, since the natural motion dies out eventually with time. The homogeneous solution (free vibration) to the above equation was discussed in the previous section. The steady state response is the topic of this section.

The particular solution (forced vibration) for the harmonic excitation has the same frequency as that of the excitation and can be written as:

$$x_p(t) = X \cos(\Omega t - \phi) \tag{1.5-2}$$

where X is the amplitude of the steady state response and ϕ is the phase angle of the steady state response relative to the excitation. The phase lag is due to the presence of the damping term. The velocity and acceleration of the steady state solution are:

$$\dot{x}_p(t) = -\Omega X \sin(\Omega t - \phi) \tag{1.5-3}$$

$$\ddot{x}_p(t) = -\Omega^2 X \cos(\Omega t - \phi) \tag{1.5-4}$$

The steady state response amplitude and phase angle can be obtained by substituting the assumed solutions (Eq. 1.5-2, 1.5-3, 1.5-4) into the differential equation (Eq. 1.5-1). Since the phases of the velocity and acceleration of a harmonic motion are ahead of the displacement by 90 and 180 degrees respectively, the terms of the differential equation can also be shown graphically in the vector form. The amplitude and phase can be easily determined from the force vector diagram as shown in Figure 1.5-1:

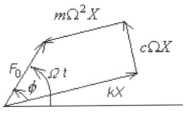

reference   **Figure 1.5-1** Force vector diagram

From the force diagram, it is easily seen that

$$F_0^2 = \left(kX - m\Omega^2 X\right)^2 + \left(c\Omega X\right)^2 \tag{1.5-5}$$

Therefore, the response amplitude is:

$$X = \frac{F_0}{\sqrt{\left(k - m\Omega^2\right)^2 + \left(c\Omega\right)^2}} \tag{1.5-6}$$

and the phase lag is:

$$\phi = \tan^{-1}\left(\frac{c\Omega}{k - m\Omega^2}\right) = \tan^{-1}\left(\frac{2\xi\gamma}{1 - \gamma^2}\right) \qquad (1.5\text{-}7)$$

where

$$\gamma = \frac{\Omega}{\omega_n} \qquad (1.5\text{-}8)$$

is called the *frequency ratio*. At resonance ($\gamma = 1$), that is when the excitation frequency coincides with the system undamped natural frequency, the steady state amplitude is limited only by the damping force, and the response lags behind the excitation by 90 degrees. The amplitudes and phase angle at resonance are:

$$X_r = \frac{F_0}{c\omega_n} = \frac{F_0}{2\xi k} \quad \text{and} \quad \phi = 90^0 \quad \text{when} \quad \gamma = 1 \ (\Omega = \omega_n) \qquad (1.5\text{-}9)$$

For the undamped systems, the steady-state response amplitude increases as the excitation frequency approaches the system undamped natural frequency, and becomes infinity at resonance. For the damped systems, the steady-state response amplitude at resonance is limited by the damping force, as shown in Eq. 1.5-9. For the unbalance response, the unbalance forcing amplitude, $F_0 = me\Omega^2$, is dependent on the square of the speed. At resonance the response amplitude can be obtained by substituting the forcing amplitude into Eq. (1.5-9):

$$X_r = \frac{e}{2\xi} \quad \text{and} \quad \phi = 90^0 \quad \text{when} \quad \gamma = 1 \ (\Omega = \omega_n) \qquad (1.5\text{-}10)$$

However, the maximum steady-state peak amplitude due to unbalance excitation occurs at an excitation frequency (rotor speed) **higher** than the undamped natural frequency and can be obtained through the following expression:

$$dX\!\big/\!d\Omega_p = 0 \qquad (1.5\text{-}11)$$

The maximum steady-state peak amplitude from Eq. (1.5-11) is:

$$X_p = \frac{e}{2\xi\sqrt{1 - \xi^2}} \qquad (1.5\text{-}12)$$

at rotor speed

$$\Omega_p = \frac{\omega_n}{\sqrt{1 - 2\xi^2}} \qquad (1.5\text{-}13)$$

Note that the peak amplitude speed exists only when

$$\sqrt{1-2\xi^2} > 0 \quad \rightarrow \quad \xi < \frac{1}{\sqrt{2}} = 0.707 \tag{1.5-14}$$

When $\xi > 0.707$, no apparent peak exists and the amplitude increases with the speed and asymptotically approaches the mass eccentricity. For steady state unbalance response, the response amplitude is zero at zero speed, peaks at the speed higher than the undamped natural frequency, and approaches to the mass eccentricity (e) at very high speed, as shown in Figure 1.5-2. The phase angle relative to the excitation is zero at zero speed, is 90 degrees at the undamped natural frequency, and 180 degrees at very high speed. Figure 1.5-2 shows the steady state unbalance responses for various damping levels. It is evident that the damping has a large influence on the response amplitude and phase angle near the resonance, but little influence when the excitation frequency is further away from the resonance. Figure 1.5-2 is commonly referred to as a *Bode plot* where the amplitude and phase are plotted against the frequency.

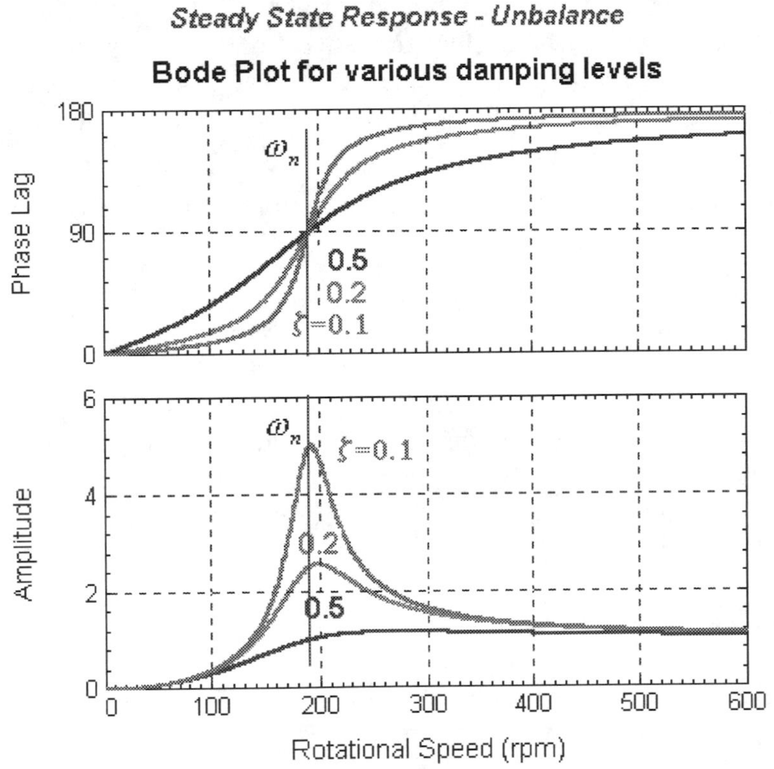

Figure 1.5-2 Steady state response due to mass unbalance

Another common synchronous harmonic excitation, which occurs in the large rotor system, is caused by the initial shaft bow. The excitation amplitude, due to shaft bow, is a constant.

$$F_0 = kq_b \tag{1.5-15}$$

Where k is the shaft stiffness, and q_b is the initial bow. At resonance, ($\gamma=1$) the response amplitude and phase angle due to initial bow from Eq. (1.5-9) becomes:

$$X_r = \frac{q_b}{2\xi} \quad \text{and} \quad \phi = 90^0 \quad \text{when} \quad \gamma = 1 \quad (\Omega = \omega_n) \tag{1.5-16}$$

For the steady state response due to the initial shaft bow, the maximum peak amplitudes occur at an excitation frequency (rotor speed), which is **lower** than the undamped natural frequency. The peak steady state amplitude and frequency (rotor speed) can be determined from Eq. (1.5-11) and they are:

$$X_p = \frac{q_b}{2\xi\sqrt{1-\xi^2}} \tag{1.5-17}$$

and

$$\Omega_p = \omega_n \sqrt{1-2\xi^2} \tag{1.5-18}$$

For steady-state response due to shaft bow, the amplitude starts at the initial shaft bow at zero speed, peaks at the speed lower than the undamped natural frequency, and approaches to zero at very high speed, as shown in Figure 1.5-3. The phase expression is the same as the phase due to unbalance excitation.

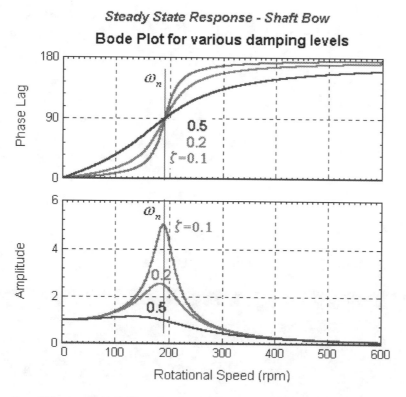

Figure 1.5-3 Steady state response due to shaft initial bow

Example 1.2: SDOF – Steady State Response to Unbalance

Following the previous example, consider the following equation of motion for a single degree-of-freedom system, subject to the unbalance excitation:

$$m\ddot{x} + c\dot{x} + kx = me\Omega^2 \cos\Omega t$$

where

$$m = 0.1, \ c = 0.8, \ k = 40, \ e = 1$$

The undamped natural frequency and damping factor are:

$$\omega_n = \sqrt{\frac{k}{m}} = 20 \ \text{rad/sec} = 190.986 \ \text{rpm}$$

$$\xi = \frac{c}{2m\omega_n} = 0.2 \qquad 20\% \ \text{damping}$$

At resonance ($\Omega = \omega_n$), the phase lag is 90 degrees and the amplitude is:

$$X = \frac{e}{2\xi} = 2.5$$

The maximum peak amplitudes occur at an excitation frequency (rotor speed), which is **higher** than the undamped natural frequency. The peak (maximum) steady state amplitude is:

$$X_p = \frac{e}{2\xi\sqrt{1-\xi^2}} = 2.5515$$

at speed

$$\Omega_p = \frac{\omega_n}{\sqrt{1-2\xi^2}} = 199 \ \text{rpm}$$

The unbalance excitation (*me*) is entered under the Unbalance tab as shown below:

Rotor Bearing System Data							☒
Axial Forces	Static Loads	Constraints	Misalignments	Shaft Bow	Time Forcing	Torsional/Axial	
Units / Description	Material	Shaft Elements	Disks	Unbalance	Bearings	Supports	User's Elements

	Ele	Sub	Left Unb.	Left Ang.	Right Unb.	Right Ang.	Comments	
1	1	1	0.1	0	0	0	me=0.1	
2								
3								

The analysis input for the steady state unbalance response is shown below:

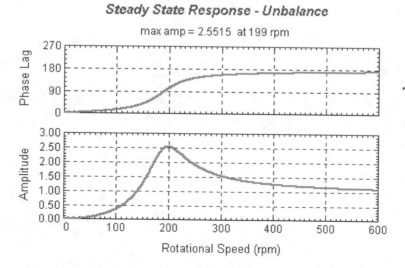

The amplitude and phase of steady state unbalance response are plotted in the following Bode Plot.

Figure 1.5-4 Bode plot for steady state unbalance response

Polar plot is also commonly used to present the steady state response. Figure 1.5-5 shows amplitude and phase in the polar coordinates, over a speed range for a given finite element station. The shaft rotational direction is counterclockwise.

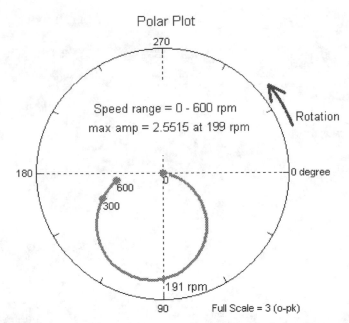

Figure 1.5-5 Polar plot for steady state unbalance response

Both plots show that the steady state unbalance response amplitude is zero at zero speed, and increases as the speed approaches resonance. At resonance (191 rpm), the amplitude is 2.5, and the maximum amplitude is 2.55 at the speed (199 rpm) that is higher than the resonance speed. After the peak, the amplitude decreases and approaches to the mass eccentricity ($e=1$) at very high speed. The phase angle relative to the excitation is zero at zero speed, 90 degrees at resonance, and 180 degrees at very high speed. The steady state response due to mass imbalance for various damping levels is shown in Figure 1.5-2. The numerical results from **DyRoBeS** are in agreement with the analytical results.

Example 1.3: SDOF – Steady State Response to Shaft Bow

Again, following the previous example, consider a single degree of freedom system subject to a harmonic excitation with constant excitation amplitude:

$$m\ddot{x} + c\dot{x} + kx = kq_b \cos\Omega t$$

where q_b is the initial shaft bow and k is the shaft bending stiffness.

$$m = 0.1, \ c = 0.8, \ k = 40, \ q_b = 1$$

The undamped natural frequency and damping factor are:

$$\omega_n = \sqrt{\frac{k}{m}} = 20 \text{ rad/sec} = 190.986 \text{ rpm}$$

$$\xi = \frac{c}{2m\omega_n} = 0.2 \qquad 20\% \text{ damping}$$

At resonance ($\Omega = \omega_n$), the phase lag is 90 degrees and the amplitude is:

$$X = \frac{q_b}{2\xi} = 2.5$$

The maximum peak amplitudes occur at an excitation frequency (rotor speed), which is **lower** than the undamped natural frequency.   The peak (maximum) steady state amplitude is:

$$X_p = \frac{q_b}{2\xi\sqrt{1-\xi^2}} = 2.5515$$

at speed

$$\Omega_p = \omega_n\sqrt{1-2\xi^2} = 183 \text{ rpm}$$

Since the stiffness k for the shaft bow excitation is the *shaft bending stiffness*, the input data must be modified from the previous example. The damping coefficient c is still entered under the *Bearing* tab, however the stiffness k must be entered under the *User's Element* tab. The following steps are used to build and analyze this system:

1. Recall the previous example file
2. Change the material number to zero in the *Shaft Elements* tab for User's Element
3. Remove the bearing stiffness, since k is the shaft bending stiffness in this case, and will be entered from the *User's Element* tab
4. Enter $K_{xx}=40$ for element number 1 under the *User's Element* tab.
5. Enter $q_b=1$ for the initial shaft bow under the *Shaft Bow* tab.
6. Run the steady-state synchronous response with the shaft bow effect checked and other effects unchecked.

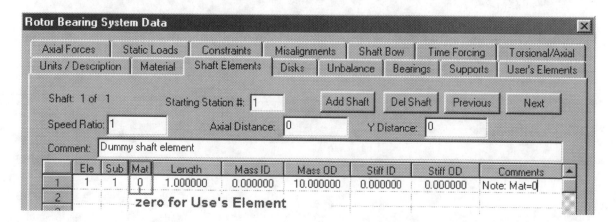

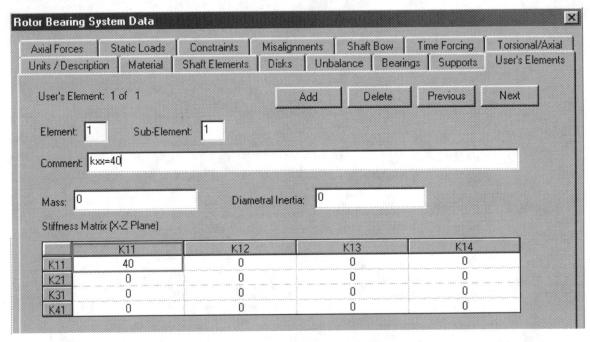

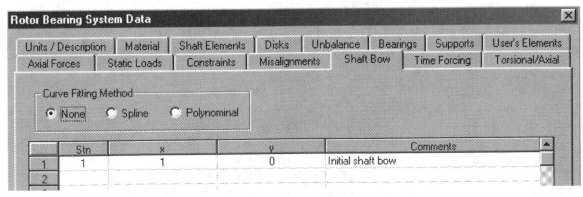

The amplitude and phase of steady-state response due to initial shaft bow are plotted in the following Bode Plot and Polar Plot.

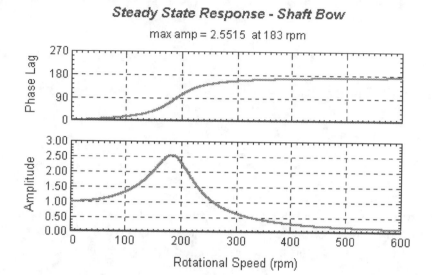

Figure 1.5-6 Bode plot for steady state shaft bow response

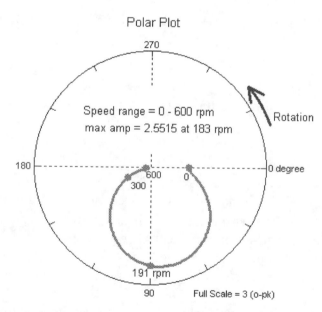

Figure 1.5-7 Polar plot for steady state shaft bow response

They show that the amplitude of the steady-state response, due to shaft bow, starts at the initial shaft bow at zero speed, peaks at the speed (183 rpm) that is lower than the undamped frequency (191 rpm), and approaches zero at very high speed. The phase angle relative to the excitation is zero at zero speed, 90 degrees at resonance, and 180 degrees at very high speed. The phase expression due to shaft bow is the same as the phase due to unbalance excitation. The steady state response due to shaft initial bow for various damping levels is shown in Figure 1.5-3. Again, the numerical results from *DyRoBeS* are agreed with the analytical results.

1.6 Total Motion

The governing equation of motion for a single-degree-of-freedom linear system as presented before, is a second order ordinary differential equation.

$$m\ddot{x}(t) + c\dot{x}(t) + kx(t) = f(t) \qquad (1.6\text{-}1)$$

with two initial conditions at $t = 0$

$$\begin{aligned} x(0) &= x_0 \\ \dot{x}(0) &= \dot{x}_0 \end{aligned} \qquad (1.6\text{-}2)$$

The complete solution of this second-order ordinary differential equation is the sum of the homogeneous solution and the particular solution. The homogeneous solution and steady-state solution to the above equation have been discussed in the previous sections. The total response is the sum of the free vibration and the forced response.

$$x(t) = x_h(t) + x_p(t) \qquad (1.6\text{-}3)$$

The total response for a single-degree-of-freedom system under harmonic excitation is as follows:

$$\begin{aligned} x(t) &= x_h(t) + x_p(t) \\ &= e^{-\xi\omega_n t}\left(A_1\cos\omega_d t + A_2\sin\omega_d t\right) + X\cos(\Omega t - \phi) \end{aligned} \qquad (1.6\text{-}4)$$

where the constants A_1 and A_2 are determined from the initial conditions; they are also dependent on the steady-state solution at $t = 0$. Several numerical integration methods are included in **DyRoBeS-Rotor**. Gear's and Runge-Kutta methods solve the first-order differential equation. The original second-order equations need to be rearranged to be in a state-space form, in order to use these methods. Newmark-β and Wilson-θ methods are developed for the second-order differential equations. Details of the numerical integration schemes are not described here and will be presented in later Chapters. Readers can find these materials in most of numerical analysis and finite element textbooks.

The equation of motion for a single degree-of-freedom system subject to the static loading and harmonic excitation is given by:

$$m\ddot{x} + c\dot{x} + kx = W_0 + F_0\cos\Omega t \qquad (1.6\text{-}5)$$

Note that this is still a linear differential equation. The nonlinear systems will be discussed in the later chapters. Since this is a linear system, the complete solution is the sum of the homogeneous solution and the particular solutions. The homogeneous solution and the particular solution for a harmonic excitation have been discussed before. The particular solution for static loading is simply the static displacement:

$$x_{ps} = \frac{W_0}{k}$$

(1.6-6)

The complete solution is:

$$x(t) = x_h(t) + x_p(t)$$

$$= e^{-\xi \omega_n t}\left(A_1 \cos \omega_d t + A_2 \sin \omega_d t\right) + X \cos(\Omega t - \phi) + \frac{W_0}{k}$$

(1.6-7)

where the constants A_1 and A_2 are determined from the initial conditions. For damped systems, the first term (homogeneous response) dies out with time, and the steady-state response is the sum of a constant (static displacement), which is due to the static load, and a harmonic response, which is due to the harmonic excitation.

Example 1.4: SDOF – Total Response to Harmonic Excitation

Following the previous example, consider the following equation of motion for a single-degree-of-freedom system subject to harmonic excitation:

$$m\ddot{x} + c\dot{x} + kx = F_0 \cos \Omega t$$

The system is initially at rest, that is:

$$x(0) = 0$$

$$\dot{x}(0) = 0$$

where $m = 0.1, k = 40, F_0 = 10,$ and $\Omega = 10 \ rad/\sec$. Three different damping levels are considered in this example and they are

$\xi = 0,$ $c = 0$ undamped system, and
$\xi = 0.05,$ $c = 0.2$ 5% damping
$\xi = 0.2,$ $c = 0.8$ 20% damping

The system undamped natural frequency is $\omega_n = 20 \ rad/\sec$ and the natural period is 0.314 sec. The steady state response has the same frequency as the excitation frequency (10 rad/sec) and the period is 0.628 sec. The total responses for the three different dampings are plotted in Figure 1.6-1. For the undamped case, a *beating phenomenon* occurs between the natural motion with a period of 0.314 seconds (frequency of 20 rad/sec), and the forced motion with a period of 0.628 seconds (frequency of 10 rad/sec). The total response is a combination of two harmonic motions. It also should be noted that although the system is at rest initially, the natural motion still exists, due to the non-zero steady state response at t=0. As the damping increases, the natural motion, also referred to as *starting-transient-motion*, quickly dies out with time and only the steady state motion continues with time.

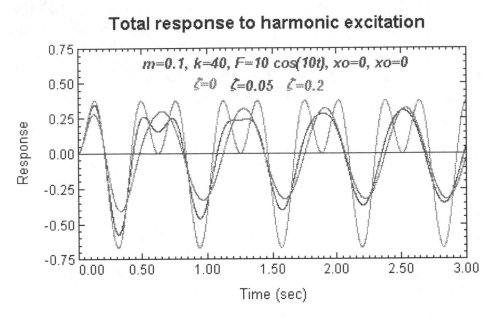

Figure 1.6-1 Total response to harmonic excitation for various damping factors

The above results are obtained from **DyRoBeS**. The model is the same as the previous example except for the excitation force. The harmonic excitation, with constant amplitude, can be entered under the *Time Forcing* tab as shown below:

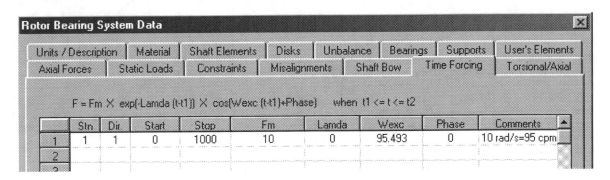

Since the numerical integration will be performed for the first several seconds, the stop time (t_2) should be large enough to include this excitation during the integration. The excitation frequency required in the data field is in unit of cpm (cycles per minute). Therefore the rad/sec must be converted into cpm. The system is at rest initially. The initial conditions can be either specified by using the (*.ics) file with zero displacement and velocity as explained in an earlier example or by using default initial conditions (No) as shown in the analysis input. The Time Forcing effect must be Checked (included) in the analysis as shown below.

Lateral Analysis Option & Run Time Data   ✕

Analysis: | Time Transient Analysis ▾ | Gravity (g) - X: | 0 | Y: | -386.088 |

Shaft Element Effects
☑ Rotatory Inertia   ☑ Shear Deformation   ☑ Gyroscopic

Transient Analysis
RPM: | 0 |
Time-Start: | 0 |
Ending: | 3 |
Increment: | 0.001 |

StartUp
ShutDown

Constant
Speed

Static Deflection
☐ Constrained Bearing Stations

Critical Speed Map
Spin/Whirl Ratio: | 1 |
Bearing K - Min: | 1000 |
Npts: | 0 | Max: | 1e+009 |
Stiffness to be varied at
Bearings: | All |

Critical Speed Analysis
Spin/Whirl Ratio: | 1 |
No. of Modes: | 3 |
Stiffness: | Kxx ▾ |

Solution Method
○ Runge-Kutta
● Newmark
○ Wilson-theta
Initial Cs: | No ▾ |

Effects:
☐ Unbalance
☐ Shaft Bow
☐ Disk Skew
☐ Gravity
☐ Static Loads
☑ Time Forcing

Whirl Speed and Stability Analysis
RPM-Starting: | 0 |
Ending: | 0 |
Increment: | 0 |
No. of Modes: | 4 |

Steady State Synchronous Response Analysis
RPM-Starting: | 0 |
Ending: | 0 |
Increment: | 0 |
Excitation Shaft: | 1 |

Effects:
☑ Unbalance
☑ Shaft Bow
☑ Disk Skew
☑ All Synchronized

Run

Cancel

The analytical solution can easily be obtained by substituting the initial conditions into the total response expression. Readers are encouraged to verify the numerical results with the analytical solutions. The analytical solutions for two damping factors given below are:

For $\xi = 0$ (undamped case)

$$x(t) = -\frac{1}{3}\cos(20t) + \frac{1}{3}\cos(10t)$$

For $\xi = 0.2$ (damped case)

$$x(t) = -e^{-4t}\left[0.31\cos(19.6t) + 0.11\sin(19.6t)\right]$$
$$+ 0.32\cos(10t - 0.26)$$

For the undamped case, the total response is a combination of two harmonic motions. One is the natural motion and the other is the forced motion. It is shown that for the damped system, the natural motion dies out quickly, and only the steady-state solution is left in the response time history. For instance, when $\xi = 0.2$, the transient response dies out within the first second. Therefore, we can also study the steady-state response by using the time transient analysis, although the steady-state response is normally performed in the frequency domain.

Assuming the damping factor is a constant (ξ=0.2) and three excitation frequencies are studied:

$$\gamma = 0.50 \qquad \Omega =10 \text{ rad/sec} = \quad 95.493 \text{ cpm}$$
$$\gamma = 0.75 \qquad \Omega =15 \text{ rad/sec} = 143.239 \text{ cpm}$$
$$\gamma = 1.0 \qquad \Omega =20 \text{ rad/sec} = 190.986 \text{ cpm}$$

The total responses for three different frequency ratios are plotted below:

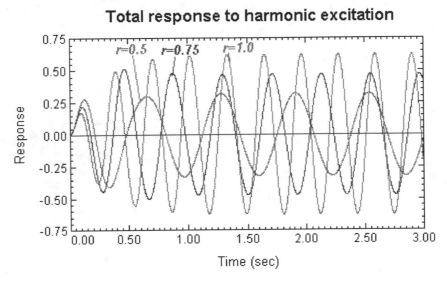

Figure 1.6-2 Total response to harmonic excitation for various excitation frequencies

It shows that the amplitude of steady-state response increases as the frequency ratio approaches unity (resonance). At resonance (γ=1), the amplitude of steady-state response is found to be of 0.625, which can be verified by the theoretical value:

$$X = \frac{F_0}{2\xi k}$$

For γ=0.5, the steady state amplitude is found to be of 0.322, which can also be verified through the amplitude equation. At steady state, the motion oscillates with a frequency equal to the excitation frequency. It also shows that the phase angle is different with different frequency ratios. Since the steady state response is an important subject in rotordynamics, it will be discussed in greater detail in the upcoming chapters.

Example 1.5: SDOF – Total Response

Using the previous example, consider the following equation of motion for a single-degree-of-freedom system subject to a static load and a harmonic excitation:

$$m\ddot{x} + c\dot{x} + kx = W_0 + F_0 \cos\Omega t$$

Again, let $m = 0.1$, $k = 40$, $F_0 = 10$, and $\Omega = 10 \ rad/\sec$. Set damping factor to be of 0.2, that is, $\xi = 0.2$, $c = 0.8$, and the system is initially at rest, $x(0) = 0$, $\dot{x}(0) = 0$. For simplicity, let $W_0 = 40$, the static deflection is:

$$X_s = \frac{W_0}{k} = 1$$

The amplitude of steady state harmonic response is:

$$X = \frac{F_0}{\sqrt{\left(k - m\Omega^2\right)^2 + \left(c\Omega\right)^2}} = 0.322$$

To add the additional static load from the previous example, the static load is entered as shown below:

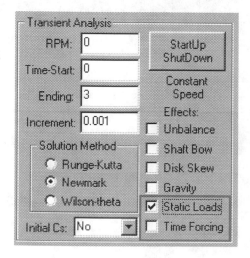

Since the response to a harmonic excitation was studied in the previous example, let us study the response to a static load first. To simulate this effect, the "Static Loads" needs to be Checked and the "Time Forcing" needs to be Unchecked in the analysis input as shown below:

The total response for the step loading is calculated by the numerical integration and shown in Figure 1.6-3:

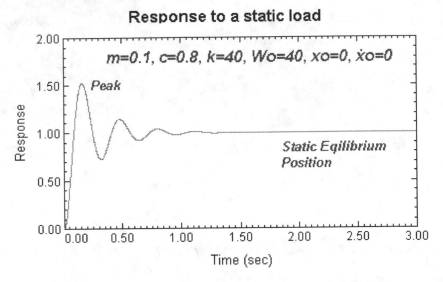

Figure 1.6-3 Total response to static load

It shows that the natural motion due to initial conditions dies out with time, and the mass element stays still at the static equilibrium position without any motion. The analytical solution is available in most vibration textbooks (Thomson, 1981) and is given as:

$$x(t) = \frac{W_0}{k}\left\{1 - e^{-\xi\omega_n t}\left[\cos(\omega_d t) + \left(\frac{\xi\omega_n}{\omega_d}\right)\sin(\omega_d t)\right]\right\}$$

(1.6-8)

The peak response for the above equation is:

$$X_{peak} = \frac{W_0}{k}\left[1 + \exp\left(\frac{-\xi\pi}{\sqrt{1-\xi^2}}\right)\right] = 1.526$$

(1.6-9)

which occurs at

$$t_{peak} = \frac{\pi}{\omega_n\sqrt{1-\xi^2}} = 0.16 \text{ sec.}$$

(1.6-10)

The computational results from **DyRoBeS** are in agreement with the analytical results. Now let us include the harmonic excitation effect into the analysis. Again, the total response caused by the step loading and time forcing harmonic excitation is calculated by numerical integration with both effects Checked in the analysis input. The result is shown in Figure 1.6-4.

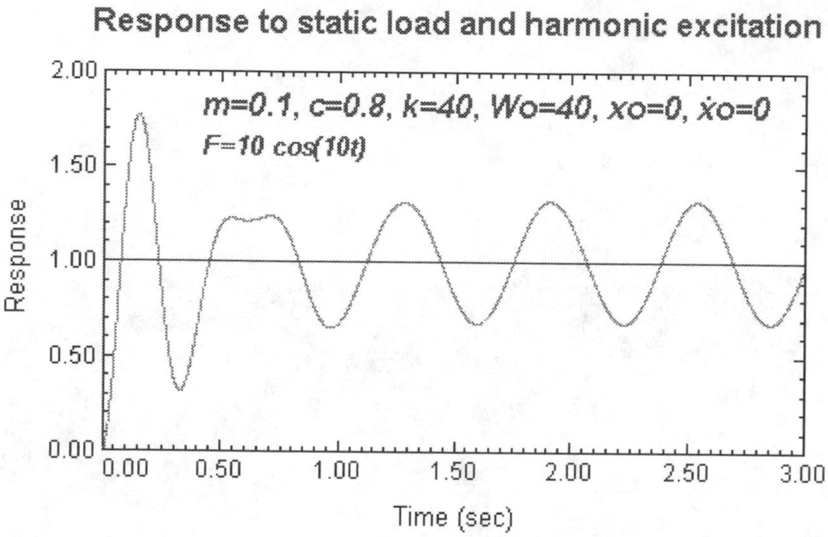

Response to static load and harmonic excitation

Figure 1.6-4 Total response to static load and harmonic excitation

It is evident that at steady-state condition, the mass element oscillates about its static equilibrium position, with the steady-state amplitude and frequency caused by the steady state harmonic excitation. The governing differential equation is *linear* and thus subject to the principal of superposition. At steady-state condition, the total response is the sum of the steady state solutions of the static load and the harmonic excitation.

F100-PW-220/F100-PW-220E Turbofan Engine

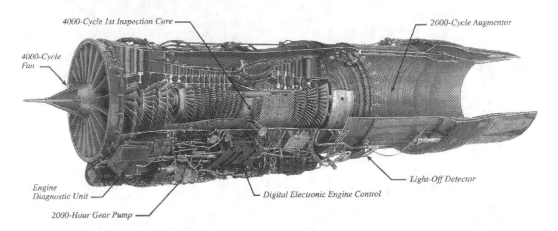

4000-Cycle 1st Inspection Core — 2000-Cycle Augmentor

4000-Cycle Fan

Engine Diagnostic Unit

2000-Hour Gear Pump

Digital Electronic Engine Control

Light-Off Detector

Vital Statistics

Maximum Thrust (Full Augmentation)	23,770 pounds (105.7 kN)
Intermediate Thrust (Nonaugmented)	14,590 pounds (64.9 kN)
Weight	3,234 pounds (1467 kg)
Length	191 inches (4.85 m)
Inlet Diameter	34.8 inches (0.88 m)
Maximum Diameter	46.5 inches (1.18 m)
Bypass Ratio	0.6
Overall Pressure Ratio	25 to 1

Schedule

Qualification Complete	March 1985
Production Introduction	November 1985
Operational Introduction	June 1986

The F100-PW-220E engine is an upgraded F100-PW-100 or F100-PW-200 with the operability, durability and maintainability features of the F100-PW-220.

UNITED TECHNOLOGIES PRATT&WHITNEY

© UTC Printed in USA

Photograph courtesy of United Technologies, Pratt & Whitney Company

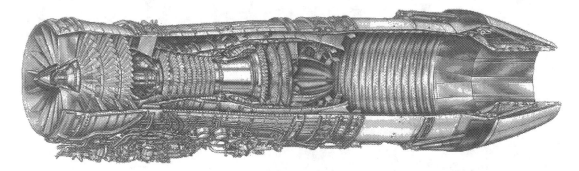

General Electric F110-GE-100 Augmented Turbofan Engine

Photograph courtesy of General Electric Company

Coordinate Systems and Rotor Motion 2

2.1 Coordinate Systems

A finite element station along the shaft, as shown in Figure 2.1-1, has a total of six (6) degrees-of-freedom; three (3) translational displacements (x,y,z) along (X,Y,Z) axes and three (3) rotational displacements $(\theta_x,\theta_y,\theta_z)$ about (X,Y,Z) axes. The (X,Y,Z) axes describe a fixed global right-hand-ruled Cartesian coordinate system, with the Z axis being collinear and coincident with the static equilibrium position of the undeformed shaft centerline. In the study of rotordynamics, the lateral, torsional, and axial vibrations are generally considered to be de-coupled and can be studied separately.

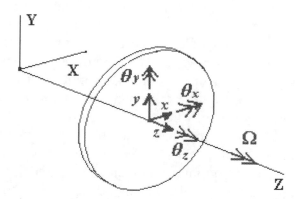

Figure 2.1-1 Coordinate system and degrees of freedom

For torsional vibration, the motion of each finite element station is described by a rotational displacement (θ_z) about the spinning axis (Z). For axial vibration, the motion of each finite element station is described by a translational displacement (z) along the spinning axis (Z). For lateral vibration, the motion of each finite element station is described by two translational displacements (x,y) in the X and Y directions respectively, and two rotational (angular) displacements (θ_x,θ_y) about the X and Y axes respectively. In the rigorous definition, the rotational displacements depend upon the translational displacements, and they cannot be treated as generalized coordinates in the Lagrange's equations. However, the lateral vibration displacements are very small in comparison to the rotor diameter, i.e., in the order of 10^{-3}. The small displacement assumption leads to the linearized equations of motion for the rotating disks and shafts.   For large displacements, Eulerian angles are normally used, that lead to non-linear equations of motion for the spinning disk. The disadvantage of using the Eulerian angles is that they cannot be directly expressed in the shaft element strain energy formulation without making an assumption of small deformations.

For lateral vibration, a rotating reference frame (X',Y',Z'), as illustrated in Figure 2.1-2, is sometimes used to formulate the equations of motion for isotropic systems. The Z'-axis coincides with the Z axis and (X',Y') rotates relative to the fixed (X,Y) by a single rotation ωt about the Z axis.

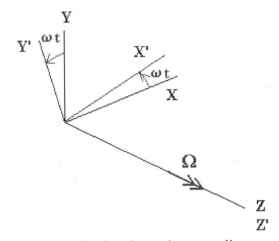

Figure 2.1-2 Fixed and rotating coordinate systems

The fixed global coordinate system (X,Y,Z) provides generality for handling problems with non-symmetric bearing stiffness and damping, flexible supports, and all other interconnecting components, since their equations of motion are normally defined in the fixed coordinate system. Also, all the vibration measurements are taken and described in the fixed reference frame. However, when analyzing systems with isotropic supports for natural frequencies of whirl, or steady-state synchronous response with circular orbits, it is convenient to utilize a rotating reference frame. Two advantages of using such a rotating reference frame for an isotropic system are: 1) the undamped critical speeds may be determined directly from a single plane of motion that leads to a reduced eigenvalue problem, and 2) the nonlinear effect of squeeze film dampers and nonlinear isotropic bearings/supports on steady-state centered circular response, can be taken into account by a relatively simple iterative procedure.

The coordinate transformation for the translational displacements between the fixed reference frame and the rotating reference frame, as illustrated in Figure 2.1-2, is:

$$q = R\,p \tag{2.1-1}$$

where

$$q = \begin{Bmatrix} x \\ y \end{Bmatrix}, \quad R = \begin{bmatrix} \cos\omega t & -\sin\omega t \\ \sin\omega t & \cos\omega t \end{bmatrix}, \quad p = \begin{Bmatrix} x' \\ y' \end{Bmatrix} \tag{2.1-2}$$

The derivative of a matrix is equal to the matrix of the derivatives of the elements, and for later use, the first two time derivatives of Eq. (2.1-1) are in the form:

$$\dot{q} = R\dot{p} + \dot{R}p = R\dot{p} + \omega\, S p \tag{2.1-3}$$

$$\ddot{q} = R\ddot{p} + 2\dot{R}\dot{p} + \ddot{R}p = R\ddot{p} + 2\omega\, S\, \dot{p} - \omega^2\, Rp \tag{2.1-4}$$

where

$$S = \begin{bmatrix} -\sin\omega t & -\cos\omega t \\ \cos\omega t & -\sin\omega t \end{bmatrix} \tag{2.1-5}$$

The above expression can be expanded to the rotational displacements. At any finite element station, the displacements in the fixed reference frame and rotating reference frame are related by:

$$\begin{Bmatrix} x \\ y \\ \theta_x \\ \theta_y \end{Bmatrix} = \begin{bmatrix} \cos\omega t & -\sin\omega t & 0 & 0 \\ \sin\omega t & \cos\omega t & 0 & 0 \\ 0 & 0 & \cos\omega t & -\sin\omega t \\ 0 & 0 & \sin\omega t & \cos\omega t \end{bmatrix} \begin{Bmatrix} x' \\ y' \\ \theta'_x \\ \theta'_y \end{Bmatrix} \tag{2.1-6}$$

The cosine and sine terms can be factored out as follows:

$$\begin{Bmatrix} x \\ y \\ \theta_x \\ \theta_y \end{Bmatrix} = \begin{Bmatrix} x' \\ y' \\ \theta'_x \\ \theta'_y \end{Bmatrix} \cos\omega t + \begin{Bmatrix} -y' \\ x' \\ -\theta'_y \\ \theta'_x \end{Bmatrix} \sin\omega t \tag{2.1-7}$$

2.2 Steady State Rotor Motion

The lateral motion of a rotor system is more complicated than the motions of torsional and axial vibration. The following sections are used to describe the lateral motion of a rotor system. As described in an earlier section, the lateral motion of a finite element station along the shaft is defined by two translational displacements and two rotational displacements (slopes). There are many types of excitations acting on a rotor system and they are frequently periodic, or can be approximated closely by the summation of periodic forces. The steady-state response to a periodic excitation is a periodic motion.

From Fourier expansion, any periodic motion can be represented by a series of harmonic motions:

$$x(t) = x_0 + \sum_{i=1}^{\infty} \left(X_i \cos(\omega_i t - \phi_i) \right)$$

$$= x_0 + \sum_{i=1}^{\infty} \left(X_{c,i} \cos(\omega_i t) + X_{s,i} \sin(\omega_i t) \right)$$

(2.2-1)

At steady state, the rotor centerline whirls (oscillates) around its equilibrium position. Typical rotor linear motion for a propylene refrigeration compressor is presented in Figure 2.2-1:

Rotor System Configuration

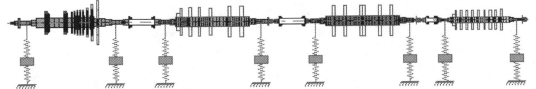

Steady State Rotor Motion

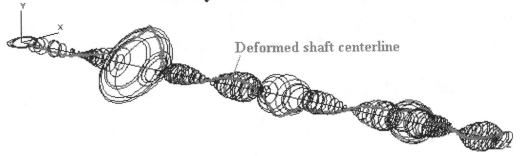

Figure 2.2-1 Rotor steady state Motion (courtesy of Malcolm E. Leader)

A non-linear rotor motion at instability region is shown in Figure 2.2-2.

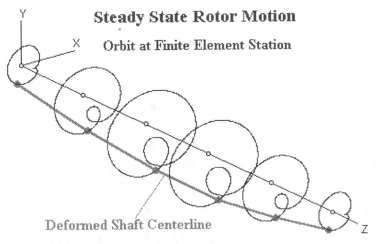

Figure 2.2-2 Rotor limit cycle motion

The rotor motion at each finite element station can be any form. Some typical rotor motions are shown in Figure 2.2-3.

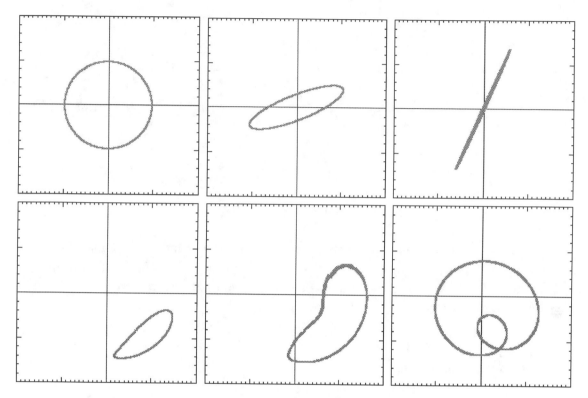

Figure 2.2-3 Typical rotor motions

2.3 Elliptical Motion

The simplest steady rotor motion is a harmonic motion with a whirl frequency of ω. At each finite element station, the rotor translational motions are described as:

$$x(t) = x_c \cos \omega t + x_s \sin \omega t = |x| \cos(\omega t - \phi_x) \qquad (2.3\text{-}1)$$

$$y(t) = y_c \cos \omega t + y_s \sin \omega t = |y| \cos(\omega t - \phi_y) \qquad (2.3\text{-}2)$$

where

$$|x| = \sqrt{x_c^2 + x_s^2} \ , \qquad \phi_x = \arctan\left(\frac{x_s}{x_c}\right) \qquad (2.3\text{-}3)$$

$$|y| = \sqrt{y_c^2 + y_s^2} \ , \qquad \phi_y = \arctan\left(\frac{y_s}{y_c}\right) \qquad (2.3\text{-}4)$$

Note that x and y displacements oscillate at the same frequency ω with different amplitudes and phase angles. This harmonic motion describes an *elliptical orbit*, as shown in Figure 2.3-1.

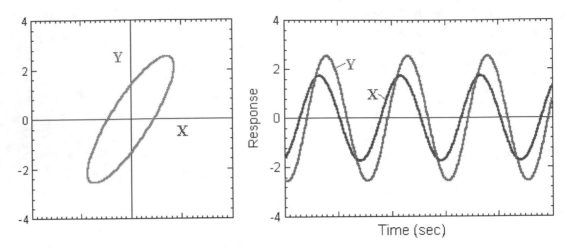

Figure 2.3-1 Rotor elliptical orbit

This simple expression can represent a steady-state synchronous response with whirl frequency ω, equal to the rotor rotational speed Ω, or can be an undamped precessional mode orbit with whirl frequency, equal to the associated natural frequency of that mode. The rotational displacements have the same forms:

$$\theta_x(t) = \theta_{xc} \cos \omega t + \theta_{xs} \sin \omega t \qquad (2.3\text{-}5)$$

$$\theta_y(t) = \theta_{yc} \cos \omega t + \theta_{ys} \sin \omega t \qquad (2.3\text{-}6)$$

The complete rotor elliptical motion at each finite element station is either forward or backward precession in relation to the direction of rotation (spin) of the rotor as shown in Figure 2.3-2. The motion is defined as a *forward precession* if it is whirling in the same direction as the rotor spinning direction, and as a *backward precession* if it is whirling opposite to the rotor spinning direction.

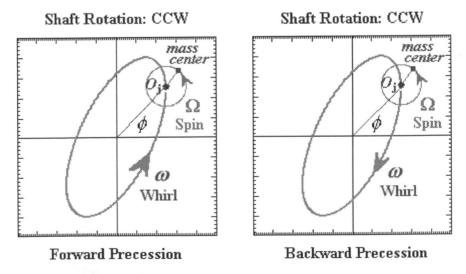

Figure 2.3-2 Forward and backward precessions

To determine the ellipse properties, i.e., semi-major axis a, semi-minor axis b, and attitude angle ϕ_a as illustrated in Figure 2.3-3, it is convenient to introduce the radius displacement vector at shaft center:

$$\hat{r}(t) = x(t) + j\, y(t) \tag{2.3-7}$$

The maximum and minimum amplitudes of displacement vector can be determined by:

$$\frac{d\left|\hat{r}(t)\right|}{d(\omega t)} = 0 \tag{2.3-8}$$

Thus, the semi-major and semi-minor axes of the elliptical orbit are:

$$a,b = \left\{ \; \frac{1}{2}\left(x_c^2 + y_c^2 + x_s^2 + y_s^2\right) \right.$$
$$\left. \pm \frac{1}{2}\left[\left(x_c^2 + y_c^2 - x_s^2 - y_s^2\right)^2 + 4\left(x_c x_s + y_c y_s\right)\right]^{\frac{1}{2}} \; \right\}^{\frac{1}{2}} \tag{2.3-9}$$

and the attitude angle of the ellipse is:

$$\tan 2\phi_a = \frac{2\left(x_c y_c + x_s y_s\right)}{x_c^2 + x_s^2 - y_c^2 - y_s^2} \tag{2.3-10}$$

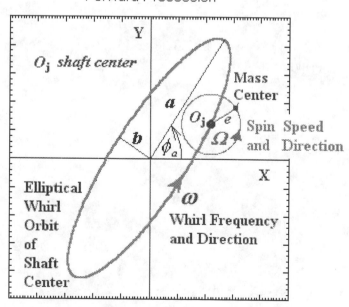

Figure 2.3-3 Elliptical orbit properties

The precessional angle is defined as:

$$\phi(t) = \arctan \frac{y(t)}{x(t)} \tag{2.3-11}$$

The direction of precession (whirling) is determined by the rate of precessional angle

$$sign(\dot{\phi}) = sign(x_c y_s - x_s y_c) \tag{2.3-12}$$

when $\dot{\phi}$ is positive value, the orbit motion is a *forward whirl* (precession). When $\dot{\phi}$ is negative value, the orbit motion is a *backward whirl* (precession). When $\dot{\phi}$ is zero, the orbit motion degenerates to a *straight-line path*.

In general, the entire rotor whirls either forward or backward. However, the complete rotor can also have mixed precession, i.e., the rotor can possess forward precession and backward precession simultaneously at different sections of the rotor. Hence, the direction of whirling should be evaluated at all the finite element stations, to determine the rotor direction of precession. Mixed precession occurs commonly in the long flexible rotors, supported by the fluid film bearings. In addition to the direction of whirling, the size, shape, and orientation of the elliptical orbits change from station to station as illustrated in Figure 2.2-1.

With the aid of complex vector, introduced before, and the complex exponential forms for the sine and cosine functions:

$$e^{j\omega t} = \cos \omega t + j \sin \omega t \quad , \quad e^{-j\omega t} = \cos \omega t - j \sin \omega t \tag{2.3-13}$$

or

$$\sin(\omega t) = \frac{-j}{2}\left(e^{j\omega t} - e^{-j\omega t}\right) \quad , \quad \cos(\omega t) = \frac{1}{2}\left(e^{j\omega t} + e^{-j\omega t}\right) \tag{2.3-14}$$

Substitution of Eq. (2.3-14) into Eqs. (2.3-1) and (2.3-2), the complex displacement can be written in the following form:

$$
\begin{aligned}
\hat{r}(t) &= x(t) + j\, y(t) \\
&= \hat{r}_f e^{j\omega t} + \hat{r}_b e^{-j\omega t} \\
&= \left|\hat{r}_f\right| e^{j(\omega t + \phi_f)} + \left|\hat{r}_b\right| e^{-j(\omega t - \phi_b)}
\end{aligned}
\tag{2.3-15}
$$

where

$$
\begin{aligned}
\hat{r}_f &= \frac{1}{2}(x_c + y_s) + \frac{j}{2}(y_c - x_s) \\
\hat{r}_b &= \frac{1}{2}(x_c - y_s) + \frac{j}{2}(y_c + x_s)
\end{aligned}
\tag{2.3-16}
$$

and

$$\left|\hat{r}_f\right| = \frac{1}{2}\left[(x_c + y_s)^2 + (y_c - x_s)^2\right]^{\frac{1}{2}}$$

$$\left|\hat{r}_b\right| = \frac{1}{2}\left[(x_c - y_s)^2 + (y_c + x_s)^2\right]^{\frac{1}{2}} \qquad (2.3\text{-}17)$$

$$\phi_f = \arctan\left(\frac{y_c - x_s}{x_c + y_s}\right)$$

$$\phi_b = \arctan\left(\frac{y_c + x_s}{x_c - y_s}\right) \qquad (2.3\text{-}18)$$

Eq. (2.3-15) shows that the elliptical orbit consists of the summation of two rotating vectors: One is a forward circular motion (progression) with an amplitude of $\left|\hat{r}_f\right|$, whirling in the same direction of rotor spinning, and the other is a backward circular motion (regression) with an amplitude of $\left|\hat{r}_b\right|$, whirling in the opposite direction of rotor spinning. Both circular motions, as shown in Figure 2.3-4, have the same angular velocity (whirl frequency) of ω. At time equals zero (t=0), these two rotating vectors start at angles ϕ_f and ϕ_b relative to the X-axis, respectively. When the forward amplitude is greater than the backward amplitude $\left|\hat{r}_f\right| > \left|\hat{r}_b\right|$, the overall motion is *forward*. When the forward amplitude is smaller than the backward amplitude $\left|\hat{r}_f\right| < \left|\hat{r}_b\right|$, the overall motion is *backward*. When the forward amplitude is equal to the backward amplitude $\left|\hat{r}_f\right| = \left|\hat{r}_b\right|$, the overall motion degenerates to a *straight line*.

Shaft Rotation: CCW
Forward Precession

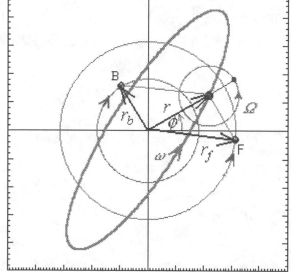

Figure 2.3-4 Elliptical orbit consists of two circular motions

The relationship between the ellipse properties and the two circular motion parameters are:

$$a = \left|\hat{r}_f\right| + \left|\hat{r}_b\right|$$

$$b = \left|\hat{r}_f\right| - \left|\hat{r}_b\right| = \frac{(x_c y_s - x_s y_c)}{a} \tag{2.3-19}$$

$$\phi_a = \frac{1}{2}\left(\phi_f + \phi_b\right)$$

or

$$\left|\hat{r}_f\right| = \frac{1}{2}(a+b) \qquad \text{and} \qquad \left|\hat{r}_b\right| = \frac{1}{2}(a-b) \tag{2.3-20}$$

By using Eq. (2.3-19), the semi-major axis a, is always positive and the semi-minor axis b, can be either positive or negative, which indicates the forward or backward precessional motion, respectively. The direction of precession of a rotor orbit can be determined from the following expressions:

Forward precession: $(x_c y_s - x_s y_c) > 0$, $b > 0$, $\left|\hat{r}_f\right| > \left|\hat{r}_b\right|$, $\pi > (\phi_y - \phi_x) > 0$

Backward precession: $(x_c y_s - x_s y_c) < 0$, $b < 0$, $\left|\hat{r}_f\right| < \left|\hat{r}_b\right|$, $2\pi > (\phi_y - \phi_x) > \pi$

Straight-Line Motion: $(x_c y_s - x_s y_c) = 0$, $b = 0$, $\left|\hat{r}_f\right| = \left|\hat{r}_b\right|$, $(\phi_y - \phi_x) = 0, \pi, -\pi$

The elliptical orbit is due to bearing/support asymmetric properties. For an isotropic system, the elliptical orbit motion becomes purely circular. Three special cases of elliptical orbit are studied below:

1. Forward Circular Precession

When $b = a$, or $\left|\hat{r}_b\right| = 0$ (i.e., $x_c = y_s$ and $y_c = -x_s$), the orbit is a purely forward circular. Forward circular whirl is the common motion of an isotropic system, since the unbalance excitation is a forward precessional force.

2. Backward Circular Precession

When $b = -a$, or $\left|\hat{r}_f\right| = 0$ (i.e., $x_c = -y_s$ and $y_c = x_s$), the orbit is a purely backward circular.

3. Straight Line Motion

When $b = 0$, or $\left|\hat{r}_f\right| = \left|\hat{r}_b\right|$ (i.e., $x_c y_s = x_s y_c$), the elliptical orbit degenerates to a straight-line path. Straight-line motion occurs when the motion changes the direction of precession, i.e., from forward to backward whirl or from backward to forward whirl as the rotor speed varies. For systems with damping, the straight line will not lie in the X or

Y axes, due to the damping effect, and the attitude angle will not be either 0 or $\pi/2$. Straight-line motion also occurs when the rotor is in a mixed precession. I.e., forward whirls in some portions of the rotor and backward whirls in other portions. At the rotor section where the change occurs, the rotor section moves along a straight line.

The elliptical orbit can be decomposed into two circular orbits. One is a forward circular motion (progression) and the other is a backward circular motion (regression). From Eqs. (2.3-1) and (2.3-2), the displacements can also be written as:

$$\begin{Bmatrix} x(t) \\ y(t) \end{Bmatrix} = \begin{Bmatrix} x_c \\ y_c \end{Bmatrix} \cos \omega t + \begin{Bmatrix} x_s \\ y_s \end{Bmatrix} \sin \omega t = \begin{Bmatrix} x_f(t) \\ y_f(t) \end{Bmatrix} + \begin{Bmatrix} x_b(t) \\ y_b(t) \end{Bmatrix} \qquad (2.3\text{-}21)$$

where

$$\begin{Bmatrix} x_f(t) \\ y_f(t) \end{Bmatrix} = \begin{Bmatrix} u_f \\ v_f \end{Bmatrix} \cos \omega t + \begin{Bmatrix} -v_f \\ u_f \end{Bmatrix} \sin \omega t \qquad (2.3\text{-}22)$$

$$\begin{Bmatrix} x_b(t) \\ y_b(t) \end{Bmatrix} = \begin{Bmatrix} u_b \\ v_b \end{Bmatrix} \cos \omega t + \begin{Bmatrix} v_b \\ -u_b \end{Bmatrix} \sin \omega t \qquad (2.3\text{-}23)$$

and

$$u_f = \frac{1}{2}(x_c + y_s) \qquad v_f = \frac{1}{2}(y_c - x_s) \qquad (2.3\text{-}24)$$

$$u_b = \frac{1}{2}(x_c - y_s) \qquad v_b = \frac{1}{2}(y_c + x_s) \qquad (2.3\text{-}25)$$

2.4 Orbit Simulator

DyRoBeS-Rotor provides two *Orbit Simulation Tools* under the main menu as shown below. *Elliptical Orbit Analysis* is for a single harmonic motion analysis and *Total Orbit Motion* is for the total steady state motion. These simulation tools allow you to visualize the whirling motion of the orbit and the spinning motion of the rotor. In additional to the following examples, readers are encouraged to experiment with this tool using other data.

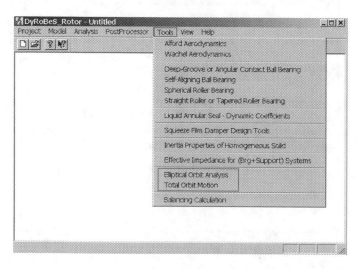

Example 2.1: Elliptical Orbit – Forward Precession

Enter $x_c = 0.5$, $x_s = -1$, and $y_c = 1.15$, $y_s = 2$ into the *Elliptical Orbit Analysis* Tool and click *Display*. **DyRoBeS-Rotor** calculates the elliptical orbit data and presents the results in the same window as shown in Figure 2.4-1. It displays the orbit in another graphic window, as illustrated in the graphic Figure 2.4-2.

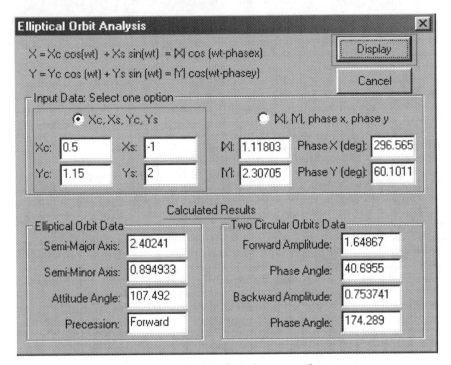

Figure 2.4-1 Orbit data input and output

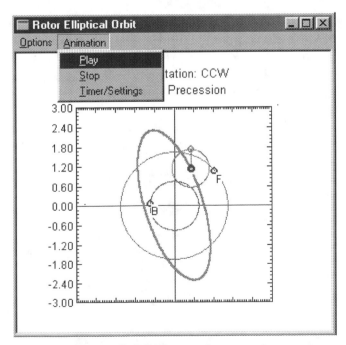

Figure 2.4-2 Forward precession

Readers are strongly encouraged to play the *Animation* feature, to see the spinning and whirling motions and also to adjust different *Settings* in the Options menu to see their effects. In ***DyRoBeS-Rotor***, the spinning direction is defined as counter-clockwise (CCW). Visualization is a very good tool in understanding the rotor motion. In this case, both spinning and whirling are in the same direction (counter-clockwise). Therefore, it is a forward precession. As expected, the circular forward amplitude (F) is greater than the circular backward amplitude (B) and the overall motion is forward. Note that the numerical results of the ellipse properties and of the two circular orbits are also confirmed by the graphic presentation.

Example 2.2: Elliptical Orbit – Backward Precession

Enter $x_c = 3$, $x_s = -1.25$, and $y_c = 1$, $y_s = -1.75$ into the *Elliptical Orbit Analysis* Tool and click *Display*. In this case, the whirling direction is CW and the spinning direction is always CCW in ***DyRoBeS-Rotor***. The rotor motion is a backward precession. The circular forward amplitude is smaller than the circular backward amplitude and the overall motion is a backward precession. The results are presented below:

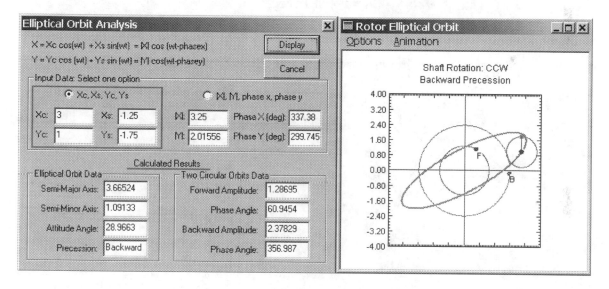

Figure 2.4-3 Backward precession

Again, readers are strongly encouraged to play the *Animation* feature, to see the spinning and whirling motions and also to adjust different *Settings* in the Options menu to see their effects. In this case, spinning and whirling are in the opposite directions. Therefore, it is a backward precession.

Example 2.3: Circular Orbit – Forward Precession

In this case, we have $x_c = y_s = 1$ and $y_c = -x_s = 1$. The motion is a purely circular forward precession. The circular backward amplitude equals zero.

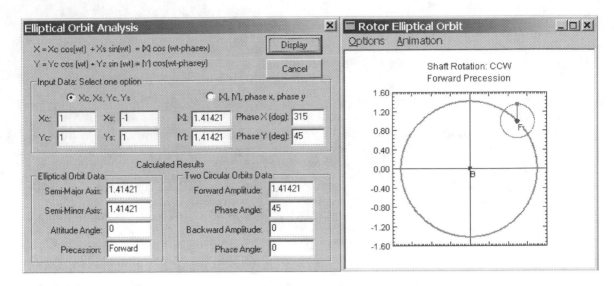

Figure 2.4-4 Circular forward precession

Example 2.4: Circular Orbit – Backward Precession

In this case, we have $x_c = -y_s = 1$ and $y_c = x_s = 1$. The motion is a purely circular backward precession. The circular forward amplitude equals zero.

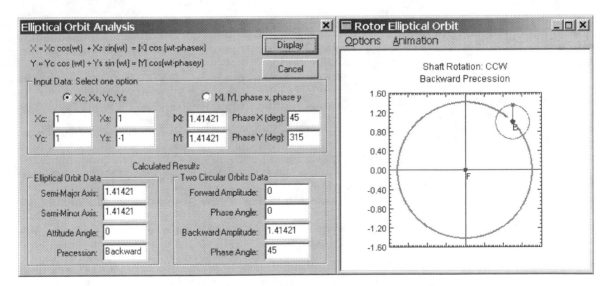

Figure 2.4-5 Circular backward precession

Example 2.5: Straight Line Motion

In this case, we have $b = 0$, $|\hat{r}_f| = |\hat{r}_b|$, $x_c y_s = x_s y_c$, the elliptical orbit degenerates to a straight-line path.

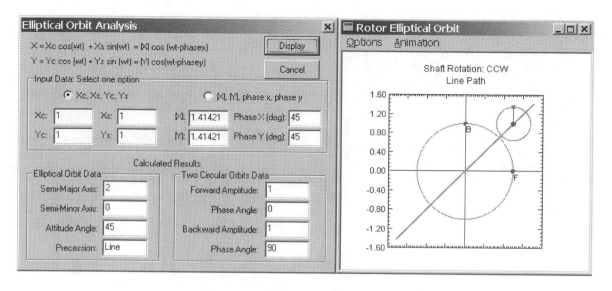

Figure 2.4-6 Straight line motion

Example 2.6: Total Motion – Forward Sub-Harmonic

Now, let us examine total motion by using the *Total Orbit Motion* option. Total steady-state motion is the sum of the static equilibrium position and of all the frequency components.

$$x(t) = x_0 + \sum_{i=1}^{\infty}\left(X_i \cos\left(\omega_i t - \phi_{x,i}\right)\right) = x_0 + \sum_{i=1}^{\infty}\left(x_{c,i}\cos(\omega_i t) + x_{s,i}\sin(\omega_i t)\right)$$

$$y(t) = y_0 + \sum_{i=1}^{\infty}\left(Y_i \cos\left(\omega_i t - \phi_{y,i}\right)\right) = y_0 + \sum_{i=1}^{\infty}\left(y_{c,i}\cos(\omega_i t) + y_{s,i}\sin(\omega_i t)\right)$$

In this example, total motion is a superposition of two forward circular orbits. Both orbits have an amplitude of 1.0, one with a whirl frequency $\omega=\Omega$, and the other with a frequency of $\omega/2$. The total motion is:

$$x(t) = X_1 \cos\left(\omega_1 t - \phi_{x,1}\right) + X_2 \cos\left(\omega_2 t - \phi_{x,2}\right)$$
$$y(t) = Y_1 \cos\left(\omega_1 t - \phi_{y,1}\right) + Y_2 \cos\left(\omega_2 t - \phi_{y,2}\right)$$

where

$$X_1 = Y_1 = X_2 = Y_2 = 1$$
$$\phi_{x,1} = \phi_{x,2} = 0$$
$$\phi_{y,1} = \phi_{y,2} = 90^0$$
$$\omega_1 = \Omega$$
$$\omega_2 = \omega_1 / 2$$

Enter the above data into the *Total Orbit Motion* as shown in the following data input and click *Display* to view the total motion. In this plot, the speed (frequency) ratios are more important than their absolute values. Therefore, the shaft speed is arbitrarily set to be 1.

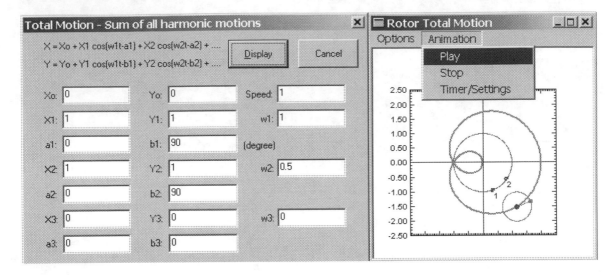

Figure 2.4-7 Sub-harmonic component is forward

Again, readers are strongly encouraged to use the *Animation* and *Settings* options to visualize the total motion. In the orbit display, "1" represents the first harmonics and "2" represents the second harmonics.

Example 2.7: Total Motion – Backward Sub-Harmonic

This example is similar to the previous example, with the exception that the sub-harmonic orbit is a backward orbit.

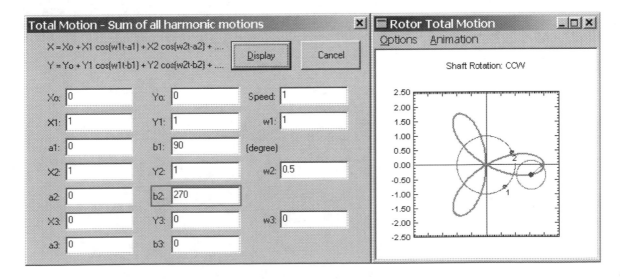

Figure 2.4-8 Sub-harmonic component is backward

Example 2.8: Total Motion – Two Forward Harmonics

In this case, $\omega_1 = \Omega$ and $\omega_2 = 2\Omega$. The input data and results are shown below:

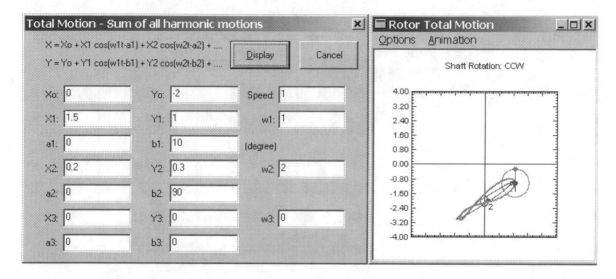

Figure 2.4-9 Two forward harmonic components

Shaft Rotation: CCW

Forward Precession

Shaft Rotation: CCW

Backward Precession

Shaft Rotation: CCW
Forward Precession

The Laval-Jeffcott Rotor Model **3**

3.1 The Two-Degrees-of-Freedom Rotor System

The fundamental dynamic characteristics of a rotor system can be studied and analyzed by using simple rotor models. Two types of rotor models are discussed in this chapter: one is the flexible rotor with rigid bearings and the other is the rigid rotor with flexible bearings. Two-degrees-of-freedom (2 DOF) systems are considered in these models and many assumptions made here are not practical and do not correspond with reality, but they can simplify the solution and allow for parametric studies be performed. This allows us to understand the effects of each parameter on rotor dynamics behaviors. They also provide many valuable physical insights into more complicated systems.

3.1.1 The Flexible Rotor with Rigid Bearings

A single disk centrally mounted on a uniform, flexible, and massless shaft, which is supported by two identical bearings, as illustrated in Figure 3.1-1, is most widely utilized by researchers to study and understand basic rotordynamics phenomena. If the bearings are infinitely stiff (rigid bearings), this model is normally referred to as the *Laval Rotor* in Europe, and *Jeffcott Rotor* in other parts of the world.

The Laval-Jeffcott Rotor

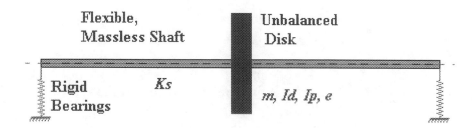

Figure 3.1-1 A simple Laval-Jeffcott rotor

For the centrally mounted disk, the system is symmetric and the first two fundamental translational and rotational motions are decoupled and can be considered separately, as shown in Figure 3.1-2. For the pure translational motion, as shown in Figure 3.1-2 (a), the disk has maximum translational displacement and zero rotation (slope). For the pure rotational motion, as shown in Figure 3.1-2 (b), the disk has maximum rotation (slope) and zero translational displacement.

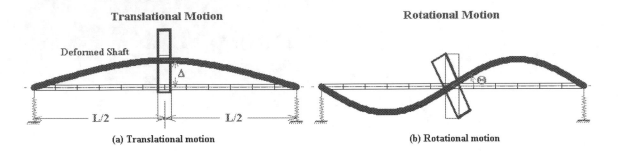

Figure 3.1-2 Two fundamental motions for a symmetric flexible rotor

Since the translational and rotational motions are decoupled, the translational and rotational stiffness of the elastic shaft can be obtained from the basic beam deflection equations. For the determination of translational shaft bending stiffness, consider a simply supported beam with a concentrated load at midspan (disk location). At midspan, the slope is zero and the defection due to the point load (F) is:

$$\Delta = \frac{FL^3}{48EI} \qquad (3.1\text{-}1)$$

From the linear force-displacement relationship, the shaft bending stiffness for the fundamental translational motion is:

$$k_T = \frac{48EI}{L^3} \qquad (3.1\text{-}2)$$

For the determination of rotational shaft bending stiffness, consider a simply supported beam with a moment load at midspan (disk location). At midspan, the deflection is zero and the slope due to the moment (M) load is:

$$\Theta = \frac{ML}{12EI} \qquad (3.1\text{-}3)$$

Again, the shaft rotation stiffness for the fundamental rotational motion is:

$$k_R = \frac{12EI}{L} \qquad (3.1\text{-}4)$$

For the Laval-Jeffcott rotor system, with flexible rotor and rigid bearings, the translational and rotational motions correspond to the first two bending modes.

3.1.2 The Rigid Rotor with Flexible Bearings

Another simple rotor model, commonly used to study the fundamental rotordynamics, is a symmetric rigid rotor on two identical flexible supports, as illustrated in Figure 3.1-3. The rotor is considered rigid and symmetric. Again, the translational and rotational motions at the center of mass are decoupled and can be studied separately, as shown in Figure 3.1-4.

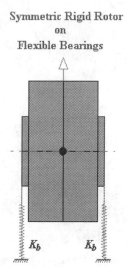

Figure 3.1-3 A symmetric rigid rotor with flexible bearings

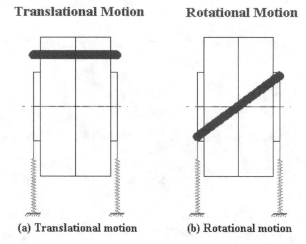

Figure 3.1-4 Two fundamental motions for a symmetric rigid rotor

Since the rotor is rigid and symmetric, and the two bearings are identical with a stiffness of K_b, the translational and rotational stiffnesses can be determined from the force and moment equations as:

$$k_T = 2k_b \qquad \text{Translational stiffness} \qquad (3.1\text{-}5)$$

$$k_R = \frac{1}{2}k_b L^2 \qquad \text{Rotational stiffness} \qquad (3.1\text{-}6)$$

For the symmetric rigid rotor with flexible bearings, the translational and rotational motions are commonly referred as two fundamental *rigid body modes*, i.e. *translatory* and *conical* modes.

3.2 Translational Motion

For a pure translational motion, consider a more generalized Laval-Jeffcott rotor system with flexible supports, as illustrated in Figure 3.2-1. When the bearings are flexible and each bearing has a stiffness of K_b, the equivalent stiffness of the system K, combining shaft stiffness K_s and bearing stiffness K_b is:

$$\frac{1}{K} = \frac{1}{K_s} + \frac{1}{2K_b} \quad \Rightarrow \quad K = \frac{2K_b K_s}{2K_b + K_s} = \frac{K_s}{1 + \dfrac{K_s}{2K_b}} = \frac{2K_b}{\dfrac{2K_b}{K_s} + 1} \tag{3.2-1}$$

It is evident from Eq. (3.2-1) that when the bearing stiffness is much larger than the shaft bending stiffness ($K_b \gg K_s$), the equivalent stiffness reduces to be K_s. This is a typical Laval-Jeffcott rotor system with elastic shaft and rigid supports. When the shaft bending stiffness is much larger than the bearing stiffness ($K_s \gg K_b$), the equivalent stiffness reduces to be $2K_b$ and it becomes a symmetric rigid rotor supported by flexible bearings.

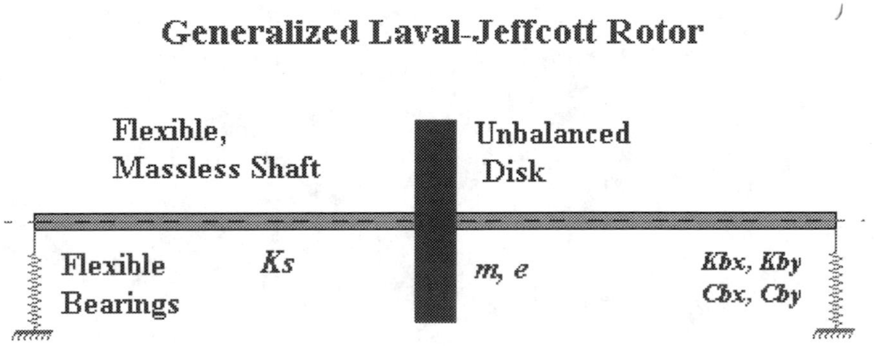

Generalized Laval-Jeffcott Rotor

Figure 3.2-1 A generalized Laval-Jeffcott rotor

When the stiffnesses and dampings in both X and Y directions are the same, the system is referred to as an *isotropic system*. In general, the equivalent stiffnesses (K_x, K_y) and viscous dampings (C_x, C_y) in both X and Y directions, are not the same due to the asymmetric properties of the bearings, even though the shaft is axisymmetric (isotropic). This system is thus referred to as *anisotropic system*.

Consider a generalized Laval-Jeffcott rotor system with equivalent support stiffnesses of K_x and K_y and associated viscous damping C_x and C_y in the X and Y direction, as illustrated in Figure 3.2-2. The disk has a mass of m and the center of gravity is offset from the shaft geometric center by an eccentricity of e. The motion at the disk center is described by two translational displacements (x, y), as illustrated in Figure 3.2-2.

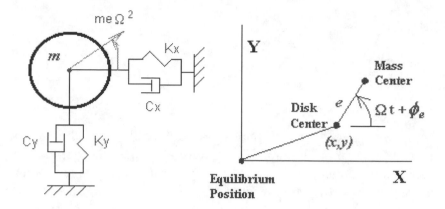

Figure 3.2-2 A two-degrees-of-freedom model

For the case of constant angular speed of rotation, Ω, the equations of motion for the mass center, can be derived from Newton's Laws of motion:

$$m\frac{d^2}{dt^2}\left(x + e\cos\left(\Omega t + \phi_e\right)\right) = -C_x\dot{x} - K_x x \tag{3.2-2}$$

$$m\frac{d^2}{dt^2}\left(y + e\sin\left(\Omega t + \phi_e\right)\right) = -C_y\dot{y} - K_y y \tag{3.2-3}$$

It can be rewritten as

$$m\ddot{x} + C_x\dot{x} + K_x x = me\Omega^2 \cos\left(\Omega t + \phi_e\right) \tag{3.2-4}$$

$$m\ddot{y} + C_y\dot{y} + K_y y = me\Omega^2 \sin\left(\Omega t + \phi_e\right) \tag{3.2-5}$$

where ϕ_e is the phase angle for the mass unbalance position. For single unbalance force, as in this case, ϕ_e can be set to zero without loss of generality. The equations of motion show that the motions in the X and Y directions are both dynamically (inertially) and statically (elastically) decoupled in this simple model. Therefore, they can be solved separately. Since there are no cross-coupling stiffnesses in this model, it is sometimes referred to as an *orthotropic system*. For the isotropic systems, the equations of motion for the x and y displacements are identical, except for the 90 degrees phase difference in unbalance excitations.

3.3 Natural Frequencies and Natural Modes

The undamped natural frequency, viscous damping factor (ratio), and damped natural frequency for each direction are:

$$\omega_{nx} = \sqrt{\frac{K_x}{m}}, \qquad \zeta_x = \frac{C_x}{2m\omega_{nx}}, \qquad \omega_x = \omega_{nx}\sqrt{1-\zeta_x^2}$$

$$\omega_{ny} = \sqrt{\frac{K_y}{m}}, \qquad \zeta_y = \frac{C_y}{2m\omega_{ny}}, \qquad \omega_y = \omega_{ny}\sqrt{1-\zeta_y^2}$$

$$(3.3\text{-}1)$$

For this simple rotor system with purely translational motion, the natural frequencies are independent of the rotor spin-speed. In rotordynamics, the natural frequencies are commonly referred to as the *whirl speeds*. For each natural frequency (eigenvalue), there is an associated modal orbit (eigenvector). Since the motions in both directions are decoupled, the modal orbit at each natural frequency degenerates into a straight line. Hence, the modal orbits (eigenvectors) for the natural frequencies are:

$$\omega = \omega_x : \quad x = |x|, \quad y = 0$$

$$\omega = \omega_y : \quad x = 0, \quad y = |y|$$

$$(3.3\text{-}2)$$

Following the discussion from Chapter 2, the straight line path on the other hand, can be considered as being made up of two circular orbits of equal amplitude and whirling in opposite directions, with the same frequency. Therefore, each natural frequency may be treated as two natural modes with the same whirl frequency, where one is a purely forward circular mode and the other is a purely backward circular mode.

When the excitation frequency of a periodic force applied to a rotor system coincides with a natural frequency of that system, the rotor system may be in a state of *resonance* (or critical condition). The most common excitation in rotating machinery is the unbalance excitation that is synchronized with the rotor spin speed. There are other synchronous excitations, such as excitations due to shaft bow and disk skew. Non-synchronous excitations include aerodynamic excitations for compressors, gear mesh excitation for geared systems, etc. In rotating machinery, the excitation frequencies are commonly related to the rotor spin speed with a constant multiple or fraction. Traditionally, when the rotor spin speed coincides with one of the natural frequencies, the spin speed is referred to as **critical speed**, since the unbalance excitation is the most common excitation. In a more general definition, when a rotor spin speed coincides with a constant multiple or fraction of one of the natural frequencies of the rotor system, then the spin speed is defined as a critical speed. That is, at the critical speed, one of the natural frequencies coincides with the excitation frequency.

To determine the damped critical speeds, a "*Whirl Speed Map*" is normally needed. A whirl speed map, also called "Campbell Diagram" or "Frequency Interference Diagram", is a plot of damped natural frequencies (whirl speeds) of the rotor system vs. the rotor spin speed. The damped critical speeds and any excitation resonance speeds are determined by noting the coincidence of the shaft speed with the system natural frequencies for a given excitation frequency line ($\omega_{exc} = \alpha\Omega$) in the whirl speed map. A value of one (1) in the excitation slope is associated with the synchronous excitation that is of main interest, because rotating unbalance is always present no matter how well the rotor is balanced. For the translational motion of this simple Laval-Jeffcott rotor system, the natural frequencies are independent of the spin speed. The whirl speed map with synchronous excitation is presented in Figure 3.3-1. It shows that the first critical speed is the first natural frequency and the second critical speed is the second natural frequency in this case.

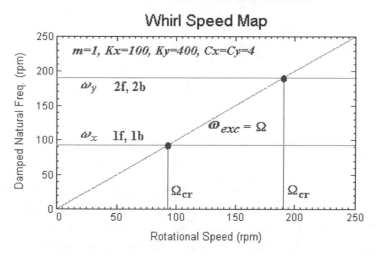

Figure 3.3-1 Whirl speed map for a 2DOF system

The modal motions at X and Y directions are decoupled in this case. There is one natural frequency for each direction of motion. Also, for each natural frequency, the associated modal orbit is a straight line in each direction. However, a straight line path can be considered as being made up of two circular orbits of equal amplitude and whirling in opposite directions, with the same natural frequency. Therefore, each natural frequency may be treated as two natural modes with the same whirl speed, where one is a purely forward (f) circular mode and the other is a purely backward (b) circular mode, as noted in Figure 3.3-1 by (1f, 1b) and (2f, 2b). Since the natural frequencies are independent of the spin speed for this simple rotor system, the natural frequencies are the critical speeds for synchronous unbalance excitation.

3.4 Steady State Response to Unbalance

Following the discussion in Chapter 1, the steady-state responses, due to mass unbalance excitation are:

$$x(t) = x_c \cos \Omega t + x_s \sin \Omega t = |x| \cos(\Omega t - \phi_x) \qquad (3.4\text{-}1)$$

$$y(t) = y_c \cos \Omega t + y_s \sin \Omega t = |y| \cos(\Omega t - \phi_y) \qquad (3.4\text{-}2)$$

where

$$x_c = \frac{\left(K_x - \Omega^2 m\right)\left(me\Omega^2\right)}{\left(K_x - \Omega^2 m\right)^2 + \left(\Omega C_x\right)^2} \; , \qquad x_s = \frac{\left(\Omega C_x\right)\left(me\Omega^2\right)}{\left(K_x - \Omega^2 m\right)^2 + \left(\Omega C_x\right)^2} \qquad (3.4\text{-}3)$$

$$|x| = \frac{me\Omega^2}{\left[\left(K_x - \Omega^2 m\right)^2 + \left(\Omega C_x\right)^2\right]^{1/2}} \quad , \quad \phi_x = \arctan\left(\frac{\Omega C_x}{K_x - \Omega^2 m}\right) \qquad (3.4\text{-}4)$$

and

$$y_c = \frac{\left(-\Omega C_y\right)\left(me\Omega^2\right)}{\left(K_y - \Omega^2 m\right)^2 + \left(\Omega C_y\right)^2} \quad , \qquad y_s = \frac{\left(K_y - \Omega^2 m\right)\left(me\Omega^2\right)}{\left(K_y - \Omega^2 m\right)^2 + \left(\Omega C_y\right)^2} \qquad (3.4\text{-}5)$$

$$|y| = \frac{me\Omega^2}{\left[\left(K_y - \Omega^2 m\right)^2 + \left(\Omega C_y\right)^2\right]^{1/2}} \quad , \quad \phi_y = \arctan\left(\frac{K_y - \Omega^2 m}{-\Omega C_y}\right) \qquad (3.4\text{-}6)$$

Note that x and y displacements oscillate at the same frequency, Ω (same as rotor spin speed) with different amplitudes and phase angles. Following the discussion in Chapter 2, the steady-state unbalance response of Eqs. (3.4-1) and (3.4-2) define an elliptical motion. The ellipse properties, such as semi-major (a), semi-minor axes (b), and attitude angle (ϕ_a), may be calculated from equations given in Chapter 2. The direction of precession (whirling) of the steady state orbit, is determined by the sign of the rate of precession ($\dot{\phi}$):

$$\begin{aligned} sign(\dot{\phi}) &= sign(x_c y_s - x_s y_c) \\ &= sign\left[\left(K_x - \Omega^2 m\right)\left(K_y - \Omega^2 m\right) + \left(\Omega^2 C_x C_y\right)\right] \\ &= sign\left[\left(\omega_{nx}^2 - \Omega^2\right)\left(\omega_{ny}^2 - \Omega^2\right) + \left(\Omega^2 4\zeta_x \zeta_y \omega_{nx} \omega_{ny}\right)\right] \end{aligned} \qquad (3.4\text{-}7)$$

That is,

$$\left(\omega_{nx}^2 - \Omega^2\right)\left(\omega_{ny}^2 - \Omega^2\right) + \left(\Omega^2 4\zeta_x \zeta_y \omega_{nx} \omega_{ny}\right) > 0 \quad \text{Forward Precession} \qquad (3.4\text{-}8a)$$

$$\left(\omega_{nx}^2 - \Omega^2\right)\left(\omega_{ny}^2 - \Omega^2\right) + \left(\Omega^2 4\zeta_x \zeta_y \omega_{nx} \omega_{ny}\right) < 0 \quad \text{Backward Precession} \qquad (3.4\text{-}8b)$$

$$\left(\omega_{nx}^2 - \Omega^2\right)\left(\omega_{ny}^2 - \Omega^2\right) + \left(\Omega^2 4\zeta_x \zeta_y \omega_{nx} \omega_{ny}\right) = 0 \quad \text{Straight Line Motion} \qquad (3.4\text{-}8c)$$

From the above equations, steady-state unbalance orbit motion is a forward precession when the rotor speed is below the first resonance frequency (critical speed) and above the second resonance frequency (critical speed). The orbit motion is a backward precession between the two resonance frequencies (critical speeds) if damping does not exist. However, the backward whirl zone decreases as the damping increases, and when there is enough damping, the backward whirl does not occur at all, as illustrated in Eq. (3.4-8). Since for the isotropic systems, there is only one resonance, backward whirl does not exist. For anisotropic systems, backward whirl may exist, depending on the amount of damping present in the systems.

To gain some insights into orbit motion, an undamped orthotropic system ($C_x=C_y=0$) with $K_x < K_y$ is used to illustrate orbit behavior. There are two distinct resonance frequencies (critical speeds):

$$\Omega_{cr,1} = \omega_x \ < \ \Omega_{cr,2} = \omega_y \tag{3.4-9}$$

From previous equations for the steady state response, we have:

$$x_c = \frac{me\Omega^2}{K_x - \Omega^2 m} = \frac{e\Omega^2}{\omega_x^2 - \Omega^2} \quad , \quad x_s = 0 \tag{3.4-10}$$

$$y_s = \frac{me\Omega^2}{K_y - \Omega^2 m} = \frac{e\Omega^2}{\omega_y^2 - \Omega^2} \quad , \quad y_c = 0 \tag{3.4-11}$$

and

$$sign(x_c y_s - x_s y_c) = sign\left[\left(\omega_x^2 - \Omega^2\right)\left(\omega_y^2 - \Omega^2\right)\right] \tag{3.4-12}$$

Thus, the steady-state response becomes:

$$x(t) = x_c \cos\Omega t = \frac{e\Omega^2}{\omega_x^2 - \Omega^2}\cos\Omega t \tag{3.4-13}$$

$$y(t) = y_s \sin\Omega t = \frac{e\Omega^2}{\omega_y^2 - \Omega^2}\sin\Omega t \tag{3.4-14}$$

The above expressions satisfy the following relationship:

$$\left(\frac{x(t)}{x_c}\right)^2 + \left(\frac{y(t)}{y_s}\right)^2 = 1 \tag{3.4-15}$$

As expected, the above equation defines an elliptical orbit. The steady-state unbalance response for this undamped orthotropic 2DOF system is presented in Figure 3.4-1. The pertinent parameters are listed in the plot for reference. The orbit motions for various rotor speeds are analyzed and summarized below:

Case 1: $\Omega < \omega_x$ below the first critical speed

$$x_c > 0, \quad y_s > 0, \quad x_c > y_s$$
$$a = x_c, \quad b = y_s, \quad \phi_a = 0, \quad \dot{\phi} > 0$$

The motion is a forward elliptical orbit where the semi-major axis is aligned with the X axis.

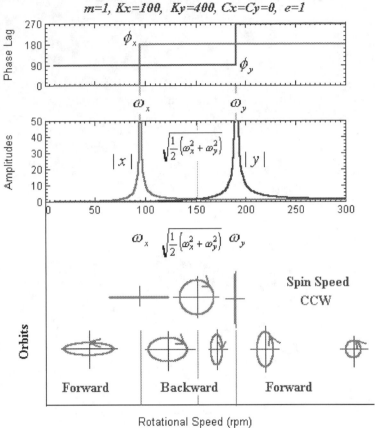

Figure 3.4-1 The steady state unbalance response and orbit analysis

Case 2: $\Omega = \omega_x$ on the first critical speed (at resonance)

$$x_c = \infty, \quad y_s > 0, \quad x_c > y_s$$
$$a = \infty, \quad b = y_s, \quad \phi_a = 0, \quad \dot{\phi} = 0$$

The motion is an unbounded straight line (resonance) aligned with the X axis. In practice, the amplitude is limited by the damping and geometric constraints.

Case 3: $\omega_x < \Omega < \sqrt{\dfrac{1}{2}\left(\omega_x^2 + \omega_y^2\right)}$ between the critical speeds

$$x_c < 0, \quad y_s > 0, \quad |x_c| > y_s$$
$$a = |x_c|, \quad b = -y_s, \quad \phi_a = 0, \quad \dot{\phi} < 0$$

The motion is a backward elliptical orbit where the semi-major axis is aligned with the X axis.

Case 4: $\Omega = \sqrt{\dfrac{1}{2}\left(\omega_x^2 + \omega_y^2\right)}$ between the critical speeds

$$x_c < 0, \quad y_s > 0, \quad |x_c| = y_s$$
$$a = |x_c|, \quad b = -a, \quad x_c = -y_s, \quad \phi_a = 0, \quad \dot{\phi} < 0$$

The motion is a purely backward **circular** orbit.

Case 5: $\sqrt{\dfrac{1}{2}\left(\omega_x^2 + \omega_y^2\right)} < \Omega < \omega_y$ between the critical speeds

$$x_c < 0, \quad y_s > 0, \quad |x_c| < y_s$$
$$a = y_s, \quad b = x_c, \quad \phi_a = 90, \quad \dot{\phi} < 0$$

The motion is a backward elliptical orbit where the semi-major axis is aligned with the Y axis.

Case 6: $\Omega = \omega_y$ on the second critical speed (at resonance)

$$x_c < 0, \quad y_s = \infty, \quad |x_c| < y_s$$
$$a = \infty, \quad b = x_c, \quad \phi_a = 90, \quad \dot{\phi} = 0$$

The motion is an unbounded straight line (resonance) aligned with the Y axis. In practice, the amplitude is limited by the damping and geometric constraints.

Case 7: $\Omega > \omega_y$ above the second critical speed

$$a = |y_s|, \quad b = x_c, \quad \phi_a = 90, \quad \dot{\phi} > 0$$

The motion is a forward elliptical orbit where the semi-major axis is aligned with the Y axis.

Case 8: $\Omega = \infty$ far above the second critical speed

$$a = b = |x_c| = |y_s| = e, \quad x_c = y_s = -e, \quad \phi_a = 0, \quad \dot{\phi} > 0$$

The motion is a purely forward **circular** orbit where the radius is equal to the mass eccentricity.

It is evident from Figure 3.4-1 that:

1. There are two distinct natural frequencies (critical speeds), ω_x and ω_y for X and Y directions, respectively. The motions become infinite straight lines when the rotor spin speeds are at critical speeds $(\Omega = \omega_x,\ \Omega = \omega_y)$.

2. Phase angle in the X displacement, ϕ_x, is zero before the first critical speed, ω_x, and becomes 180 degrees after the first critical speed. The phase angle changes from zero to 180 degrees at critical speed. Phase angle in the Y displacement, ϕ_y, behaves the same, except it has a 90 degrees phase lag due to the 90 degrees lag in the unbalance excitation.

3. The direction of precession can be easily identified by using the phase angle relationship when the phase angles are available:

When $\Omega < \omega_x$, $\pi > (\phi_y - \phi_x) > 0$ Forward precession

When $\omega_x < \Omega < \omega_y$, $\pi > (\phi_x - \phi_y) > 0$ Backward precession

When $\Omega = \omega_x,\ \Omega = \omega_y$, $\phi_y = \phi_x$ Straight line

For the isotropic systems ($K_x = K_y$ and $C_x = C_y$), the natural frequencies are the same for both directions. From previous equations, we have:

$$\omega_n = \omega_{nx} = \omega_{ny}$$

$$x_c = y_s \quad \text{and} \quad y_c = -x_s$$

and

$$b = a \qquad\qquad\qquad \text{(circular orbit)}$$

$$sign(x_c y_s - x_s y_c) = sign(x_c^2 + y_c^2) > 0 \qquad \text{(forward whirl)}$$

Therefore, for the *isotropic* systems the steady-state unbalance response orbit motion is a purely *forward circular* whirl, with the whirl speed equal to the shaft spinning speed. That is, the steady-state unbalance response is a *forward synchronous circular motion* for the isotropic systems without any backward whirl.

Complex Notation

It is sometimes desirable and convenient to analyze the equations of motion and solutions by using complex displacements. The harmonic excitation can be represented by complex vector, and the response orbit motion can be represented by a combination of the forward and backward circular precessions of the motion. The equations of motion for a simple 2DOF Laval-Jeffcott rotor system in complex notation becomes

$$m\ddot{\hat{r}} + C_p \dot{\hat{r}} + C_m \dot{\hat{r}}^* + K_p \hat{r} + K_m \hat{r}^* = me\Omega^2 e^{j\Omega t} \qquad (3.4\text{-}16)$$

where

$$\hat{r}(t) = x(t) + j\, y(t)$$
$$= \hat{r}_f e^{j\Omega t} + \hat{r}_b e^{-j\Omega t} \qquad (3.4\text{-}17)$$

and the complex conjugate vector is

$$\hat{r}^*(t) = x(t) - j\, y(t)$$
$$= \hat{r}_f^{\,*} e^{-j\Omega t} + \hat{r}_b^{\,*} e^{j\Omega t} \qquad (3.4\text{-}18)$$

also,

$$C_p = \frac{1}{2}\left(C_x + C_y\right), \qquad C_m = \frac{1}{2}\left(C_x - C_y\right) \qquad (3.4\text{-}19)$$

$$K_p = \frac{1}{2}\left(K_x + K_y\right), \qquad K_m = \frac{1}{2}\left(K_x - K_y\right) \qquad (3.4\text{-}20)$$

Differentiating Equations (3.4-17) and (3.4-18), and substituting into Equation (3.4-16), yields:

$$\hat{r}_f = \frac{1}{\Delta}\left(K_p - m\Omega^2 + j\Omega C_p\right)me\Omega^2 \qquad (3.4\text{-}21)$$

$$\hat{r}_b^{\,*} = \frac{-1}{\Delta}\left(K_m + j\Omega C_m\right)me\Omega^2 \qquad (3.4\text{-}22)$$

where

$$\Delta = \left(K_p - m\Omega^2 + j\Omega C_p\right)^2 - \left(K_m + j\Omega C_m\right)^2 \qquad (3.4\text{-}23)$$

For isotropic systems, $K_m = 0, C_m = 0$, we then have $\hat{r}_b = \hat{r}_b^{\,*} = 0$. The response orbit is a purely forward circular motion.

For anisotropic undamped systems, we have:

$$\hat{r}_f = \frac{\left(\frac{1}{2}\left(\omega_x^2 + \omega_y^2\right) - \Omega^2\right)e\Omega^2}{\left(\omega_x^2 - \Omega^2\right)\left(\omega_y^2 - \Omega^2\right)} \qquad (3.4\text{-}24)$$

$$\hat{r}_b = \frac{\frac{-1}{2}\left(\omega_x^2 - \omega_y^2\right)e\Omega^2}{\left(\omega_x^2 - \Omega^2\right)\left(\omega_x^2 - \Omega^2\right)} \qquad (3.4\text{-}25)$$

From the above equations, the response can be divided into several spinning speed zones:

$$\Omega < \omega_x \quad \text{and} \quad \Omega > \omega_y \quad \Rightarrow |\hat{r}_f| > |\hat{r}_b| \qquad \text{Forward elliptical whirl}$$

$$\omega_x < \Omega < \omega_y \qquad\qquad \Rightarrow |\hat{r}_f| < |\hat{r}_b| \qquad \text{Backward elliptical whirl}$$

$$\Omega = \sqrt{\frac{1}{2}\left(\omega_x^2 + \omega_y^2\right)} \qquad \Rightarrow |\hat{r}_f| = 0 \qquad \text{Purely backward circular whirl}$$

These conclusions are in agreement with the previous discussion.

Example 3.1: 2DOF System – Steady State Unbalance Response

Before we move on to more practical and general examples, a generalized 2DOF orthotropic system, subject to unbalance excitation, is presented in this example. The pertinent parameters are listed below:

$$m = 1, \ K_x = 100, \ K_y = 400, \ C_x = C_y = C = 4, \ e = 1$$

The equations of motion are:

$$m\ddot{x} + C_x \dot{x} + K_x x = me\Omega^2 \cos(\Omega t)$$

$$m\ddot{y} + C_y \dot{y} + K_y y = me\Omega^2 \sin(\Omega t) = me\Omega^2 \cos\left(\Omega t - \frac{\pi}{2}\right)$$

There are several ways to model this simple 2DOF system in **DyRoBeS**. Since we are going to perform some parametric studies, the simplest way to construct the rotor system is demonstrated in this example. The rotor system contains a dummy shaft element with a zero mass density and a zero elastic modulus, as shown in Figure 3.4-2. The rotational degrees-of-freedom at station 1, and all four degrees-of-freedom (translational and rotational) at station 2 are constrained. A concentrated mass with an unbalance and a bearing with translational properties, are placed at station 1. They are the only active components in this case.

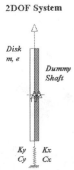

Figure 3.4-2 2DOF model

Formulas for the basic parameters and results have been presented before and are listed below for reference.

$$\omega_n = \sqrt{\frac{K}{m}} \qquad \text{Undamped natural frequency}$$

$$\xi = \frac{C}{2m\omega_n} \qquad \text{Damping factor}$$

$$\omega_d = \omega_n\sqrt{1-\xi^2} \qquad \text{Damped natural frequency}$$

The maximum steady-state response due to mass unbalance occurs at

$$\Omega_{peak} = \frac{\omega_n}{\sqrt{1-2\xi^2}} \qquad \text{with amplitude} = \frac{e}{2\xi\sqrt{1-\xi^2}}$$

The numerical results are summarized below:

Numerical results	X direction	Y direction
Undamped Natural Frequency (R/S)	10	20
Undamped Natural Frequency (RPM)	95	191
For C=4		
Damping Factor (Zeta)	0.2	0.10
Damped Natural Frequency (RPM)	94	190
Max. Amplitude at Speed (RPM)	100	193
Max. Amplitude	2.552	5.025

The whirl speed map for this simple system was shown in Figure 3.3-1. The steady-state unbalance response for speeds from 0 to 300 rpm with an increment of 1 rpm was analyzed using **DyRoBeS**. The Bode Plot for the x and y displacements are presented in Figure 3.4-3.

As expected, there is a peak response at each direction due to the bearing asymmetry. The maximum amplitudes and speeds calculated using **DyRoBeS**, are in agreement with the analytical solutions. The y displacement phase angle (ϕ_y) equals the x displacement phase angle (ϕ_x) at the rotor speed of 98 rpm that is greater than the first undamped natural frequency of 95 rpm, and at the rotor speed of 186 rpm that is less than the second undamped natural frequency of 191 rpm. That is:

$$\pi > \left(\phi_y - \phi_x\right) > 0 \quad \text{when } \Omega < 98\,\text{rpm and } \Omega > 186\,\text{rpm}$$

$$\pi > \left(\phi_x - \phi_y\right) > 0 \quad \text{when } 98\,\text{rpm} < \Omega < 186\,\text{rpm}$$

$$\phi_x = \phi_y \qquad\qquad \text{when } \Omega = 98\,\text{rpm and } \Omega = 186\,\text{rpm}$$

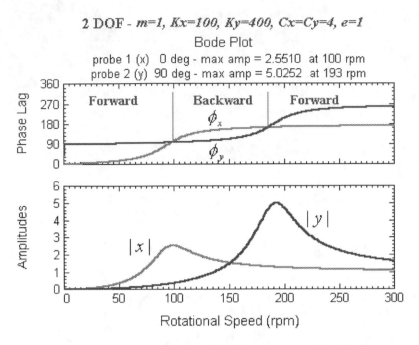

Figure 3.4-3 Bode plot for 2DOF system with unbalance

It indicates that the rotor whirls in forward precession when the rotor speed is below 98 rpm and oscillates along a straight line at 98 rpm. Then it reverses the direction of precession and becomes a backward precession until the rotor speed reaches 186 rpm, where the rotor oscillates along a straight line again. When the rotor speed is above 186 rpm, the direction of precession is reversed from a backward to a forward precession. This phenomenon can be easily observed from the phase angle data shown in the Bode plot and also from the major and minor axes of the elliptical orbit plot, as shown in Figure 3.4-4. The negative semi-minor axis indicates the backward precession. The orbit changes its direction of precession by going through the straight-line motion.

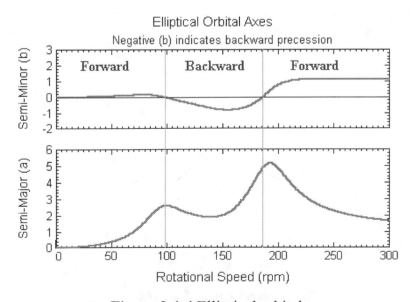

Figure 3.4-4 Elliptical orbital axes

Several orbits at different rotor speeds are shown in Figure 3.4-5. For systems with damping, the straight-line motions will not be aligned in the X or Y axes, due to the damping effect.

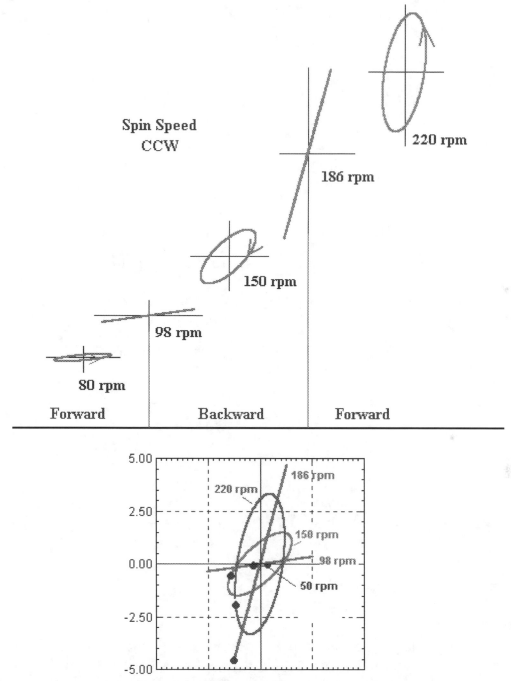

Figure 3.4-5 Steady state rotor orbits at different rotor speeds

This phenomenon can also easily be observed by utilizing the *Displacement Orbit Animation* feature provided by **DyRoBeS**. Readers are encouraged to run the *Startup/Shutdown Animation*, to visualize the orbit changing sizes and directions as the rotor speed increases or decreases, as illustrated in Figure 3.4-6.

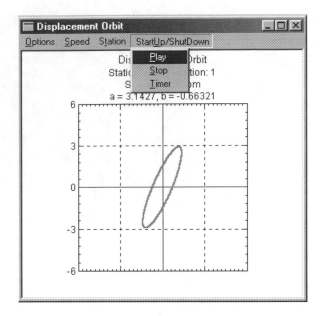

Figure 3.4-6 Animation for the orbits startup and shutdown

Typically, the vibration probes (displacement pick-up) are not in the X and Y directions, as illustrated in Figure 3.4-7. The Bode plot at the probes locations can be quite different from the x and y displacements. Figure 3.4-8 is the Bode plot for probes located at 45 and 135 degrees, measured from the X axis, respectively.

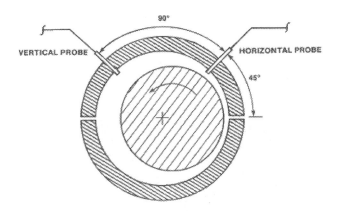

Figure 3.4-7 Two vibration probes

Unlike the Bode plot for the *x* and *y* displacement in Figure 3.4-3, there are two peak responses at each probe in Figure 3.4-8. Each peak corresponds to a resonance condition: one for each natural frequency. One can still identify the forward/backward precessions and straight-line motion by the difference in the phase angles.

Forward precession: $\pi > (\phi_y - \phi_x) > 0$, or $3\pi > (\phi_y - \phi_x) > 2\pi$, and so on.

Backward precession: $2\pi > (\phi_y - \phi_x) > \pi$, or $0 > (\phi_y - \phi_x) > -\pi$, and so on.

Straight-Line Motion: $(\phi_y - \phi_x) = 0, \pi, 2\pi, 3\pi$, and so on.

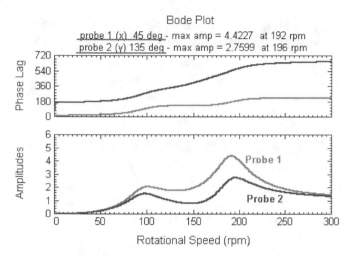

Figure 3.4-8 Bode plot for two probes

It is common in many Industry Standards that utilize the peak amplitude measurements at the vibration probes, to locate the critical speeds instead of using the natural frequencies to define the critical speeds.   The above figure shows that for different probe angles, the result can be significantly different.   Caution must be taken when utilizing the response measurements to define critical speeds.

It is also noted that for this damped system, the range between two transition speeds where the straight-line motion occurred, is smaller than those in the undamped system.   It indicates that the speed range for the backward precession gets smaller as damping increases.   In this example, when $C=Cx=Cy=10$ ($\xi_x = 0.5$ and $\xi_y = 0.25$), there is only one speed (135 rpm), which has a straight-line motion.   When C>10, there is no straight-line motion and backward precession does not exist.   The major and minor axes of the elliptical orbits for various damping are plotted below:

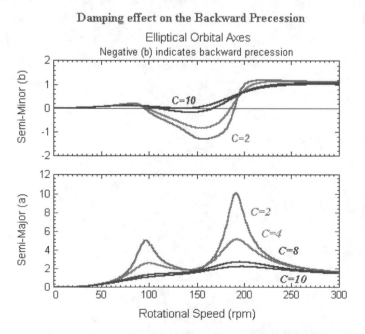

Figure 3.4-9 Elliptical orbital axes for various damping levels

Since this is a simple 2DOF model, the speeds for the straight-line motion can be obtained mathematically from the equations derived before. For the orbit to be a straight-line motion, we have:

$$b = 0, \quad \text{or} \quad \phi_x = \phi_y \tag{3.4-26}$$

By substitution phase equations (3.4-4) and (3.4-6) into Equation (3.4-26), yields

$$\tan^{-1}\left(\frac{C_x\Omega}{K_x - M\Omega^2}\right) = \tan^{-1}\left(\frac{C_y\Omega}{K_y - M\Omega^2}\right) + \frac{\pi}{2} \tag{3.4-27}$$

By using the following relationships:

$$\tan^{-1}\alpha = \frac{\pi}{2} - \tan^{-1}\frac{1}{\alpha} \tag{3.4-28}$$

and

$$\tan^{-1}\alpha = -\tan(-\alpha) \tag{3.4-29}$$

We have

$$M^2\Omega^4 + \left(C_x C_y - M\left(K_x + K_y\right)\right)\Omega^2 + K_x K_y = 0 \tag{3.4-30}$$

This draws the same conclusion as presented before, by using the semi-minor axis magnitude equation or the rate of precession. For a given simple Laval-Jeffcott rotor system as presented in this example, the above equation can be used to determine the speeds where the straight-line motion occurs. The speeds where straight-line motions occurred, by using the above equation, can be graphed versus damping ($C=Cx=Cy$) in Figure 3.4-10.

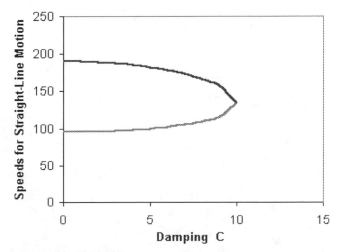

Figure 3.4-10 Speeds for straight-line motion vs. damping

However, for the above equation to have real roots, the following equation must be satisfied:

$$\left(C_x C_y - M\left(K_x + K_y\right)\right)^2 - 4M^2 K_x K_y \geq 0 \qquad (3.4\text{-}31)$$

The equal sign can also be used to determine if the change in the direction of precession exists or not. The results are identical to those obtained from the steady-state response analysis, by varying the damping for every analysis as shown in Figure 3.4-9. It is evident from Eq. (3.4-31) that for the undamped system, the two straight-line motions occur at the two undamped natural frequencies (95, 191 rpm). As the damping increases, these two speeds approach each other with different approaching rates. When damping reaches a certain value (C=10 in this case), the backward precession does not exist anymore, thus there will be no straight-line motion.

The amount of damping required to eliminate the backward precession is also dependent upon the bearing asymmetry. Figure 3.4-11 shows the damping required to eliminate the backward precession for a range of bearing asymmetry.

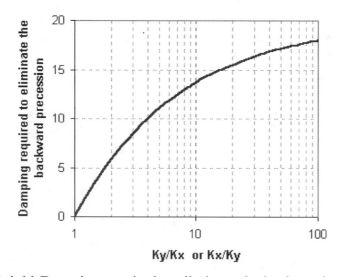

Figure 3.4-11 Damping required to eliminate the backward precession

It shows that for an isotropic system (Ky/Kx=1), the displacement orbit is forward circular and no straight-line motion exists. As the bearing asymmetry increases, more damping is required to eliminate the backward precession.

3.5 Steady State Response to Shaft Bow

The residual shaft bow may be present in the rotor-bearing systems due to many various reasons, including assembly tolerances and uneven thermal distribution. When the residual shaft bow exists in a rotor system, a constant magnitude rotating force synchronized with the shaft spin speed acts on the rotor system. The excitation force caused by the residual shaft bow is of the form:

$$F = K_s q_b \qquad\qquad (3.5\text{-}1)$$

where K_s is the shaft bending stiffness and q_b is the amount of shaft initial displacements in fixed reference frame, due to residual shaft bow. The shaft bow rotates with the rotating reference and is specified in the rotating reference frame. By utilizing the coordinate transformation between fixed and rotating reference frames, presented in Chapter 2, the synchronous excitation force due to the residual shaft bow in the fixed reference frame becomes:

$$\begin{Bmatrix} F_x \\ F_y \end{Bmatrix} = K_s \left(\begin{Bmatrix} x' \\ y' \end{Bmatrix} \cos\Omega t + \begin{Bmatrix} -y' \\ x' \end{Bmatrix} \sin\Omega t \right) \qquad\qquad (3.5\text{-}2)$$

The synchronous excitation force due to residual shaft bow is similar to the synchronous excitation due to mass unbalance. However, the excitation amplitude of a shaft bow is a constant, while the excitation amplitude of an unbalance is a function of square of spin speed.

Example 3.2: Laval-Jeffcott Rotor– Residual Shaft Bow and Unbalance Response

A Laval-Jeffcott rotor system supported by rigid bearings is presented in this example to demonstrate the steady-state response due to shaft bow, mass imbalance, and their combined effects. As shown in the following Figure 3.5-1, a rigid disk is located at the midspan of a flexible shaft with residual shaft bow. Since only the fundamental translational motion is considered, the rigid disk has a mass of 4 Lbm (0.01036 Lbf-s^2/in) and zero mass moment of inertia (I_d=I_p=0). The flexible shaft with elastic modulus of 3.0E07 psi has a length of 12 inches and a diameter of 0.275 inches. The shaft mass is negligible in this case, otherwise, the shaft modal mass ($0.4857\rho AL$) needs to be added into the disk to form an effective (modal) mass for this translational motion, as demonstrated in Chapter 1. A linear isotropic viscous damping with a value of C=0.3114 Lbf-s/in, is applied at the disk location. For demonstrative purposes, the pertinent parameters for this Laval-Jeffcott rotor are summarized below:

L = 12.0 in (total length)
D = 0.275 in (shaft diameter)
E = 3.0E07 psi (Young's modulus)
M = 4.0 Lbm (0.01036 Lbf-s^2/in) (disk mass)
C = 0.3114 Lbf-s/in (viscous damping at disk)
e = 0.001 in (mass eccentricity)
q_b= 0.001 in (shaft bow at disk)

The translational stiffness at the midspan of the shaft due to bending is:

$$K = \frac{48EI}{L^3} = 233.95 \,\text{Lbf/in} \quad \text{where} \quad I = \frac{\pi D^4}{64}$$

The Laval-Jeffcott Rotor

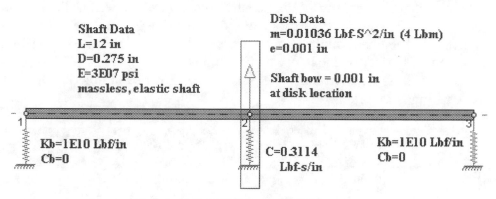

Figure 3.5-1 The Laval-Jeffcott rotor

This simple 2DOF system can be modeled by using geometric constraints, that is, constrain all the degrees-of-freedom at stations 1 and 3, and rotational degrees-of-freedom at station 2, as demonstrated in the earlier examples. However, in this example a bearing stiffness of 1E10 Lbf/in is applied at both ends, instead of constraints as used in earlier examples. This bearing stiffness is much larger than the shaft bending stiffness. Therefore, they act like rigid bearings. Without loss of generality, the residual bow is assumed to be:

$$x' = q_b, \quad \text{and} \quad y' = 0$$

The above assumption is similar to assuming the phase angle of unbalance eccentricity be zero in the unbalance response analysis, which was presented in the previous sections. However, the phase angle of the unbalance force in this example is a parameter that will be studied. The steps used to build this 2-elements (3 stations) rotor system, are listed below for reference:

1. Select *Consistent Units*, and input descriptions under the *Units/Description*.
2. Enter the material properties under the *Material*. Note that a small density value is used to avoid the numerical singularity. Technically speaking, this 3-stations rotor system has 12 degrees-of-freedom. A small positive density does not affect the fundamental translational motion, however, it will ensure that the mass matrix is positive definite and a large separation between the fundamental translational mode and the second rotational mode. Therefore, the fundamental translational motion can be studied without any influence from the higher frequency modes. The shear modulus is entered here, however, the shear deformation effect can be neglected in the analysis.
3. Enter shaft data, 2 elements with equal length, under the *Shaft Elements*. The equal length is important to ensure the disk will be centrally located on the shaft.
4. Input disk data under the *Disks*. Note that the disk is located at station 2.
5. Input 3 bearings data under the *Bearings*. The first two bearings at both ends of the shaft (stations 1 and 3) are identical with very large translational stiffness

compared to the shaft bending stiffness. The third bearing at station 2 only has viscous damping.

6. Enter unbalance moment (*me*) under the *Unbalance*. Note that the unbalance at disk (station 2) can be applied at the left side of element 2, or the right side of element 1. The phase angle of this unbalance force will be varied to demonstrate the combined effect of unbalance and shaft bow.

7. Enter residual shaft bow under the *Shaft Bow*. For this simple 2DOF system no curve fitting is required.

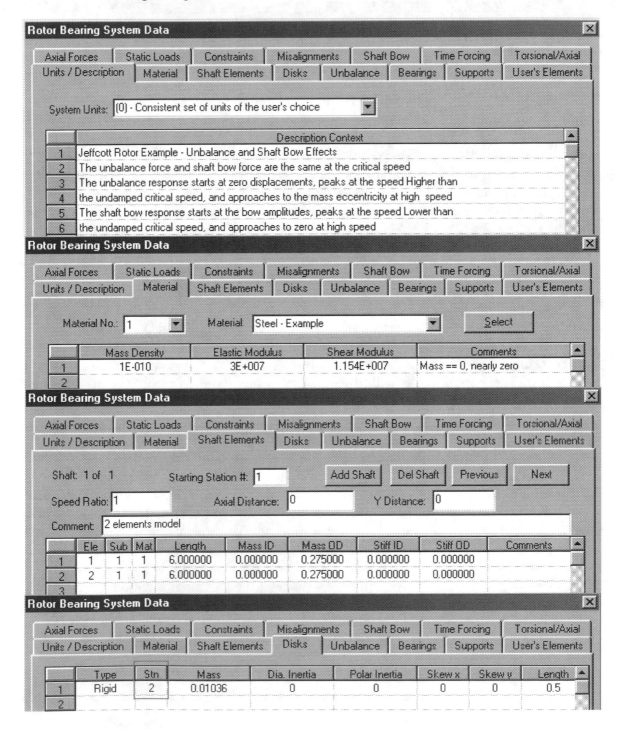

Rotor Bearing System Data ☒

| Axial Forces | Static Loads | Constraints | Misalignments | Shaft Bow | Time Forcing | Torsional/Axial |
| Units / Description | Material | Shaft Elements | Disks | Unbalance | Bearings | Supports | User's Elements |

Bearing: 1 of 3   [Add Brg] [Del Brg] [Previous] [Next]

3 for Bearing #2.

Station I: 1   J: 0   Angle: 0

Type: 0-Linear Constant Bearing

Comment: Rigid bearing, Kb >> Ks

Translational Bearing Properties

Kxx: 10000000000   Kxy: 0   Cxx: 0   Cxy: 0

Kyx: 0   Kyy: 10000000000   Cyx: 0   Cyy: 0

Rotor Bearing System Data ☒

| Axial Forces | Static Loads | Constraints | Misalignments | Shaft Bow | Time Forcing | Torsional/Axial |
| Units / Description | Material | Shaft Elements | Disks | Unbalance | Bearings | Supports | User's Elements |

Bearing: 3 of 3   [Add Brg] [Del Brg] [Previous] [Next]

Station I: 2   J: 0   Angle: 0

Type: 0-Linear Constant Bearing

Comment: viscous damping at the disk

Translational Bearing Properties

Kxx: 0   Kxy: 0   Cxx: 0.3114   Cxy: 0

Kyx: 0   Kyy: 0   Cyx: 0   Cyy: 0.3114

Rotor Bearing System Data ☒

| Axial Forces | Static Loads | Constraints | Misalignments | Shaft Bow | Time Forcing | Torsional/Axial |
| Units / Description | Material | Shaft Elements | Disks | Unbalance | Bearings | Supports | User's Elements |

	Ele	Sub	Left Unb.	Left Ang.	Right Unb.	Right Ang.	Comments	▲
1	2	1	1.036E-005	0	0	0	Me	
2	Same as below - Only one is needed							
3	1	1	0	0	1.036E-005	0		

Rotor Bearing System Data ☒

| Units / Description | Material | Shaft Elements | Disks | Unbalance | Bearings | Supports | User's Elements |
| Axial Forces | Static Loads | Constraints | Misalignments | Shaft Bow | Time Forcing | Torsional/Axial |

Curve Fitting Method
◉ None   ○ Spline   ○ Polynominal

	Stn	x	y	Comments	▲
1	2	0.001	0		
2					

The governing equations for translational motions at the disk (station 2) are:

$$m\ddot{x} + C\dot{x} + Kx = Kq_b \cos\Omega t + me\Omega^2 \cos(\Omega t + \phi_e)$$

$$m\ddot{y} + C\dot{y} + Ky = Kq_b \sin\Omega t + me\Omega^2 \sin(\Omega t + \phi_e)$$

(3.5-3)

where m is the disk mass, C is the linear viscous damping at the disk, and K is the shaft bending stiffness for this simple Laval-Jeffcott rotor. Since the damping and stiffness in both the X and Y directions are the same, the system is referred to as an *isotropic system*. The equations of motion in the X and Y directions are also decoupled in this example. For this isotropic 2DOF system, the two natural frequencies with the same value (one for each direction) are equal to two distinct natural modes. That is, the system has a repeated eigenvalue and two independent eigenvectors.

The undamped natural frequency, damping factor, logarithm decrement and damped natural frequency are:

$$\omega_n = \sqrt{\frac{K}{M}} = \sqrt{\frac{233.95}{0.01036}} = 150.3 \, rad/\sec = 1435 \text{ rpm}$$

$$\xi = \frac{C}{C_c} = \frac{C}{2M\omega_n} = 0.1 \qquad \text{i.e. 10 percent critical damping}$$

$$\delta = \frac{2\pi\xi}{\sqrt{1-\xi^2}} = 0.631$$

$$\omega_d = \omega_n\sqrt{1-\xi^2} = 149.5 \, rad/\sec = 1428 \text{ rpm}$$

The natural frequencies are constant and independent of the spin speed in this example. If the phase angle in the unbalance excitation is zero ($\phi_e = 0$), the unbalance excitation is in-phase with the shaft bow. If the phase angle in the unbalance excitation is 180 degrees ($\phi_e = \pi$), the unbalance excitation is out-of-phase with the shaft bow. The steady-state synchronous responses, due to unbalance and shaft bow, may add or subtract with each other depending upon the magnitudes and phases of the unbalance and shaft bow distributions. Note that the force magnitude of a shaft bow is constant, and the force magnitude of an unbalance is a function of square of spin speed, as illustrated in Figure 3.5-2. In this example, the unbalance and shaft bow were chosen, such that the force amplitudes are the same at the resonance ($\Omega = \omega_n$). Since this is an isotropic system ($C=C_x=C_y$, $K=K_x=K_y$), the steady-state response amplitudes at the X and Y directions are the same, and the response orbits are forward circular at any speed as discussed previously.

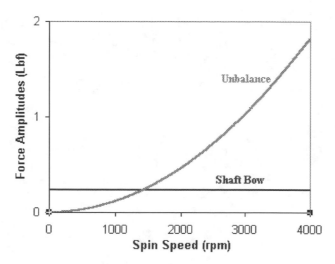

Figure 3.5-2 Force magnitudes for unbalance and shaft bow

For purpose of comparison, the response amplitudes at various speeds are summarized below:

1. For steady-state **unbalance** response:

$$\Omega = 0 \qquad\qquad |x| = |y| = 0$$

$$\Omega = \omega_n = 1435 \text{ rpm} \qquad |x| = |y| = \frac{e}{2\xi} = 0.005$$

$$\Omega = \frac{\omega_n}{\sqrt{1 - 2\xi^2}} = 1450 \text{ rpm} \qquad |x| = |y| = \frac{e}{2\xi\sqrt{1 - \xi^2}} = 0.005025 \text{ Peak Amplitude}$$

2. For steady-state **shaft bow** response:

$$\Omega = 0 \qquad\qquad |x| = |y| = q_b$$

$$\Omega = \omega_n = 1435 \text{ rpm} \qquad |x| = |y| = \frac{q_b}{2\xi} = 0.005$$

$$\Omega = \omega_n\sqrt{1 - 2\xi^2} = 1420 \text{ rpm} \qquad |x| = |y| = \frac{q_b}{2\xi\sqrt{1 - \xi^2}} = 0.005025 \text{ Peak Amplitude}$$

The steady-state responses due to unbalance and shaft bow, are analyzed separately and presented in the following Figure 3.5-3. In this case, the unbalance angle is zero and is in-phase with the shaft bow.

For steady-state **unbalance** response, the response amplitude is zero at zero spin speed, peaks at the spin speed that is slightly higher than the undamped natural frequency, and approaches the mass eccentricity (e) at very high speed. For steady-state **shaft bow** response, the amplitude starts at the initial shaft bow at zero spin speed, peaks at the spin speed that is lower than the undamped natural frequency, and approaches zero at very high speed.

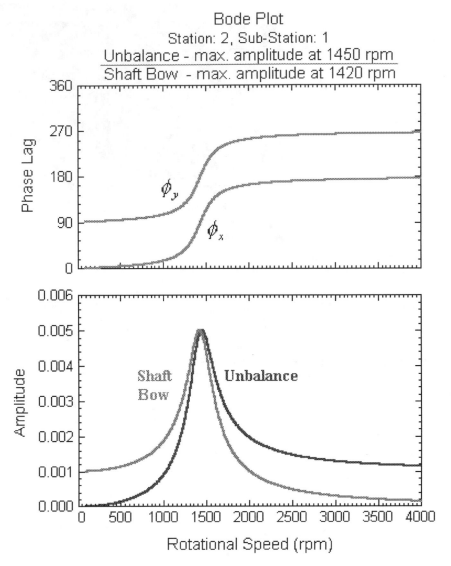

Figure 3.5-3 Comparison of unbalance response and shaft bow response

The phase angles due to unbalance and shaft bow excitations are the same. The phase angle of the x displacement relative to the excitation is zero at zero spin speed. The phase angle changes rapidly near the resonant speed. It is 90 degrees at the undamped natural frequency, and is 180 degrees at very high speed. The phase angle of the y displacement lags 90 degrees behind the phase angle of the x displacement.

$$\phi_y = \phi_x + 90$$

From the previous discussion, the above equation also indicates that the displacement orbits are forward precession for all rotor speeds.

Figure 3.5-4 shows the total responses due to the combined excitations of unbalance and shaft bow. Both results for in-phase and out-of-phase are presented. The numerical results from **DyRoBeS** are in agreement with the analytical solutions.

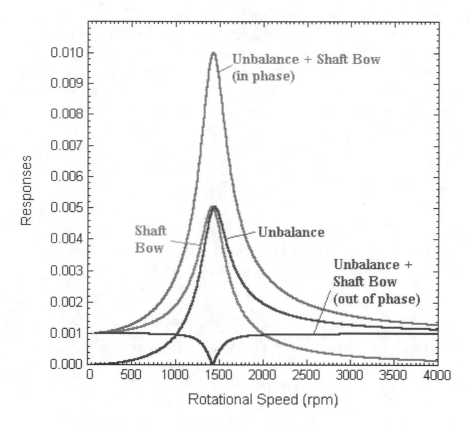

Figure 3.5-4 Combined effect of unbalance and shaft bow

3.6 Internal and External Viscous Dampings

Consider a classical Laval-Jeffcott rotor system with rigid bearings. The forces acting upon the disk include elastic restoring force, mass unbalance force, as well as both external and internal viscous damping forces. The external damping force opposing to the motion, is proportional to the velocity of the disk, relative to the fixed reference frame. The internal damping force is proportional to the velocity of the disk, relative to the rotating shaft. The internal damping forces in the rotating reference frame which rotates with a constant spin speed of Ω are:

$$\begin{Bmatrix} F_{x'} \\ F_{y'} \end{Bmatrix} = \begin{bmatrix} c_i & 0 \\ 0 & c_i \end{bmatrix} \begin{Bmatrix} \dot{x}' \\ \dot{y}' \end{Bmatrix} \tag{3.6-1}$$

By using the coordinate transformation presented in Chapter 2, the internal damping forces in the fixed reference frame become:

$$\begin{Bmatrix} F_x \\ F_y \end{Bmatrix} = \begin{bmatrix} c_i & 0 \\ 0 & c_i \end{bmatrix} \begin{Bmatrix} \dot{x} \\ \dot{y} \end{Bmatrix} + \begin{bmatrix} 0 & c_i\Omega \\ -c_i\Omega & 0 \end{bmatrix} \begin{Bmatrix} x \\ y \end{Bmatrix} \tag{3.6-2}$$

It shows that the internal damping contributes not only to the conventional velocity dependent *dissipative force*, but also to the speed and displacement dependent *circulatory force*. The coefficients in the stiffness matrix that produce the circulatory force are also called *cross-coupled stiffness* terms. One unique feature in rotordynamics study is the destabilizing effect, due to the circulatory forces commonly caused by fluid film bearings, liquid seals, and other fluid interactions. The equations of motion of the disk in the fixed reference frame are:

$$\begin{bmatrix} m & 0 \\ 0 & m \end{bmatrix}\begin{Bmatrix} \ddot{x} \\ \ddot{y} \end{Bmatrix} + \begin{bmatrix} c_e + c_i & 0 \\ 0 & c_e + c_i \end{bmatrix}\begin{Bmatrix} \dot{x} \\ \dot{y} \end{Bmatrix} + \begin{bmatrix} k & c_i\Omega \\ -c_i\Omega & k \end{bmatrix}\begin{Bmatrix} x \\ y \end{Bmatrix} = me\Omega^2\begin{Bmatrix} \cos\Omega t \\ \sin\Omega t \end{Bmatrix} \quad (3.6\text{-}3)$$

where c_i and c_e are the internal and external viscous damping coefficients, and k is the shaft bending stiffness. It is usually convenient, although not necessary, to analyze this isotropic system by introducing the complex notation:

$$\hat{r} = x + jy \tag{3.6-4}$$

By multiplying j in the second row, and adding into the first row of Eq. (3.6-3), we have:

$$m\ddot{\hat{r}} + \left(c_e + c_i\right)\dot{\hat{r}} + \left(k - j\Omega c_i\right)\hat{r} = me\Omega^2 e^{j\Omega t} \tag{3.6-5}$$

The steady-state unbalance response in complex form, was discussed in the previous section and it has the form:

$$\hat{r}(t) = \hat{r}_f e^{j\Omega t} + \hat{r}_b e^{-j\Omega t} \tag{3.6-6}$$

By differentiating and substituting the solution into the equation of motion, we can obtain:

$$\hat{r}_f = \frac{me\Omega^2}{\left(k - m\Omega^2\right) + j\Omega c_e} \tag{3.6-7}$$

$$\hat{r}_b = 0 \tag{3.6-8}$$

It is evident from Eqs. (3.6-7) and (3.6-8) that the steady-state unbalance response orbit is a purely forward circular motion for this isotropic system, and that internal damping has no effect on the circular response.

The stability of the rotor system is determined by the transient (homogeneous) solution. The real parts of the eigenvalues are used to determine system stability in the linear sense. A positive value indicates that the homogeneous solution is an exponentially growing function of time. However, when the amplitude increases with time, the linear model becomes invalid, and the nonlinear theory should be used. The eigenvalues are the roots of the characteristic equation. The characteristic equation is obtained by substituting the solution $(r = |r| e^{\lambda t})$ into the homogeneous equation of motion, and is:

$$m\lambda^2 + (c_e + c_i)\lambda + (k - j\Omega c_i) = 0 \qquad (3.6\text{-}9)$$

The roots of the characteristic equation or eigenvalues are:

$$\lambda_{1,2} = \frac{-(c_e + c_i)}{2m} \pm \sqrt{\left[\frac{(c_e + c_i)}{2m}\right]^2 - \omega_n^2 + j\frac{\Omega c_i}{m}} = \sigma_{1,2} \pm j\omega_{1,2} \qquad (3.6\text{-}10)$$

where

$$\omega_n = \sqrt{\frac{k}{m}} \qquad \text{is the undamped natural frequency}$$

and $\omega_{1,2}$ are the damped natural frequencies or whirl speeds. At the onset of instability, the shaft spin speed is Ω_0 and one eigenvalue has zero real part:

$$\lambda_0 = j\omega_0 \qquad (3.6\text{-}11)$$

At the instability threshold, the corresponding frequency (ω_0) is referred to as unstable whirl frequency. In this case, the characteristic equation becomes:

$$-m\omega_0^2 + j(c_e + c_i)\omega_0 + (k - j\Omega_0 c_i) = 0 \qquad (3.6\text{-}12)$$

When separating the real and imaginary parts, we get:

$$k - m\omega_0^2 = 0 \qquad (3.6\text{-}13a)$$

and

$$(c_e + c_i)\omega_0 - \Omega_0 c_i = 0 \qquad (3.6\text{-}13b)$$

Therefore, at the instability threshold, the whirl frequency is the undamped natural frequency:

$$\omega_0 = \sqrt{\frac{k}{m}} = \omega_n \qquad (3.6\text{-}14)$$

and the shaft speed is

$$\Omega_0 = \omega_n \left(\frac{c_e + c_i}{c_i}\right) \qquad (3.6\text{-}15)$$

The condition for instability is when the rotor spin speed is greater than the instability threshold:

$$\Omega > \omega_n \left(\frac{c_e + c_i}{c_i} \right) \tag{3.6-16}$$

It shows that the dissipative force due to external damping always stabilizes the rotor system, and the circulatory force due to internal damping destabilizes the system when the spin speed is greater than the system natural frequency. If the external damping does not present, the system becomes unstable when the spin speed is greater than the system undamped natural frequency. The forces acting on the disk, at the instability threshold, can be obtained by multiplying the displacement r with the Eq. (3.6-12) and are shown in Figure 3.6-1. It shows that the unstable motion is a forward precession.

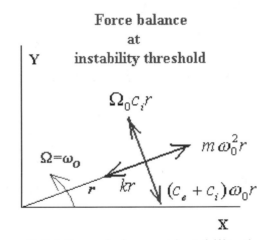

Figure 3.6-1 Force balance at instability threshold

Example 3.3: Laval-Jeffcott Rotor– Internal Damping Effects

An internal viscous damping is included in the Laval-Jeffcott rotor system, presented in the previous example. The internal damping coefficient of 0.3 is acting on the disk (station 2). An additional speed-dependent bearing at the disk station, as shown in the following data input table, is added to include this internal damping effect. Since the cross-coupled stiffness due to the internal damping is linearly proportional to the speed, the speed-dependent bearing type is used to model this internal damping mechanism. There two speed points are used to specify the direct damping and cross-coupled stiffness. In **DyRoBeS**, if only two speed points are given, linear interpolation is used, and if more than three points are given, spline function is used to interpolate the data. Note that the cross-coupled stiffness is Ωc_i where Ω is in rad/sec.

The results of steady-state unbalance response with internal damping, as shown in Figure 3.6-2, are identical as the results presented in previous example without internal damping. The internal damping has no effect on the steady state forced response of this isotropic system.

Internal damping input:

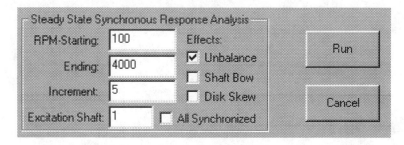

	rpm	Kxx	Kxy	Kyx	Kyy	Cxx	Cxy	Cyx	Cyy
1	1000	0	31.416	-31.416	0	0.3	0	0	0.3
2	3000	0	94.248	-94.248	0	0.3	0	0	0.3
3			Ωc_i rad/sec						
4									

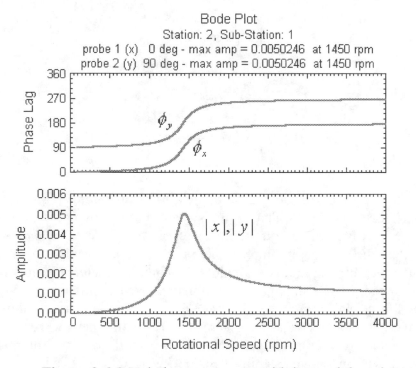

Figure 3.6-2 Unbalance response with internal damping

For the stability analysis, the eigenvalue problem is solved. A complex eigenvalue is given by:

$$\lambda = \sigma + j\omega_d \tag{3.6-17}$$

The imaginary parts of the eigenvalues, ω_d, are the system *damped natural frequencies*. These can be used to determine the damped critical speeds. The real parts of the eigenvalues, σ, are the system *damping exponents*, and they are used to determine the system stability. A positive damping exponent indicates system instability. Very often, the logarithmic decrements or damping factors are used to determine the system stability. A negative logarithmic decrement or damping factor indicates system instability. The eigenvectors define the mode shapes. If the damped natural frequency is a non-zero value, this mode is a *precessional mode* with an oscillating frequency equal to the damped natural frequency. If the damped natural frequency equals zero, this mode is a *real mode* or non-oscillating mode.

The logarithmic decrement and damping factor for a precessional mode is defined to be:

$$\delta = \frac{-2\pi\sigma}{\omega_d} = \frac{2\pi\xi}{\sqrt{1-\xi^2}} \tag{3.6-18}$$

and

$$\xi = \frac{\delta}{\sqrt{(2\pi)^2 + \delta^2}} \tag{3.6-19}$$

In this example, a Whirl Speed and Stability Analysis is performed as shown below.

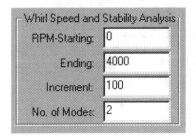

The results are presented in both the Stability Map in Figure 3.6-3 and the Whirl Speed Map in Figure 3.6-4. Note that in this system the first two eigenvalues have the same imaginary parts (damped natural frequencies), but two different real parts (damping exponents). Unlike the previous examples, there is a non-conservative force (circulatory force) due to internal damping in this example. The damping factors are not constant, but vary with rotor speed. The damping factors for the first two modes (eigenvalues) are plotted in the Stability Map as shown in Figure 3.6-3. At zero spin speed, the two modes have the same damping factors (repeated eigenvalues). As the speed increases, the cross-coupled stiffness due to internal damping increases. It shows that internal damping destabilizes the forward mode, yet, stabilizes the backward mode. At a spin speed around

2925 rpm, the forward mode becomes unstable in the linear sense. This spin speed is referred to as the *onset of instability* or *instability threshold*.

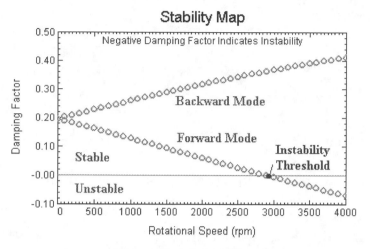

Figure 3.6-3 Stability map

The first two damped natural frequencies are plotted in the Whirl Speed Map as shown in Figure 3.6-4. Since they are identical, only one frequency curve is shown. The natural frequencies are not constant as in the previous example, but slowly increase with speed when internal damping is present. There are two excitation lines that are superimposed on the whirl speed map. One is the conventional synchronous excitation line, which determines damped critical speed due to synchronous excitation; such as unbalance, shaft bow, and disk skew. Damped critical speed is determined at around 1414 rpm. The second excitation line is the self-excitation due to internal damping, which is used to determine the instability threshold. The instability threshold, determined from the Whirl Speed Map, is in agreement with the results from the Stability Map. Since the damped natural frequencies are dependent upon the spin speed, and the damping factors are not a constant, the maximum steady-state synchronous response occurs at a speed, which is slightly higher than the damped critical speed determined from the Whirl Speed Map.

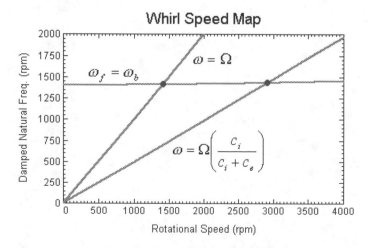

Figure 3.6-4 Whirl speed map

At zero spin speed, the results can be verified analytically:

$$\omega_n = \sqrt{\frac{k}{m}} = \sqrt{\frac{233.95}{0.01036}} = 150.3 \, \text{rad/sec} = 1435 \, \text{rpm}$$

$$\xi = \frac{c_e + c_i}{2m\omega_n} = 0.196$$

$$\omega_d = \omega_n \sqrt{1 - \xi^2} = 1407 \, \text{rpm}$$

Again, the computational results are in agreement with the analytical results.

3.7 Rotational Motion and Gyroscopic Effects

Because the translational and rotational motions of the centrally mounted disk are decoupled, the translational motions have been discussed in previous sections. This section will discuss the rotational motions. The spinning rigid disk has two angular (rotational) displacements, θ_x, θ_y, which represent small rotations of the disk about the X and Y axes, respectively as shown in Figure 3.7-1,

$$\theta_x = -\left(\frac{dy}{dz}\right), \qquad \theta_y = \frac{dx}{dz} \tag{3.7-1}$$

where z is the coordinate along the shaft axis, and x and y are the lateral displacements. Again, in the rigorous definition, these angles depend upon the translational displacements, and they cannot be treated as independent generalized coordinates. However, they are very small and they lead to linearized equations of motion for a spinning disk. In addition, they can be used directly in the shaft element strain energy formulation.

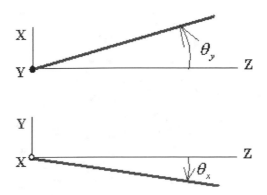

Figure 3.7-1 Rotational displacements

The axisymmetric disk has a polar mass moment of inertia I_p and a transverse (diametral) mass moment of inertia I_d, spinning at a constant rotational speed Ω about the Z axis. Since we only discuss the dynamics, I_p, I_d are often simply referred to as polar and transverse moments of inertia. The angular momentum for small angular displacements, as illustrated in Figure 3.7-2, disregarding the higher order terms are:

$$H_x = I_d \dot{\theta}_x + I_p \Omega \theta_y \qquad (3.7\text{-}2a)$$

$$H_y = I_d \dot{\theta}_y - I_p \Omega \theta_x \qquad (3.7\text{-}2b)$$

$$H_z = I_p \Omega \qquad (3.7\text{-}2c)$$

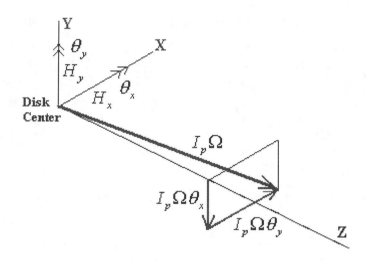

Figure 3.7-2 Angular momentum for a spinning disk

The terms $I_p \Omega \theta_y$, and $-I_p \Omega \theta_x$ in Eq. (3.7-2) represent the gyroscopic moments. The angular restoring moment from the shaft is represented by the shaft stiffness k_R. When small angles are assumed, the moment equations are linear and the equations of motion are:

$$M_x = \frac{dH_x}{dt} = -k_R \theta_x \qquad (3.7\text{-}3a)$$

$$M_y = \frac{dH_y}{dt} = -k_R \theta_y \qquad (3.7\text{-}3b)$$

The equations of motion for rotational motions of the disk become:

$$\begin{bmatrix} I_d & 0 \\ 0 & I_d \end{bmatrix} \begin{Bmatrix} \ddot{\theta}_x \\ \ddot{\theta}_y \end{Bmatrix} + \Omega \begin{bmatrix} 0 & I_p \\ -I_p & 0 \end{bmatrix} \begin{Bmatrix} \dot{\theta}_x \\ \dot{\theta}_y \end{Bmatrix} + \begin{bmatrix} k_R & 0 \\ 0 & k_R \end{bmatrix} \begin{Bmatrix} \theta_x \\ \theta_y \end{Bmatrix} = \begin{Bmatrix} 0 \\ 0 \end{Bmatrix} \qquad (3.7\text{-}4)$$

The gyroscopic moments are cross-coupled at both coordinates. Since the gyroscopic moments are proportional to angular velocities, they are commonly grouped with the viscous damping terms in the formulation of the equations of motion. However, it should be noted that the gyroscopic matrix is a skew symmetric matrix and a function of rotor spin speed. It is derived from the kinetic energy and dissipates no energy. Therefore it is a conservative force (moment) unlike the viscous damping terms. It will be shown that because of the gyroscopic effect, each natural frequency of whirl (mode) is split into two frequencies (modes) when rotor speed is not zero. One mode is a forward whirl and the other is a backward whirl. It will also show that the gyroscopic effect raises the forward whirl frequencies, and lowers the backward whirl frequencies.

Since this is a conservative gyroscopic system with 2 degrees-of-freedom, there are two natural frequencies $(\omega_i, i = 1,2)$, and two corresponding natural (normal) modes. The eigenvalues of Eq. (3.7-4) consist of 2 pairs of pure imaginary complex conjugates, $\lambda_i = \pm j\omega_i$ $(i = 1,2)$. The corresponding eigenvectors also occur in pairs of complex conjugates. The natural frequencies of whirl (eigenvalues) and associated modes (eigenvectors) are determined by assuming the solution with the exponential form of $\hat{\theta} e^{j\omega t}$. The eigenvalue problem becomes:

$$\begin{bmatrix} k_R - \omega^2 I_d & j\omega\Omega I_p \\ -j\omega\Omega I_p & k_R - \omega^2 I_d \end{bmatrix} \begin{Bmatrix} \hat{\theta}_x \\ \hat{\theta}_y \end{Bmatrix} = \begin{Bmatrix} 0 \\ 0 \end{Bmatrix} \tag{3.7-5}$$

From the determinant of the matrix equation, the two positive natural frequencies are:

$$\omega_1 = -\left(\frac{\Omega I_p}{2 I_d}\right) + \sqrt{\left(\frac{\Omega I_p}{2 I_d}\right)^2 + \left(\frac{k_R}{I_d}\right)} \tag{3.7-6}$$

$$\omega_2 = \left(\frac{\Omega I_p}{2 I_d}\right) + \sqrt{\left(\frac{\Omega I_p}{2 I_d}\right)^2 + \left(\frac{k_R}{I_d}\right)} \tag{3.7-7}$$

The natural frequencies for this rotational motion are dependent on the rotor spin speed Ω. Note that Ω has a unit of rad/sec here. At $\Omega = 0$, $\omega_1 = \omega_2 = \sqrt{\dfrac{k_R}{I_d}}$, the vibration modes are planar. As the rotor speed increases, these two frequencies split. One frequency (backward mode) decreases with the rotor speed while the other (forward mode) increases with the rotor speed. The direction of precession is determined from the natural modes. Again, it is convenient to introduce the complex rotation vector $\hat{r} = \theta_x + j\theta_y$. By multiplying the second row of Eq. (3.7-4) by j and adding to the first row, the equations of motion become:

$$I_d \ddot{\hat{r}} - j\Omega I_p \dot{\hat{r}} + k_R \hat{r} = 0 \tag{3.7-8}$$

For this conservative gyroscopic system, the natural vibration is a harmonic motion. The harmonic motion can be decomposed into two components: forward and backward circular precessions, as discussed in Chapter 2:

$$\hat{r} = \theta_x + j\theta_y = \hat{r}_f e^{j\omega t} + \hat{r}_b e^{-j\omega t} \tag{3.7-9}$$

Substituting Eq. (3.7-9) into Eq. (3.7-8), and separating the forward and backward motions, yields:

$$\left(-\omega^2 I_d + \Omega I_p \omega + k_R \right) \hat{r}_f e^{j\omega t} = 0 \tag{3.7-10}$$

$$\left(-\omega^2 I_d - \Omega I_p \omega + k_R \right) \hat{r}_b e^{-j\omega t} = 0 \tag{3.7-11}$$

Hence, the natural frequencies associated with the forward precession $(\hat{r}_f e^{j\omega t})$ are:

$$\omega_f = \left(\frac{\Omega I_p}{2I_d} \right) + \sqrt{\left(\frac{\Omega I_p}{2I_d} \right)^2 + \left(\frac{k_R}{I_d} \right)} \tag{3.7-12}$$

and the natural frequencies associated with the backward precession $(\hat{r}_b e^{-j\omega t})$ are:

$$\omega_b = -\left(\frac{\Omega I_p}{2I_d} \right) + \sqrt{\left(\frac{\Omega I_p}{2I_d} \right)^2 + \left(\frac{k_R}{I_d} \right)} \tag{3.7-13}$$

These results are identical with the results obtained from the real domain analysis. However, it shows that the forward natural frequency increases with spin speed due to the "*gyroscopic stiffening*" effect, and backward natural frequency decreases as the spin speed increases due to the "*gyroscopic softening*" effect. It should also be noted that some authors like to use negative natural frequency to represent the backward whirl for a progressive precession. That is, a backward mode can be a mode with a backward precession with a positive frequency, or a forward precession but a negative frequency. However, it will be easier to interpret the physics if only the positive frequencies are remained. Under the conditions of positive spin speed and positive frequencies, a forward mode is a mode with progressive (forward) precession, and a backward mode is a mode with retrograde (backward) precession.

The synchronous critical speeds are determined from Eqs. (3.7-12) and (3.7-13) by setting $\omega = \Omega$. The forward and backward critical speeds are:

$$(\Omega_{cr})_{forward} = \sqrt{\frac{k_R}{I_d - I_p}} \tag{3.7-14}$$

$$(\Omega_{cr})_{backward} = \sqrt{\frac{k_R}{I_d + I_p}} \tag{3.7-15}$$

When $I_p > I_d$ (thin disk), the forward critical speed does not exist in this pure rotational motion. In general, when the mass of the shaft is included in the practical rotors, the polar mass moment of inertia will be less than the transverse (diametral) mass moment of inertia. Although the backward critical speed always exists, it cannot be excited by unbalance, or by residual shaft bow for isotropic rotor systems, as explained earlier.

One unique feature makes the rotordynamics study so fascinating and different from the non-rotating structural dynamics, is the gyroscopic effect previously described. Now we will examine the gyroscopic effect from energy point of view. The gyroscopic moment is a conservative force. It can however, be deceiving due to its entry into the equation of motion. The skew-symmetric gyroscopic matrix that couples two planes of motion is contributed by the part of rotational kinetic energy caused by the gyroscopic moments. The complete rotational kinetic energy of a spinning disk including the gyroscopic energy is:

$$T_{rot} = \frac{1}{2}\left(H_x \dot{\theta}_x + H_y \dot{\theta}_y + H_z \Omega\right)$$
$$= \frac{1}{2}I_d\left(\dot{\theta}_x^2 + \dot{\theta}_y^2\right) + \frac{1}{2}\Omega I_p\left(\dot{\theta}_x \theta_y - \theta_x \dot{\theta}_y\right) + \frac{1}{2}I_p \Omega^2$$

$$(3.7\text{-}16)$$

The first term is due to rotatory inertia, the second term is due to the gyroscopic moments, and the third term is due to the pure spinning effect.

For a harmonic motion with a whirl frequency of ω, the kinetic energy generated by the gyroscopic effect becomes:

$$T_{gyro} = \frac{1}{2}\Omega I_p\left(\dot{\theta}_x \theta_y - \theta_x \dot{\theta}_y\right) = \frac{1}{2}\Omega I_p \omega\left(\theta_{xs}\theta_{yc} - \theta_{xc}\theta_{ys}\right)$$
$$= \frac{1}{2}\Omega I_p \omega\left(-a_\theta \cdot b_\theta\right) = \frac{1}{2}\Omega I_p \omega\left(-\left|\hat{r}_{\theta,f}\right|^2 + \left|\hat{r}_{\theta,b}\right|^2\right)$$

$$(3.7\text{-}17)$$

where a_θ, b_θ, and $\left|\hat{r}_{\theta,f}\right|, \left|\hat{r}_{\theta,b}\right|$ are the semi-major axis, semi-minor axis, forward circular amplitude, and backward circular amplitude, respectively, for the rotational displacement orbit. The expression shows that the kinetic energy due to the gyroscopic moments is linearly proportional to the polar moment of inertia, rotor spin speed, whirl frequency, and area of rotational displacement (slope) orbit. Since I_p, Ω, and ω are all positive values, the sign of Eq. (3.7-17) is determined by the value inside the parenthesis. For the forward precessional mode, $b_\theta > 0$, and $\left|\hat{r}_{\theta,f}\right| > \left|\hat{r}_{\theta,b}\right|$, the gyroscopic effect contributes negative kinetic energy (much like inertia decreased effect), and tends to raise the corresponding forward whirl frequency (gyroscopic stiffening). For the backward precessional mode, $b_\theta < 0$, and $\left|\hat{r}_{\theta,f}\right| < \left|\hat{r}_{\theta,b}\right|$, the gyroscopic effect contributes positive kinetic energy (much like inertia added effect), and tends to lower the corresponding backward whirl frequency (gyroscopic softening). The gyroscopic effect can be significant in the study of high-speed overhung rotor system, where a large wheel is mounted outside the bearing span and has large rotational displacement.

Disk Skew Effect

If the disk is mounted in a skewed position on the shaft with a small skewed angle τ, the external moments due to the skew of the disk are:

$$M_x = (I_p - I_d) \tau \, \Omega^2 \, \cos(\Omega t + \phi_\tau) \tag{3.7-18}$$

$$M_y = (I_p - I_d) \tau \, \Omega^2 \, \sin(\Omega t + \phi_\tau) \tag{3.7-19}$$

where ϕ_τ is the angle of the disk skew with respect to the shaft and reference mark. The disk skew excitation forces are similar to the unbalance forces associated with mass eccentricity. Both amplitudes are functions of square of rotor speed, and the excitation frequencies are synchronous with the rotor speed. Disk skew produces an external moment, but mass eccentricity produces an external force. The precessional motion of the axis of the disk caused by the disk skew describes a cone with an elliptical cross-section. The steady-state response due to the disk skew can be obtained by following the discussion previously used for translational motion. Therefore, it will not be discussed here.

Example 3.4: Laval-Jeffcott Rotor– Rotational Motion

The Laval-Jeffcott rotor system, used in the previous examples, is presented here to demonstrate the gyroscopic effect. As shown in the following Figure 3.3-3, a rigid disk is located at the midspan of a flexible shaft. Only the fundamental rotational motion is being considered, thus, the rigid disk has a zero mass and the diametral (transverse) mass moment of inertia (I_d) of the disk is 0.768 Lbf-s^2-in. Three different polar moments of inertia (I_p) of the disk will be used to study the gyroscopic effect. They are: 0, 0.5, and 1.5 times I_d. The flexible shaft with elastic modulus of 3.0E07 psi has a length of 12 inches and a diameter of 0.275 inches. The shaft mass is negligible, so a very small mass density of 1.0E-10 is entered in the input, to avoid the singularity of the translational degrees-of-freedom. Technically speaking, this is a 12-degrees-of-freedom system (4DOF per finite element station). However, since the total mass is extremely small, the fundamental translational motion and any higher frequency motions can be disregarded in this example. Only the fundamental rotational motion will be considered due to its much lower natural frequency compared with others. Again, a very large bearing stiffness of 1E10 Lbf/in is applied at both ends instead of constraints as used in earlier examples to simulate the rigid bearings. This bearing stiffness is much larger than the shaft bending stiffness. Therefore it acts like rigid bearing. The pertinent parameters for this rotational motion of the Laval-Jeffcott rotor are summarized below:

$L = 12.0$ in (total length)
$D = 0.275$ in
$E = 3.0E07$ psi
$m = 0$ Lbm
$I_d = 0.768$ Lbf-s^2-in (296.5 Lb-in^2)
$I_p = 0$, $I_p = 0.5I_d$, $I_p = 1.5I_d$

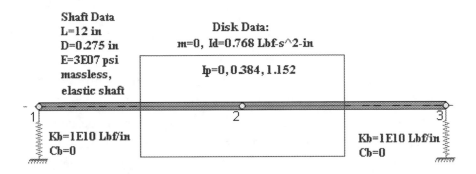

Figure 3.7-3 Rotation motion for the Laval-Jeffcott rotor

The whirl speed and stability analysis is performed from 0 rpm to 3000 rpm, with an increment of 500 rpm, using three different polar moments of inertia. The results are presented in Figure 3.7-4.

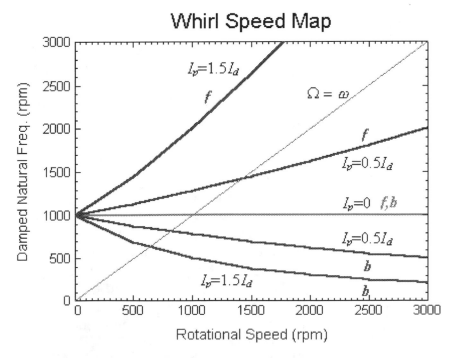

Figure 3.7-4 Whirl speed map with three different polar moments of inertia

For $I_p = 0$, there is no gyroscopic effect, and the two natural frequencies are identical with a value of 1000 cpm (repeated eigenvalues). Readers are encouraged to open the **DyRoBeS** output text file, to review the input and output data. There are 12 precessional modes calculated by **DyRoBeS**, since technically speaking, this is a 12 DOF system. However, the third and higher natural frequencies are significantly higher than the first two fundamental rotational frequencies, and their effects can be neglected. For $I_p = 0.5I_d$, the gyroscopic effect splits the two natural frequencies. The forward natural frequency increases with the spin speed, and the backward natural frequency decreases as the spin speed increases. There are two synchronous critical speeds; one is forward and one is backward. For $I_p = 1.5I_d$, the two natural frequencies split even further and the synchronous forward critical speed does not exist anymore.

The numerical results can be verified by using the analytical solutions previously presented. The fundamental rotational stiffness at the midspan of the shaft due to bending is:

$$k_R = \frac{12EI}{L} = 8422.13 \text{ Lbf-in/rad} \qquad \text{where} \quad I = \frac{\pi D^4}{64}$$

and the two natural frequencies are function of spin speed, given in Eqs. (3.7-12) and (3.7-13). At zero spin speed,

$$\omega_f = \omega_b = \sqrt{\frac{k_R}{I_d}} = \sqrt{\frac{8422.13}{0.768}} = 104.72 \text{ rad/sec} = 1000 \text{ rpm}$$

3.8 Energy Consideration

To gain insight into rotor dynamic behavior, this section will study the work done by various forces, and the significance of each effect. Consider a generalized 2-DOF linear rotor system:

$$\begin{bmatrix} m & 0 \\ 0 & m \end{bmatrix} \begin{Bmatrix} \ddot{x} \\ \ddot{y} \end{Bmatrix} + \begin{bmatrix} c_{xx} & c_{xy} \\ c_{yx} & c_{yy} \end{bmatrix} \begin{Bmatrix} \dot{x} \\ \dot{y} \end{Bmatrix} + \begin{bmatrix} k_{xx} & k_{xy} \\ k_{yx} & k_{yy} \end{bmatrix} \begin{Bmatrix} x \\ y \end{Bmatrix} = \begin{Bmatrix} me\Omega^2 \cos(\Omega t) \\ me\Omega^2 \sin(\Omega t) \end{Bmatrix} \qquad (3.8-1)$$

where the generalized damping and stiffness matrices contain all the linear forces acting on the mass (m), which include shaft bending stiffness, internal and external viscous dampings, linearized bearing forces, aerodynamics cross-coupling forces, and other linearized interactive forces. From linear algebra, any real matrix can be written as the sum of a symmetric matrix and a skew-symmetric matrix. For example, the generalized stiffness matrix can be written as:

$$\boldsymbol{K} = \boldsymbol{K}_{sy} + \boldsymbol{K}_{sk} \qquad (3.8-2)$$

where K_{sy} and K_{sk} are the symmetric and skew-symmetric matrices, respectively. They are defined as:

$$K_{sy} = \frac{1}{2}\left(K + K^T\right) = \begin{bmatrix} k_{xx} & k_m \\ k_m & k_{yy} \end{bmatrix} \qquad \text{where} \quad k_m = \frac{1}{2}\left(k_{xy} + k_{yx}\right) \tag{3.8-3}$$

and

$$K_{sk} = \frac{1}{2}\left(K - K^T\right) = \begin{bmatrix} 0 & k_d \\ -k_d & 0 \end{bmatrix} \qquad \text{where} \quad k_d = \frac{1}{2}\left(k_{xy} - k_{yx}\right) \tag{3.8-4}$$

The symmetric matrix K_{sy} is referred to as an *elastic stiffness matrix* and the skew-symmetric matrix K_{sk} is referred to as a *circulatory matrix*. Note that the circulatory matrix contains only cross-coupled stiffness terms.

Similarly, for the generalized damping matrix:

$$C = C_{sy} + C_{sk} \tag{3.8-5}$$

where C_{sy} and C_{sk} are the symmetric and skew-symmetric matrices, respectively. They are defined as:

$$C_{sy} = \frac{1}{2}\left(C + C^T\right) = \begin{bmatrix} c_{xx} & c_m \\ c_m & c_{yy} \end{bmatrix} \qquad \text{where} \quad c_m = \frac{1}{2}\left(c_{xy} + c_{yx}\right) \tag{3.8-6}$$

and

$$C_{sk} = \frac{1}{2}\left(C - C^T\right) = \begin{bmatrix} 0 & c_d \\ -c_d & 0 \end{bmatrix} \qquad \text{where} \quad c_d = \frac{1}{2}\left(c_{xy} - c_{yx}\right) \tag{3.8-7}$$

The symmetric matrix C_{sy} is referred to as a *dissipative damping matrix* and the skew-symmetric matrix C_{sk} is referred to as a *gyroscopic matrix*.

Hence, the generalized stiffness and damping matrices have been decomposed into conservative (i.e., elastic and gyroscopic), and non-conservative (i.e., circulatory and dissipative) matrices. For a harmonic motion with a whirl frequency of ω, the displacements are of the form:

$$q = \begin{Bmatrix} x(t) \\ y(t) \end{Bmatrix} = \begin{Bmatrix} x_c \\ y_c \end{Bmatrix} \cos \omega t + \begin{Bmatrix} x_s \\ y_s \end{Bmatrix} \sin \omega t \tag{3.8-8}$$

The total work done on the system is the summation of the work done by elastic force, gyroscopic force, dissipative force, and circulatory force. It is known that the elastic and gyroscopic forces are conservative forces, and their work done per cycle of motion is zero. The dissipative force and circulatory force are non-conservative forces, and their work done per cycle of motion is not zero.

The work done by dissipative force over one cycle of motion is:

$$W_{dissipative} = \int_0^{2\pi/\omega} -\mathbf{C}_{sy}\dot{\mathbf{q}}\cdot\dot{\mathbf{q}}\,dt = -\int_0^{2\pi/\omega} \begin{Bmatrix} c_{xx}\dot{x}+c_m\dot{y} \\ c_m\dot{x}+c_{yy}\dot{y} \end{Bmatrix} \cdot \begin{Bmatrix} \dot{x} \\ \dot{y} \end{Bmatrix} dt$$

$$= -\int_0^{2\pi/\omega}\left(c_{xx}\dot{x}^2 + c_{yy}\dot{y}^2 + 2c_m\dot{x}\dot{y}\right)dt \tag{3.8-9}$$

$$= -\omega\pi\left[c_{xx}\left(x_c^2+x_s^2\right)+c_{yy}\left(y_c^2+y_s^2\right)+\left(c_{xy}+c_{yx}\right)\cdot\left(x_c y_c + x_s y_s\right)\right]$$

The work done by circulatory force over one cycle of motion is:

$$W_{circulatory} = \int_0^{2\pi/\omega} -\mathbf{K}_{sk}\mathbf{q}\cdot\dot{\mathbf{q}}\,dt = -\int_0^{2\pi/\omega} \begin{Bmatrix} k_d y \\ -k_d x \end{Bmatrix} \cdot \begin{Bmatrix} \dot{x} \\ \dot{y} \end{Bmatrix} dt$$

$$= -k_d\int_0^{2\pi/\omega}\left(y\dot{x}-x\dot{y}\right)dt \tag{3.8-10}$$

$$= \pi\left(k_{xy}-k_{yx}\right)\cdot\left(x_c y_s - x_s y_c\right)$$

$$= \pi\left(k_{xy}-k_{yx}\right)\cdot\left(a\cdot b\right) = \pi\left(k_{xy}-k_{yx}\right)\cdot\left(|\hat{r}_f|^2 - |\hat{r}_b|^2\right)$$

Following the same procedure, the work done by elastic force and gyroscopic force can be shown to be zero.

$$W_{elastic} = W_{gyroscopic} = 0 \tag{3.8-11}$$

Every elliptical motion can be decomposed into two pure circular motions; one being a forward motion and the other being a backward motion with the same whirl frequency. For forward circular motion, we have ($y_c = -x_s$, $y_s = x_c$). For backward circular motion, we have ($y_c = x_s$, $y_s = -x_c$). Using these relationships, it shows that the work done on the system by the dissipative force, Eq. (3.8-9), is always negative (that is, the energy is removed from the system) for both forward and backward motions. Thus, the dissipative force always stabilizes the system. Also, the work done by the dissipative force is linear proportion to the whirl frequency. For the higher frequency modes, the dissipative force removes more energy from the system. Therefore, when the rotor system becomes unstable, in general, the lowest frequency mode becomes unstable. However, the work done on the system by the circulatory force, Eq. (3.8-10), depends on the direction of precession. With the positive ($k_{xy} - k_{yx}$) term, the work done is positive for the forward motion and negative for the backward motion. That is, the circulatory force destabilizes the forward modes and stabilizes the backward modes.

Let us review the internal and external damping effect on the stability presented before, by examining the work done by these damping forces. The total work done by the internal and external dampings per cycle of motion on the rotor system, is the sum of work done by dissipative force and circulatory force:

$$W_{total} = -\omega\pi\left[(c_e + c_i)(x_c^2 + x_s^2 + y_c^2 + y_s^2)\right]$$
$$+ \pi(2\Omega c_i)\cdot(x_c y_s - x_s y_c) \tag{3.8-12}$$

For the forward circular whirl motion, ($y_c = -x_s$, $y_s = x_c$), we have:

$$W_{total} = -2\pi(x_c^2 + x_s^2)\cdot(\omega(c_e + c_i) - \Omega c_i) \tag{3.8-13}$$

The total energy is positive (system is unstable), if

$$\omega(c_e + c_i) - \Omega c_i < 0 \quad \Rightarrow \quad \Omega > \omega\left(\frac{c_e + c_i}{c_i}\right) \tag{3.8-14}$$

It shows that internal damping can destabilize the forward whirl motion, when speed is greater than the first whirl natural frequency.
For the backward circular whirl motion, ($y_c = x_s$, $y_s = -x_c$), we have:

$$W_{total} = -2\pi(x_c^2 + x_s^2)\cdot(\omega(c_e + c_i) + \Omega c_i) \tag{3.8-15}$$

It indicates that internal damping always stabilizes the backward whirl.

The work done by the synchronous excitations, such as unbalance, shaft bow, and disk skew, over one cycle of steady state elliptical motion is:

$$W_F = \int_0^{2\pi/\omega} \left[|F|\cos\omega t \cdot \dot{x} + |F|\sin\omega t \cdot \dot{y}\right] dt \tag{3.8-16}$$
$$= \pi|F|(x_s - y_c)$$

where $|F|$ is the magnitude of the rotating force, such as $me\Omega^2$ for the unbalance excitation and $k_s q_b$ for the shaft bow. It shows that the rotating force can only do work on the forward circular component of the elliptical motion, and there is no work done on the backward circular component of the elliptical motion. Therefore, for isotropic systems, the steady-state response due to synchronous excitation is a pure forward circular motion and no backward precession exists. For anisotropic systems, the response is an elliptical orbit and it always contains a pure forward circular motion regardless whether the total elliptical motion is forward or backward. Therefore, the rotating force can always supply energy to the anisotropic systems with the elliptical motion.

3.9 Transient Startup and Shutdown

In many applications, it is necessary to study the rotor motion during startup, shutdown, going through critical speeds, or during rotor drop when magnetic bearings failure occurs. In these situations, the angular velocity (spin speed) is no longer a constant, but is a function of time. Two more terms introduced by using Lagrange's equation for the derivation of governing equations of rotating components, due to the speed variation are: circulatory matrix and forcing function. The governing equations of motion become:

$$M\ddot{q}(t) + \left[C + \dot{\varphi}\,G\right]\dot{q}(t) + \left[K + \ddot{\varphi}\,G\right]q(t)$$
$$= \dot{\varphi}^2\,Q_1(\varphi) + \ddot{\varphi}\,Q_2(\varphi) + Q_n(\dot{q}, q, \varphi, \dot{\varphi}, \ddot{\varphi}, t) \tag{3.9-1}$$

where all linearized damping and stiffness terms are in the damping and stiffness matrices. $\varphi, \dot{\varphi}, \ddot{\varphi}$ are the angular displacement, velocity, and acceleration of the system. Q_1 and Q_2 are functions of $(\varphi, \dot{\varphi})$ and $(\varphi, \ddot{\varphi})$, respectively. They are mainly due to mass unbalance and disk skew. All the nonlinear forces are included in Q_n.

Two types of speed profiles are commonly used for the startup and shutdown analysis.

1. Linear Speed Profile

Linear profile has a constant acceleration during startup and deceleration during shutdown, as illustrated in Figure 3.9-1.

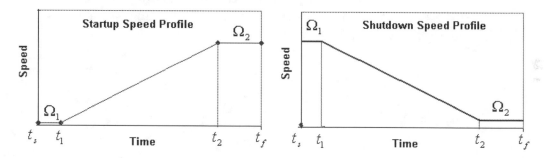

Figure 3.9-1 Linear speed profile

Note that numerical time integration starts from t_s and ends at t_f. Ω_1 is the initial speed from t_s to t_1 and Ω_2 is the final speed after t_2. The angular acceleration, velocity and displacement for three different time zones are:

When $t \le t_1$: Constant Speed

$$\ddot{\varphi}_0 = 0$$
$$\dot{\varphi}_0 = \Omega_1 \tag{3.9-2}$$
$$\varphi_0 = \Omega_1 t$$

When $t_1 \leq t \leq t_2$: Linear Acceleration or Deceleration

$$\ddot{\varphi}_1 = \frac{\Omega_2 - \Omega_1}{t_2 - t_1} = \frac{\Delta\Omega}{\Delta t}$$

$$\dot{\varphi}_1 = \Omega_1 + \ddot{\varphi}_1 \left(t - t_1\right) = \left(\frac{\Omega_1 t_2 - \Omega_2 t_1}{t_2 - t_1}\right) + \left(\frac{\Omega_2 - \Omega_1}{t_2 - t_1}\right) t$$

$$\varphi_1 = C_1 + \left(\frac{\Omega_1 t_2 - \Omega_2 t_1}{t_2 - t_1}\right) t + \frac{1}{2}\left(\frac{\Omega_2 - \Omega_1}{t_2 - t_1}\right) t^2 \qquad (3.9\text{-}3)$$

$$C_1 = \Omega_1 t_1 - \left(\frac{\Omega_1 t_2 - \Omega_2 t_1}{t_2 - t_1}\right) t_1 - \frac{1}{2}\left(\frac{\Omega_2 - \Omega_1}{t_2 - t_1}\right) t_1^2$$

When $t_2 \leq t$: Constant Speed

$$\ddot{\varphi}_2 = 0$$
$$\dot{\varphi}_2 = \Omega_2$$
$$\varphi_2 = \Omega_2 t + C_2 \qquad (3.9\text{-}4)$$
$$C_2 = \phi_1(t_2) - \Omega_2 t_2$$

2. Exponential Speed Profile

In some industrial applications, the rotor speed increases rapidly during initial startup and then gradually increases as the rotor speed approaches operating speed, as shown in Figure 3.9-2. The angular velocity can be expressed in the exponential form:

$$\dot{\varphi}(t) = A + B\,e^{-\lambda t} \qquad (3.9\text{-}5)$$

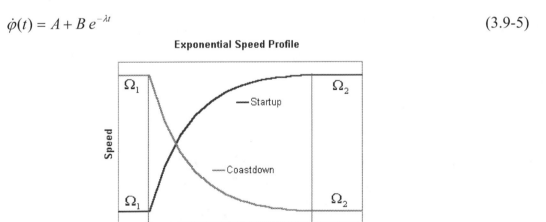

Figure 3.9-2 Exponential speed profile

Due to the characteristics of the exponent, the startup curve can be expressed as:

$$\dot{\varphi} = A\left(1 - e^{-\lambda(t - t_1)}\right) + \Omega_1 \qquad (3.9\text{-}6)$$

which satisfies the initial condition at t_1 instant. At t_2 instant,

$$\Omega_2 = A\left(1 - e^{-\lambda(t_2 - t_1)}\right) + \Omega_1 \qquad (3.9\text{-}7)$$

where A and λ can be related by a single parameter δ, let

$$\Delta\Omega = \Omega_2 - \Omega_1$$
$$\Delta t = t_2 - t_1 \qquad (3.9\text{-}8)$$
$$A = (1 + \delta)\Delta\Omega$$

then we have

$$\lambda = -\frac{\ln\left(\dfrac{\delta}{1 + \delta}\right)}{\Delta t} \qquad (3.9\text{-}10)$$

δ determines the shape of the curve. The startup speed curves for various δ are shown in Figure 3.9-3.

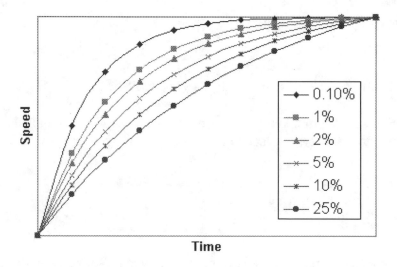

Figure 3.9-3 Exponential speed profiles for various shape parameters

The larger δ value is, the straight the curve becomes. The angular acceleration, velocity and displacement for three different time zones are:

When $t \le t_1$: Constant Speed

$$\ddot{\varphi}_0 = 0$$
$$\dot{\varphi}_0 = \Omega_1 \qquad (3.9\text{-}11)$$
$$\varphi_0 = \Omega_1 t$$

When $t_1 \leq t \leq t_2$: Exponential profile

$$\ddot{\varphi}_1 = A\lambda\, e^{-\lambda(t-t_1)}$$
$$\dot{\varphi}_1 = A\left(1 - e^{-\lambda(t-t_1)}\right) + \Omega_1$$
$$\varphi_1 = C_1 + \left(A + \Omega_1\right)t + \frac{A}{\lambda}e^{-\lambda(t-t_1)} \tag{3.9-12}$$
$$C_1 = -At_1 - \frac{A}{\lambda}$$

When $t_2 \leq t$: Constant Speed

$$\ddot{\varphi}_2 = 0$$
$$\dot{\varphi}_2 = \Omega_2$$
$$\varphi_2 = \Omega_2 t + C_2 \tag{3.9-13}$$
$$C_2 = \phi_1(t_2) - \Omega_2 t_2$$

The shutdown speed curve can be expressed in the same format as the startup curve:

$$\dot{\varphi} = B\left(1 - e^{-\lambda(t-t_1)}\right) + \Omega_1 \tag{3.9-14}$$
$$B = -A$$

Example 3.5: Transient Response through a Critical Speed

A simple 2DOF rotor system taken from Yamamoto and Ishida (2001) is used to illustrate the rotor motion though a critical speed with a constant acceleration. The equations of motion are:

$$m\ddot{x} + c\dot{x} + kx = e\dot{\varphi}^2 \cos\varphi + e\ddot{\varphi}\sin\varphi \tag{3.9-15}$$
$$m\ddot{y} + c\dot{y} + ky = e\dot{\varphi}^2 \sin\varphi - e\ddot{\varphi}\cos\varphi \tag{3.9-16}$$

Constant acceleration is used in this example. The acceleration, velocity, and angle are given by:

$$\ddot{\varphi} = \text{constant}$$
$$\dot{\varphi} = \dot{\varphi}_0 + \ddot{\varphi}\, t \tag{3.9-17}$$
$$\varphi = \varphi_0 + \dot{\varphi}_0 t + \frac{1}{2}\ddot{\varphi}\, t^2$$

This is an isotropic system and the related parameters used by Yamamoto and Ishida (2001) are:

$$m = 1,\ c = 0.02,\ k = 1,\ e = 0.02$$

$$\ddot{\varphi} = 0.002$$

$$\dot{\varphi}_0 = 0.6 \, (\text{rad/sec}) = 5.72958 \, (\text{rpm})$$

$$\varphi_0 = 0$$

Following the previous discussion, the undamped natural frequency and damping factor are:

$$\omega_n = \sqrt{\frac{k}{m}} = 1 \, \text{rad/sec} = 9.55 \, \text{rpm}$$

$$\xi = \frac{c}{2m\omega_n} = 0.01 \quad \text{i.e.} \quad 1 \, \text{percent critical damping}$$

Note that this is a very small damping. The equations of motion can be solved by numerical integration with a given initial conditions $x = y = 0$; and $\dot{x} = \dot{y} = 0$. The time transient responses verses time and speed, are shown in Figure 3.9-4 and Figure 3.9-5.

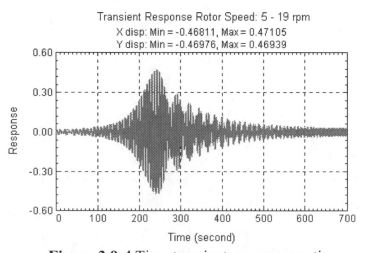

Figure 3.9-4 Time transient response vs. time

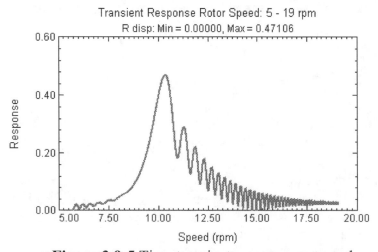

Figure 3.9-5 Time transient response vs. speed

The initial amplitude fluctuation at startup speed in Figure 3.9-5 is caused by the initial conditions in free vibration. The response amplitude increases as the rotor speed approaches resonance, and then decreases after the peak response. The peak response speed occurs at a higher speed than the undamped natural frequency. Since damping is very small, a *beating phenomenon* occurs after resonance due to the coexistence of a natural motion excited at the resonance region and a forced response due to the unbalance. When the damping increases (say $c=0.15$ in this case), the natural motion decays rapidly and the beating phenomenon disappears as illustrated in Figure 3.9-6.

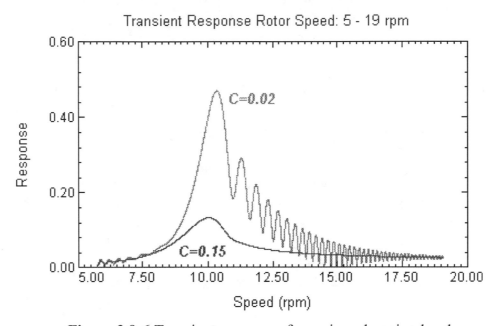

Figure 3.9-6 Transient responses for various damping levels

It is interesting to study the effect of the angular acceleration on the rotor response. For zero acceleration and constant spin speed, the steady-state synchronous response has been studied before. The steady-state maximum peak amplitude, due to mass unbalance, occurs at a rotor speed higher than the undamped natural frequency. The equations for the steady-state peak response and speed have been presented before and they are:

$$X_p = Y_p = \frac{e}{2\xi\sqrt{1-\xi^2}} = 1.0001$$

$$\Omega_p = \frac{\omega_n}{\sqrt{1-2\xi^2}} = 1.0002 \text{ rad/sec} = 9.55 \text{ rpm}$$

The Bode plot for the steady state synchronous response with a constant spin speed is shown in Figure 3.9-7. The transient responses for various accelerations are shown in Figure 3.9-8. It shows that as acceleration increases, maximum response amplitude decreases and peak response speed increases.

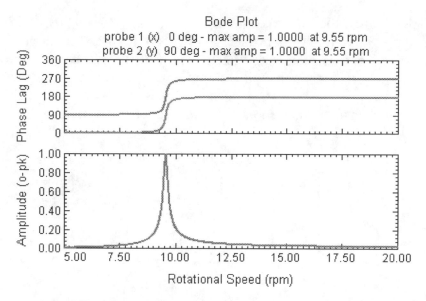

Figure 3.9-7 Bode plot for the steady state unbalance response

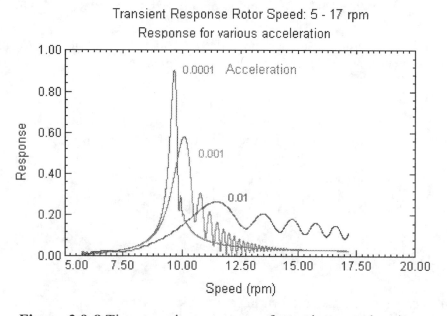

Figure 3.9-8 Time transient responses for various acceleration rates

3.10 The Effects of Flexible Support

Consider the journal mass of M_j is connected to a flexible support of mass M_s by the linearized bearing stiffness K_b and damping C_b. The flexible support is mounted to the ground by a spring of stiffness K_s and a viscous damping C_s, as illustrated in Figure 3.10-1. For simple harmonic motion with a frequency of ω, this 4 degrees-of-freedom (4DOF) journal-bearing-support system can be reduced to be a 2DOF effective journal-bearing system with the reduced equivalent bearing-support dynamic coefficients. Normally, the shaft rotational speed is selected to be the reduction frequency. Thus, the reduced bearing coefficients are called the *synchronously reduced coefficients*. This reduction procedure is commonly used to demonstrate the effects of the flexible support. However, it is not advisable to perform this reduction in the analysis of complex rotor-bearing-support systems due to its simplicity and assumptions. The reduced coefficients can be used in the steady state synchronous response analysis, but not suitable for the whirl speed/stability and transient analysis since the reduced coefficients are frequency dependent.

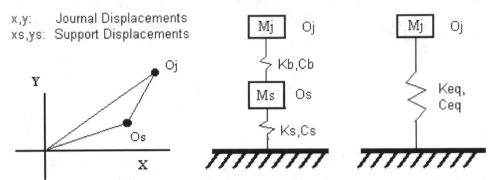

Figure 3.10-1 4DOF and equivalent 2DOF models

For the purely harmonic motions with a whirl frequency of ω, it is convenient and desirable to introduce the impedance notation in a complex form as:

$$Z = K + j\omega C \tag{3.10-1}$$

The forces acting on the journal mass for the complete 4DOF system are:

$$
\begin{aligned}
\begin{Bmatrix} F_x \\ F_y \end{Bmatrix} &= -\begin{bmatrix} K_{xx} & K_{xy} \\ K_{yx} & K_{yy} \end{bmatrix}_b \begin{Bmatrix} x - x_s \\ y - y_s \end{Bmatrix} - \begin{bmatrix} C_{xx} & C_{xy} \\ C_{yx} & C_{yy} \end{bmatrix}_b \begin{Bmatrix} \dot{x} - \dot{x}_s \\ \dot{y} - \dot{y}_s \end{Bmatrix} \\
&= -\begin{bmatrix} Z_{xx} & Z_{xy} \\ Z_{yx} & Z_{yy} \end{bmatrix}_b \begin{Bmatrix} x - x_s \\ y - y_s \end{Bmatrix} = -\mathbf{Z}_b(\mathbf{x} - \mathbf{x}_s)
\end{aligned}
\tag{3.10-2}
$$

where the subscript b represents the bearing properties. The forces acting on the journal mass for the equivalent 2DOF system are:

$$
\begin{aligned}
\begin{Bmatrix} F_x \\ F_y \end{Bmatrix} &= -\begin{bmatrix} K_{xx} & K_{xy} \\ K_{yx} & K_{yy} \end{bmatrix}_{eq} \begin{Bmatrix} x \\ y \end{Bmatrix} - \begin{bmatrix} C_{xx} & C_{xy} \\ C_{yx} & C_{yy} \end{bmatrix}_{eq} \begin{Bmatrix} \dot{x} \\ \dot{y} \end{Bmatrix} \\
&= -\begin{bmatrix} Z_{xx} & Z_{xy} \\ Z_{yx} & Z_{yy} \end{bmatrix}_{eq} \begin{Bmatrix} x \\ y \end{Bmatrix} = -\mathbf{Z}_{eq}\mathbf{x}
\end{aligned}
\tag{3.10-3}
$$

where the subscript *eq* represents the equivalent bearing-support properties. Assuming there is no external loads on the flexible support, the motion of the support mass can be described by the following equations of motion:

$$\begin{bmatrix} M_{xx} & M_{xy} \\ M_{yx} & M_{yy} \end{bmatrix}_s \begin{Bmatrix} \ddot{x}_s \\ \ddot{y}_s \end{Bmatrix} + \begin{bmatrix} C_{xx} & C_{xy} \\ C_{yx} & C_{yy} \end{bmatrix}_s \begin{Bmatrix} \dot{x}_s \\ \dot{y}_s \end{Bmatrix} + \begin{bmatrix} K_{xx} & K_{xy} \\ K_{yx} & K_{yy} \end{bmatrix}_s \begin{Bmatrix} x_s \\ y_s \end{Bmatrix} = -\begin{Bmatrix} F_x \\ F_y \end{Bmatrix} \qquad (3.10\text{-}4)$$

For harmonic motions with a whirl frequency of ω, the dynamic stiffness notation is introduced below:

$$\hat{K} = K - \omega^2 M + j\omega C \qquad (3.10\text{-}5)$$

Then, the equations of motion of the support mass may be written as:

$$\begin{bmatrix} \hat{K}_{xx} & \hat{K}_{xy} \\ \hat{K}_{yx} & \hat{K}_{yy} \end{bmatrix}_s \begin{Bmatrix} x_s \\ y_s \end{Bmatrix} = -\begin{Bmatrix} F_x \\ F_y \end{Bmatrix} \qquad \Rightarrow \qquad \hat{K}_s x_s = -F \qquad (3.10\text{-}6)$$

By inverting the dynamic stiffness matrix, the support displacements become:

$$\begin{Bmatrix} x_s \\ y_s \end{Bmatrix} = -\begin{bmatrix} \hat{K}_{xx} & \hat{K}_{xy} \\ \hat{K}_{yx} & \hat{K}_{yy} \end{bmatrix}_s^{-1} \begin{Bmatrix} F_x \\ F_y \end{Bmatrix} \qquad \Rightarrow \qquad x_s = -\hat{K}_s^{-1} F \qquad (3.10\text{-}7)$$

By substituting Eq. (3.10-7) into Eq. (3.10-2), and comparing with Eq. (3.10-3), we have

$$F = -\left(I + Z_b \hat{K}_s^{-1}\right)^{-1} Z_b x = -Z_{eq} x \qquad (3.10\text{-}8)$$

Therefore, the effective total bearing-support impedance for harmonic motion becomes:

$$Z_{eq} = \left(I + Z_b \hat{K}_s^{-1}\right)^{-1} Z_b \qquad (3.10\text{-}9)$$

To illustrate the effects of the flexible support, an isotropic and undamped system with a bearing stiffness of K_b, a support mass of M_s, and a support stiffness of K_s, is presented below. The dynamic stiffness of the flexible support for a purely harmonic motion with a whirl frequency of ω, is give as:

$$\hat{K}_s = \begin{bmatrix} K_s - \omega^2 M_s & 0 \\ 0 & K_s - \omega^2 M_s \end{bmatrix} \qquad (3.10\text{-}10)$$

and the inversion is:

$$\hat{K}_s^{-1} = \begin{bmatrix} \dfrac{1}{(K_s - \omega^2 M_s)} & 0 \\ 0 & \dfrac{1}{(K_s - \omega^2 M_s)} \end{bmatrix} \qquad (3.10\text{-}11)$$

The impedance matrix for the undamped isotropic bearing is:

$$Z_b = \begin{bmatrix} K_b & 0 \\ 0 & K_b \end{bmatrix}$$

(3.10-12)

Therefore, the total equivalent bearing-support impedance becomes:

$$Z_{eq} = \begin{bmatrix} \dfrac{K_b\left(K_s - \omega^2 M_s\right)}{\left(K_b + K_s - \omega^2 M_s\right)} & 0 \\ 0 & \dfrac{K_b\left(K_s - \omega^2 M_s\right)}{\left(K_b + K_s - \omega^2 M_s\right)} \end{bmatrix}$$

(3.10-13)

Then, the equivalent bearing-support stiffness is represented as:

$$K_{eq} = \frac{K_b\left(K_s - \omega^2 M_s\right)}{K_b + K_s - \omega^2 M_s}$$

(3.10-14)

Note that the effective bearing-support stiffness is now dependent upon the reduction frequency ω, and not a constant like individual bearing and support stiffness.

The effective bearing-support stiffness for various support conditions and frequency excitations are discussed below:

Case 1: High Support Stiffness

When K_s approaches infinity, the effective stiffness K_{eq} approaches the bearing stiffness K_b.

$$K_s \to \infty, \quad K_{eq} \to K_b$$

Case 2: High Support Mass

When $\omega^2 M_s$ approaches infinity, the effective stiffness K_{eq} approaches the bearing stiffness K_b.

$$\omega^2 M_s \to \infty, \quad K_{eq} \to K_b$$

Cases 1 and 2 show that in order to have flexible support influence on rotor behavior, the support stiffness must not be infinite, and the support mass cannot be several orders of magnitude greater than the rotor mass.

Case 3: Zero Support Mass

When the support mass is very small in comparison to the rotor mass, the effective stiffness is equivalent to two springs in series. The effective stiffness is always lower than either the bearing or the support stiffness value.

$$K_{eq} = \frac{K_b K_s}{K_b + K_s} = \frac{1}{\frac{1}{K_b} + \frac{1}{K_s}}$$

Case 4: Support Resonant Condition, $\omega^2 = \dfrac{K_s}{M_s}$

When the support is in resonance with its stiffness, the effective stiffness becomes zero.

$$K_s - \omega^2 M_s = 0, \quad \Rightarrow \quad K_{eq} = 0$$

Therefore, under these conditions, there does not appear to be a bearing acting at that bearing location, so the rotor is supported freely under the support resonance condition.

Case 5: Tuned Vibration Absorber, $\omega^2 = \dfrac{K_b + K_s}{M_s}$

When support mass is in resonance with the combination of bearing stiffness and support stiffness, the effective stiffness becomes infinite.

$$K_b + K_s - \omega^2 M_s = 0, \quad \Rightarrow \quad K_{eq} \to \infty$$

Under these conditions, the support acts as a dynamic vibration absorber, and there is zero motion at the bearing station.

Case 6: - $\omega^2 > \dfrac{K_s}{M_s}$

When the synchronous frequency is higher than the support natural frequency, the effective stiffness may be negative. The numerator of the expression for K_{eq} is given by:

$$\omega^2 > \frac{K_s}{M_s} \quad \Rightarrow \quad K_b\left(K_s - \omega^2 M_s\right) < 0$$

The denominator of K_{eq} is given by:

$$K_b + K_s - \omega^2 M_s$$

This expression may be positive for high values of bearing stiffness, depending upon the support mass and frequency. Hence, a frequency range may exist, in which the effective bearing-support stiffness value is negative.

Example 3.6: The Effects of Flexible Support

A simple undamped 2DOF system as shown in Figure 3.10-2 is used to demonstrate this reduction procedure. The journal and support displacements x_j and x_s are measured from the static equilibrium positions. The journal mass is subject to a harmonic unbalance excitation force $F(t)$.

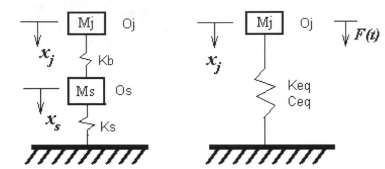

Figure 3.10-2 Two DOF system and its equivalent 1DOD model

For this undamped 2DOF system, the equations of motion are:

$$\begin{bmatrix} m_j & 0 \\ 0 & m_s \end{bmatrix} \begin{Bmatrix} \ddot{x}_j \\ \ddot{x}_s \end{Bmatrix} + \begin{bmatrix} K_b & -K_b \\ -K_b & K_b + K_s \end{bmatrix} \begin{Bmatrix} x_j \\ x_s \end{Bmatrix} = \begin{Bmatrix} F\sin\Omega t \\ 0 \end{Bmatrix} \qquad (3.10\text{-}15)$$

The natural frequencies are determined from homogeneous equations, with the assumed harmonic solutions in the following form:

$$\begin{Bmatrix} x_j \\ x_s \end{Bmatrix} = \begin{Bmatrix} A_j \\ A_s \end{Bmatrix} e^{i\omega t} \qquad (3.10\text{-}16)$$

By substituting these solutions into the homogeneous equations, the nontrivial solution is determined from the characteristic determinant of the system:

$$\begin{vmatrix} K_b - \omega^2 m_j & -K_b \\ -K_b & K_b + K_s - \omega^2 m_s \end{vmatrix} = 0 \qquad (3.10\text{-}17)$$

The above determinant leads to the characteristics equation:

$$m_j m_s \omega^4 - \left(m_s K_b + m_j (K_b + K_s) \right)\omega^2 + K_b K_s = 0 \qquad (3.10\text{-}18a)$$

or

$$\omega^4 - \left(\frac{K_b}{m_j} + \frac{K_b + K_s}{m_s} \right)\omega^2 + \frac{K_b K_s}{m_j m_s} = 0 \qquad (3.10\text{-}18b)$$

There are four values (two positive and two negative roots) of ω, which satisfy the frequency equation. Since the negative frequencies have no physical significance, only two positive values are referred to as the system natural frequencies.

The steady-state unbalance response can be determined by assuming the solutions with the same forms as the excitation:

$$\begin{Bmatrix} x_j \\ x_s \end{Bmatrix} = \begin{Bmatrix} X_j \\ X_s \end{Bmatrix} \sin \Omega t \qquad (3.10\text{-}19)$$

By substituting the solution into the equation of motion, the steady-state amplitudes are determined from the following matrix equation:

$$\begin{bmatrix} K_b - \Omega^2 m_j & -K_b \\ -K_b & K_b + K_s - \Omega^2 m_s \end{bmatrix} \begin{Bmatrix} X_j \\ X_s \end{Bmatrix} = \begin{Bmatrix} F \\ 0 \end{Bmatrix} \qquad (3.10\text{-}20)$$

From Cramer's rule, the amplitudes are:

$$X_j = \frac{\left(K_b + K_s - \Omega^2 m_s\right) F}{\left(K_b - \Omega^2 m_j\right)\left(K_b + K_s - \Omega^2 m_s\right) - K_b^2} \qquad (3.10\text{-}21)$$

$$X_s = \frac{K_b F}{\left(K_b - \Omega^2 m_j\right)\left(K_b + K_s - \Omega^2 m_s\right) - K_b^2} \qquad (3.10\text{-}22)$$

This simple undamped 2DOF system is modeled using **DyRoBeS**, as shown in Figure 3.10-3. The detailed modeling technique for this system with geometric constraints was presented in Chapter 1 and is not repeated here.

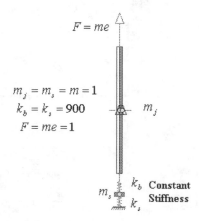

$$F = me$$

$$m_j = m_s = m = 1$$
$$k_b = k_s = 900$$
$$F = me = 1$$

m_j

k_b Constant
m_s Stiffness
k_s

Figure 3.10-3 Two DOF model

The natural frequencies can be calculated by using Whirl Speed Analysis or Critical Speed Analysis. Partial outputs from the whirl speed analysis are listed below for reference.

```
******************** Whirl Speed and Stability Analysis ********************

Shaft  1      Speed=      .00 rpm  =        .00 R/S  =        .00 Hz

              2  Precessional Modes,     0  Pure Real Modes
*************************** Precessional Modes ***************************
*********** Frequency ************    Damping      Log.   Damping
Mode   rpm          R/S         Hz     Coefficient Decrement Factor

1    177.054     18.5410     2.9509     .0000       .000     .000
2    463.533     48.5410     7.7255     .0000       .000     .000
*************************************************************************

------------------------------------------------------------------------
******************** Whirl Speed and Stability Analysis ********************

Shaft  1      Speed=    100.00 rpm  =     10.47 R/S  =      1.67 Hz

              2  Precessional Modes,     0  Pure Real Modes
*************************** Precessional Modes ***************************
*********** Frequency ************    Damping      Log.   Damping
Mode   rpm          R/S         Hz     Coefficient Decrement Factor

1    177.054     18.5410     2.9509     .0000       .000     .000
2    463.533     48.5410     7.7255     .0000       .000     .000
*************************************************************************
```

As expected, there are two natural frequencies found for this two DOF system. The natural frequencies are not dependent on the rotor speed in this case. These values can also be analytically verified from the above characteristic equation (3.10-18). The whirl speed map is shown in Figure 3.10-4. Since the system natural frequencies are not dependent on the rotor speed, the two synchronous critical speeds due to mass unbalance are the two system natural frequencies.

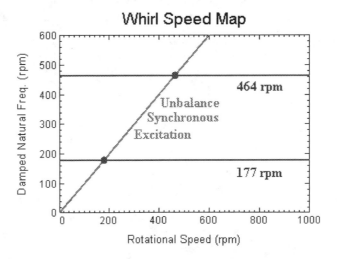

Figure 3.10-4 Whirl speed map

The steady-state unbalance response at the journal and support are shown in Figure 3.10-5 and Figure 3.10-6. It shows that the response becomes infinity when the speed approaches the natural frequencies (177 and 464 rpm). It also shows that when the speed satisfies the equation $\left(K_b + K_s - \Omega^2 m_s\right) = 0$, the journal displacement is zero. At very high speeds, the journal displacement approaches the mass eccentricity and the support displacement approaches zero. The vibration probe is commonly mounted on the support; therefore, the relative displacement (i.e., journal displacement relative to the support displacement) is measured. Figure 3.10-7 shows the journal displacement relative to the support displacement. It shows that at support resonance, $K_s - \Omega^2 M_s = 0$, the journal relative displacement is zero. Again, these results can also be analytically verified by using the equations presented above.

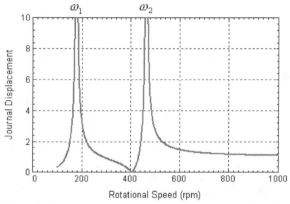

Figure 3.10-5 Steady state journal displacement

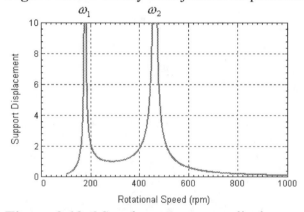

Figure 3.10-6 Steady state support displacement

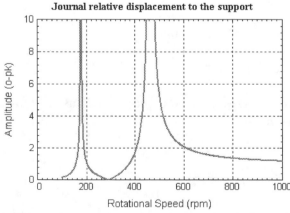

Figure 3.10-7 Steady state journal relative displacement to the support

Another purpose of this example is to demonstrate the reduction procedure. By utilizing the procedure discussed above, the equation of motion for the equivalent 1DOF system is:

$$m_j \ddot{x}_j + K_{eq} x_j = F \sin \Omega t \qquad (3.10\text{-}23)$$

where the effective stiffness previously derived is:

$$K_{eq} = \frac{K_b \left(K_s - \omega_r^2 m_s \right)}{K_b + K_s - \omega_r^2 m_s} \qquad (3.10\text{-}24)$$

Note that ω_r is the reduction frequency. The effective stiffness is not a constant anymore, but depends upon the reduction frequency. The frequency dependent effective stiffness is shown in Figure 3.10-8. It shows that the effective stiffness is zero at 286 cpm, where $K_s - \omega_r^2 m_s = 0$, and can be negative for the frequency from 286 cpm to 405 cpm, where $K_b + K_s - \omega_r^2 m_s = 0$. The value switches from negative infinity to positive infinity from 405 to 406 cpm, where the denominator of K_{eq} is zero.

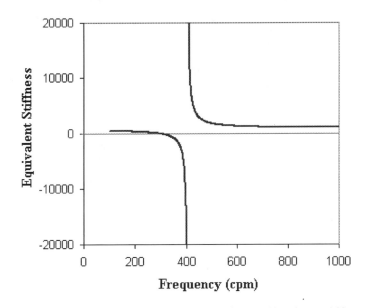

Figure 3.10-8 Frequency dependent effective stiffness

For the determination of system natural frequencies, assume a harmonic solution for the homogeneous equation, and let the reduction frequency be the natural frequency which is to be determined. The characteristic equation for this 1DOF is:

$$K_{eq} - \omega^2 m_j = 0 \qquad (3.10\text{-}25)$$

Substituting the frequency dependent stiffness K_{eq} into the above equation leads to:

$$m_j m_s \omega^4 - \left(m_s K_b + m_j (K_b + K_s)\right)\omega^2 + K_b K_s = 0 \tag{3.10-26}$$

This expression is identical to the characteristic equation (3.10-18) for the original 2DOF system. There are two system natural frequencies for this equivalent 1DOF system.

For steady-state unbalance response, the forcing frequency is the shaft rotational speed. The harmonic response has the same frequency as the excitation; therefore shaft rotational speed is selected to be the reduction frequency. By substituting the harmonic solution into the equation of motion (3.10-23), the steady-state amplitude is:

$$X_j = \frac{F}{(K_{eq} - \Omega^2 m_j)} = \frac{\left(K_b + K_s - \Omega^2 m_s\right)F}{K_b K_s - \Omega^2 \left(K_b m_s + (K_b + K_s)m_j\right) + \Omega^4 m_s m_j} \tag{3.10-27}$$

Note that this expression is identical to the expression derived for the 2DOF system. The denominator is the same as the characteristic equation, which defines the system natural frequency, with the shaft rotational speed replacing the system natural frequency. Therefore, the journal response is the same as before with two peaks due to the frequency dependent stiffness, although this is a 1DOF system.

However, since the reduction procedure is performed before knowing the system natural frequency, the effective stiffness is commonly given in a frequency table, not the analytical form as presented in equation (3.10-24). This equivalent 1DOF system is also modeled in **DyRoBeS** as shown in Figure 3.10-9.

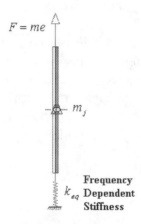

Figure 3.10-9 Equivalent 1DOF model

For the steady-state unbalance response analysis, the procedure can be straightforward, since the reduction frequency is the shaft rotational speed for the synchronous motion. The journal displacement is the same as shown in Figure 3.10-5. The support displacement is however, not available in this 1DOF model. For the determination of system natural frequency, an iterative procedure is required. The eigenvalue problem has to be solved for a range of frequencies. The system natural frequencies are the natural frequencies that coincide with the analysis frequency. For this 1DOF system, the natural frequency is plotted against the analysis frequency, as shown in Figure 3.10-10. Note that the analysis frequency is the shaft rotational speed in this case, since the bearing data is entered as speed-dependent coefficients. The system natural frequencies determined from the graph are 177 and 464 rpm, which are the same as the original 2DOF system.

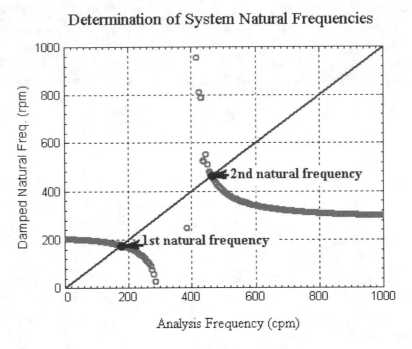

Figure 3.10-10 Determination of system natural frequencies

The reduction procedure is to eliminate the DOF at the flexible support, however it creates the frequency dependent bearing properties. The only resulting benefit is the calculation of steady-state unbalance response. It does not provide any computational advantage for whirl speed and stability analysis. Therefore, this reduction procedure is not recommended if the computational tools, such as **DyRoBeS**, can include the flexible supports into the system model.

Rotating Disk Equations and Rigid Rotor Dynamics **4**

4.1 Rigid Disk Equation

For small vibrations, the kinetic energy of a spinning axisymmetric rigid disk as illustrated in Figure 4.1-1, with translational and rotational motions, may be expressed in a fixed reference coordinate system, as the sum of the translational and rotational kinetic energies:

$$T = \frac{1}{2}m_d\left(\dot{x}^2 + \dot{y}^2\right) + \frac{1}{2}I_d\left(\dot{\theta}_x^2 + \dot{\theta}_y^2\right) + \frac{1}{2}\Omega I_p\left(\dot{\theta}_x\theta_y - \theta_x\dot{\theta}_y\right) + \frac{1}{2}\Omega^2 I_p \qquad (4.1\text{-}1)$$

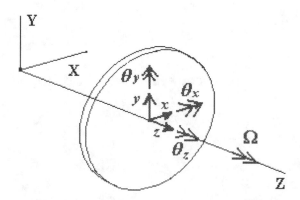

Figure 4.1-1 Displacements for a rigid disk

Where m_d, I_d, I_p are the disk mass, diametral (transverse) mass moment of inertia, and polar mass moment of inertia of the disk. The first two terms of Eq. (4.1-1) are homogeneous quadratic functions of the generalized velocities and are commonly presented in a *natural* system (Meirovitch, 1980). The first term is the translational kinetic energy contributed by the disk mass and translational displacements. The second is a fraction of the rotational kinetic energy contributed by the rotary inertia and rotational displacements. The first two terms are always positive. The third term is linear in the generalized velocities and is referred to as the *gyroscopic effect*, which is contributed by the gyroscopic moments. This term can be either positive or negative depending on the disk motion, as discussed in Chapter 3. The fourth term, caused by the

119

pure spinning of the disk, does not depend on the vibrational coordinates, and therefore can be ignored in the vibrational analysis. Due to the gyroscopic effect, the kinetic energy given in equation (4.1-1) is known as *nonnatural* (Meirovitch, 1980). The gyroscopic effect and the non-conservative nature of the fluid film bearings make the study of rotordynamics unique compared to general structural dynamics studies. The *rigid disk* possesses only kinetic energy, not potential energy. The equations of motion of a spinning rigid disk can be derived from the Lagrange's equation, as discussed in Chapter 1. The coefficient matrices in the equations of motion can also be identified directly by the observation of the energy expressed in matrix form. The kinetic energy expression, excluding the fourth term of Eq. (4.1-1), can be conveniently written in matrix form as:

$$T = \frac{1}{2}\dot{q}^T\left(M_T^d + M_R^d\right)\dot{q} + q^T\Omega g^d\,\dot{q} \qquad (4.1\text{-}2)$$

where $q^T(t) = (x, y, \theta_x, \theta_y)$ is the displacement vector of the finite element station at which the rigid disk is located. The symmetric matrices M_T^d, M_R^d, are the translational mass and rotatory inertia matrices, respectively.

$$M_T^d = \begin{bmatrix} m_d & 0 & 0 & 0 \\ 0 & m_d & 0 & 0 \\ 0 & 0 & 0 & 0 \\ 0 & 0 & 0 & 0 \end{bmatrix}, \qquad M_R^d = \begin{bmatrix} 0 & 0 & 0 & 0 \\ 0 & 0 & 0 & 0 \\ 0 & 0 & I_d & 0 \\ 0 & 0 & 0 & I_d \end{bmatrix} \qquad (4.1\text{-}3)$$

The gyroscopic matrix ΩG^d, is a skew-symmetric matrix derived from the intermediate skew-symmetric matrix Ωg^d:

$$G^d = \left(g^d\right)^T - g^d = -2g^d = \begin{bmatrix} 0 & 0 & 0 & 0 \\ 0 & 0 & 0 & 0 \\ 0 & 0 & 0 & I_p \\ 0 & 0 & -I_p & 0 \end{bmatrix} \qquad (4.1\text{-}4)$$

Note that it is convenient to put the spin speed outside the matrix for easy computational implementation. The gyroscopic matrix cross-couples the two rotational degrees-of-freedom.

For an unbalanced disk, the mass center of gravity is offset from the disk (shaft) geometric center, by an eccentricity of e and a phase angle of ϕ_e. The resulting unbalanced forces can be obtained from Newton's law:

$$F_{u,x} = me\Omega^2\cos(\Omega t + \phi_e) = F_{xc}\cos\Omega t + F_{xs}\sin\Omega t$$

$$F_{u,y} = me\Omega^2\sin(\Omega t + \phi_e) = F_{yc}\cos\Omega t + F_{ys}\sin\Omega t \qquad (4.1\text{-}5)$$

If the disk is mounted in a skew position on the shaft, with an angle of τ between the rotor axis and the disk axis, and a phase angle of ϕ_τ, the resulting moments are similar to the unbalance forces expressions:

$$
\begin{aligned}
M_{\tau,x} &= \tau(I_p - I_d)\Omega^2 \cos(\Omega t + \phi_\tau) = M_{xc}\cos\Omega t + M_{xs}\sin\Omega t \\
M_{\tau,y} &= \tau(I_p - I_d)\Omega^2 \sin(\Omega t + \phi_\tau) = M_{yc}\cos\Omega t + M_{ys}\sin\Omega t
\end{aligned}
\tag{4.1-6}
$$

If gravitational force is included in the system, the gravitational forces are:

$$
\begin{aligned}
F_{g,x} &= mg_x \\
F_{g,y} &= mg_y
\end{aligned}
\tag{4.1-7}
$$

where g_x and g_y are the gravity constants in the X and Y directions, respectively. Their values depend on the orientation of the coordinate system.

The governing equations of motion for a rotating rigid disk (4DOF) in a fixed reference frame can be summarized as:

$$
(M_T^d + M_R^d)\ddot{q} + \Omega G^d \dot{q} = Q_{(4x1)}^d
\tag{4.1-8}
$$

where the generalized force vector contains the unbalance force, skew disk moment, and gravitational force. Substituting the coordinate transformation from Chapter 2, the equations of motion in a rotating reference frame become:

$$
\begin{aligned}
(M_T^d + M_R^d)\ddot{p} &+ \omega\left[2(\hat{M}_T^d + \hat{M}_R^d) + \gamma\, G^d\right]\dot{p} \\
&- \omega^2\left[(M_T^d + M_R^d) - \gamma\,\hat{G}^d\right]p = P_{(4x1)}^d
\end{aligned}
\tag{4.1-9}
$$

where $\gamma = \dfrac{\Omega}{\omega}$ is the spin/whirl ratio and the transformed matrices are given by

$$
\hat{M}_T^d = R^T M_T^d S = \begin{bmatrix} 0 & -m_d & 0 & 0 \\ m_d & 0 & 0 & 0 \\ 0 & 0 & 0 & 0 \\ 0 & 0 & 0 & 0 \end{bmatrix}
\tag{4.1-10}
$$

$$
\hat{M}_R^d = R^T M_R^d S = \begin{bmatrix} 0 & -I_d & 0 & 0 \\ I_d & 0 & 0 & 0 \\ 0 & 0 & 0 & 0 \\ 0 & 0 & 0 & 0 \end{bmatrix}
\tag{4.1-11}
$$

and

$$\hat{G}^d = R^T G^d S = \begin{bmatrix} 0 & 0 & 0 & 0 \\ 0 & 0 & 0 & 0 \\ 0 & 0 & I_p & 0 \\ 0 & 0 & 0 & I_p \end{bmatrix} \qquad (4.1\text{-}12)$$

4.2 Rigid Rotor Dynamics

Before we discuss the flexible rotor with distributed stiffness properties, let us consider rigid rotor dynamics first. If the rotor bending stiffness is much greater than the bearing/support stiffness, and the rotor spin speed is much lower than the bending critical speed, the rotor can be treated as a *rigid rotor*. The dynamic behavior of a rigid rotor system can be described by the motion at any location along the rotor. The center of mass (gravity) is often chosen as the reference point, since it produces the diagonal mass/inertia matrix. The equation of motion for the rigid disk, as described in equation (4.1-8), can be used directly with all the mass properties of the rotor, lumped at the center of mass location. For a rigid rotor, the motion at any typical cross-section, as shown in Figure 4.2-1, located at a distance of z from the origin, is related to the motion at the center of mass at z_c by:

$$\begin{Bmatrix} x \\ y \\ \theta_x \\ \theta_y \end{Bmatrix} = \begin{bmatrix} 1 & 0 & 0 & (z-z_c) \\ 0 & 1 & -(z-z_c) & 0 \\ 0 & 0 & 1 & 0 \\ 0 & 0 & 0 & 1 \end{bmatrix} \begin{Bmatrix} x \\ y \\ \theta_x \\ \theta_y \end{Bmatrix}_c \qquad (4.2\text{-}1)$$

It can be written in matrix form as:

$$q = T q_c \qquad (4.2\text{-}2)$$

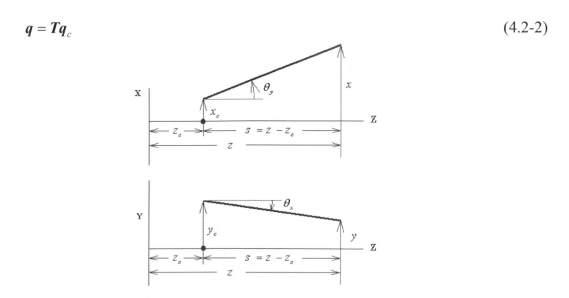

Figure 4.2-1 Displacement relationship for a rigid rotor

where z_c is the axial distance of the center of mass from the origin, and $\boldsymbol{q}_c$ is the displacement vector at the center of mass. The displacements at the center of mass are also the generalized coordinates.

The generalized force by the unbalance excitations, can be derived from the virtual work:

$$\delta W_u = \sum_{i=1}^{nu} \boldsymbol{F}_i \bullet \delta \boldsymbol{q}_i = \sum_{i=1}^{nu} \delta \boldsymbol{q}_i^T \boldsymbol{F}_i = \delta \boldsymbol{q}_c^T \sum_{i=1}^{nu} \boldsymbol{T}_i^T \boldsymbol{F}_i = \delta \boldsymbol{q}_c^T \sum_{i=1}^{nu} \boldsymbol{Q}_i \qquad (4.2\text{-}3)$$

Therefore, the generalized force for the ith unbalance excitation is:

$$\boldsymbol{Q}_i = \boldsymbol{T}_i^T \boldsymbol{F}_i \qquad (4.2\text{-}4)$$

The generalized force by the reactions of the linear and nonlinear bearings can be derived from the virtual work:

$$\delta W_b = \sum_{i=1}^{nb} \boldsymbol{F}_i \bullet \delta \boldsymbol{q}_i = \sum_{i=1}^{nb} \delta \boldsymbol{q}_i^T \boldsymbol{F}_i = \delta \boldsymbol{q}_c^T \sum_{i=1}^{nb} \boldsymbol{T}_i^T \boldsymbol{F}_i = \delta \boldsymbol{q}_c^T \sum_{i=1}^{nb} \boldsymbol{Q}_i \qquad (4.2\text{-}5)$$

Again, the generalized force for the ith bearing has the same form as equation (4.2-4), and the expression is valid for both linear and nonlinear bearings. For the linear bearing, the bearing force can be expressed in terms of the linearized stiffness and damping matrices for the ith bearing:

$$\boldsymbol{F}_i = -\boldsymbol{K}_i \boldsymbol{q}_i - \boldsymbol{C}_i \dot{\boldsymbol{q}}_i = -\boldsymbol{K}_i \boldsymbol{T}_i \boldsymbol{q}_c - \boldsymbol{C}_i \boldsymbol{T}_i \dot{\boldsymbol{q}}_c \qquad (4.2\text{-}6)$$

Therefore the generalized force due to the ith linear bearing becomes:

$$\boldsymbol{Q}_i = \boldsymbol{T}_i^T \boldsymbol{F}_i = -\left(\boldsymbol{T}_i^T \boldsymbol{K}_i \boldsymbol{T}_i\right) \boldsymbol{q}_c - \left(\boldsymbol{T}_i^T \boldsymbol{C}_i \boldsymbol{T}_i\right) \dot{\boldsymbol{q}}_c \qquad (4.2\text{-}7)$$

Thus, the effective generalized stiffness and damping matrices for the ith bearing are:

$$\boldsymbol{K}_{eq,i} = \boldsymbol{T}_i^T \boldsymbol{K}_i \boldsymbol{T}_i \qquad (4.2\text{-}8)$$

$$\boldsymbol{C}_{eq,i} = \boldsymbol{T}_i^T \boldsymbol{C}_i \boldsymbol{T}_i \qquad (4.2\text{-}9)$$

The effective generalized stiffness and damping matrices for the linear bearing can also be obtained directly by substituting the Eq. (4.2-2) into the potential energy, and the dissipation function of the ith bearing:

$$V_i = \frac{1}{2} \boldsymbol{q}_i^T \boldsymbol{K}_i \boldsymbol{q}_i = \frac{1}{2} \boldsymbol{q}_c^T \boldsymbol{T}_i^T \boldsymbol{K}_i \boldsymbol{T}_i \boldsymbol{q}_c = \frac{1}{2} \boldsymbol{q}_c^T \boldsymbol{K}_{eq,i} \boldsymbol{q}_c \qquad (4.2\text{-}10)$$

and

$$\mathfrak{I}_i = \frac{1}{2}\dot{q}_i^T C_i \dot{q}_i = \frac{1}{2}\dot{q}_c^T T_i^T C_i T_i \dot{q}_c = \frac{1}{2}\dot{q}_c^T C_{eq,i} \dot{q}_c \tag{4.2-11}$$

If the bearings are linear isotropic, with translational stiffness k_i and damping c_i, the system stiffness matrix can be reduced to be:

$$K = \sum_{i=1}^{nb} T_i^T K_i T_i = \sum_{i=1}^{nb} \begin{bmatrix} k_i & 0 & 0 & k_i s_i \\ 0 & k_i & -k_i s_i & 0 \\ 0 & -k_i s_i & k_i s_i^2 & 0 \\ k_i s_i & 0 & 0 & k_i s_i^2 \end{bmatrix} \tag{4.2-12}$$

where $s_i = z_i - z_c$ is the distance from the bearing location to the center of mass. The damping matrix has the same form with k_i being replaced by c_i.

Then the four degrees-of-freedom equations of motion for the rotating rigid rotor, with linear isotropic bearings at the center of mass are:

$$M\ddot{q}_c + (C + \Omega G)\dot{q}_c + Kq_c = Q_u \tag{4.2-13}$$

where

$$M = \begin{bmatrix} m & 0 & 0 & 0 \\ 0 & m & 0 & 0 \\ 0 & 0 & I_d & 0 \\ 0 & 0 & 0 & I_d \end{bmatrix} \qquad G = \begin{bmatrix} 0 & 0 & 0 & 0 \\ 0 & 0 & 0 & 0 \\ 0 & 0 & 0 & I_p \\ 0 & 0 & -I_p & 0 \end{bmatrix} \tag{4.2-14}$$

$$C = \sum_{i=1}^{nb} \begin{bmatrix} c_i & 0 & 0 & c_i s_i \\ 0 & c_i & -c_i s_i & 0 \\ 0 & -c_i s_i & c_i s_i^2 & 0 \\ c_i s_i & 0 & 0 & c_i s_i^2 \end{bmatrix} \tag{4.2-15}$$

$$K = \sum_{i=1}^{nb} \begin{bmatrix} k_i & 0 & 0 & k_i s_i \\ 0 & k_i & -k_i s_i & 0 \\ 0 & -k_i s_i & k_i s_i^2 & 0 \\ k_i s_i & 0 & 0 & k_i s_i^2 \end{bmatrix} \tag{4.2-16}$$

and

$$Q_u = \sum_{i=1}^{noofunbalance} T_i^T F_i \tag{4.2-17}$$

Note that if $\sum_{i=1}^{nb} k_i s_i = 0$, then the translational and rotational motions are decoupled and can be solved independently.

If the rotor is symmetric and supported by two identical isotropic bearings, with a translational stiffness of k, as shown in Figure 3.1-3, the stiffness matrix becomes:

$$K = \begin{bmatrix} 2k & 0 & 0 & 0 \\ 0 & 2k & 0 & 0 \\ 0 & 0 & \dfrac{kL^2}{2} & 0 \\ 0 & 0 & 0 & \dfrac{kL^2}{2} \end{bmatrix} \tag{4.2-18}$$

Then the translational and rotational motions are decoupled and can be analyzed separately. This result is in agreement with the results of Chapter 3, for a symmetric rigid rotor system.

Example 4.1: Rigid Rotor Dynamics

A rigid rotor system on flexible bearings, as illustrated in Figure 4.2-2, is presented in this example. The rotating shaft consists of 6-elements (7 stations) where the related geometric dimensions are given in the Figure 4.2-2. The design speed of the rotor system is 10,000 rpm. The material properties of the shaft are:

Weight density = 0.283 Lb_m/in^3
Young's Modulus = 30.0E06 Lb_f/in^2
Shear Modulus = 11.54E06 Lb_f/in^2

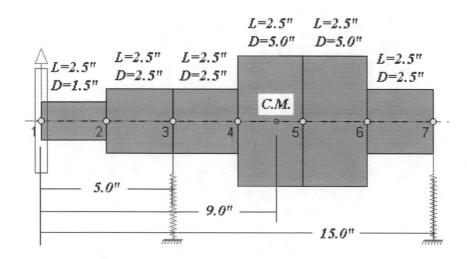

Figure 4.2-2 A rigid rotor system

An overhung disk is mounted on the shaft at station 1, with a mass of 0.755 Lb_m, a diametral moment of inertia of 1.60 $Lb_m\text{-}in^2$, and a polar moment of inertia of 3.0 $Lb_m\text{-}in^2$. The mass properties of the complete rotating assembly can be calculated and listed below for reference purposes:

Mass (m) = 40.208 Lb_m
Diametral Inertia (I_d)= 477.6 $Lb_m\text{-}in^2$
Polar Inertia (I_p) = 98.31 $Lb_m\text{-}in^2$
C.G location (z_c) = 9.0 inches from station 1

The rotor assemble is supported by two identical isotropic bearings at stations 3 and 7, with a stiffness of k=10,000 Lb_f /in, and a viscous damping coefficient of c=10 Lb_f /in. Following the discussion previously presented, the displacements at the center of mass are chosen to be the generalized coordinates. Then, the effective generalized stiffness from Eq. (4.2-16) and damping matrices from Eq. (4.2-15) for the bearings are:

$$
K = \left(\begin{bmatrix} k & 0 & 0 & -4k \\ 0 & k & 4k & 0 \\ 0 & 4k & 16k & 0 \\ -4k & 0 & 0 & 16k \end{bmatrix} + \begin{bmatrix} k & 0 & 0 & 6k \\ 0 & k & -6k & 0 \\ 0 & -6k & 36k & 0 \\ 6k & 0 & 0 & 36k \end{bmatrix} \right)
$$

$$
= \begin{bmatrix} 2k & 0 & 0 & 2k \\ 0 & 2k & -2k & 0 \\ 0 & -2k & 52k & 0 \\ 2k & 0 & 0 & 52k \end{bmatrix} (Lb_f / in)
$$

Similarly,

$$
C = \begin{bmatrix} 2c & 0 & 0 & 2c \\ 0 & 2c & -2c & 0 \\ 0 & -2c & 52c & 0 \\ 2c & 0 & 0 & 52c \end{bmatrix} (Lb_f - s/in)
$$

The rotor system is subject to an unbalance force at the disk (station 1) with me=1 $oz\text{-}in$. The generalized force due to unbalance from Eq. (4.2-4) is:

$$
Q_u = T_1^T F_1 = \begin{bmatrix} 1 & 0 & 0 & 0 \\ 0 & 1 & 0 & 0 \\ 0 & 9 & 1 & 0 \\ -9 & 0 & 0 & 1 \end{bmatrix} \begin{Bmatrix} me\Omega^2 \cos(\Omega t) \\ me\Omega^2 \sin(\Omega t) \\ 0 \\ 0 \end{Bmatrix} = \begin{Bmatrix} me\Omega^2 \cos(\Omega t) \\ me\Omega^2 \sin(\Omega t) \\ 9 \cdot me\Omega^2 \sin(\Omega t) \\ -9 \cdot me\Omega^2 \cos(\Omega t) \end{Bmatrix}
$$

As expected, the unbalance force acting at station 1 contributes not only the forces, but also the moment, at center of mass location. Then the equations of motion at the center of mass, for this rigid rotor system are:

$$
\begin{bmatrix} m & 0 & 0 & 0 \\ 0 & m & 0 & 0 \\ 0 & 0 & I_d & 0 \\ 0 & 0 & 0 & I_d \end{bmatrix} \begin{Bmatrix} \ddot{x} \\ \ddot{y} \\ \ddot{\theta}_x \\ \ddot{\theta}_y \end{Bmatrix}_c + \left(\begin{bmatrix} 2c & 0 & 0 & 2c \\ 0 & 2c & -2c & 0 \\ 0 & -2c & 52c & 0 \\ 2c & 0 & 0 & 52c \end{bmatrix} + \Omega \begin{bmatrix} 0 & 0 & 0 & 0 \\ 0 & 0 & 0 & 0 \\ 0 & 0 & 0 & I_p \\ 0 & 0 & -I_p & 0 \end{bmatrix} \right) \begin{Bmatrix} \dot{x} \\ \dot{y} \\ \dot{\theta}_x \\ \dot{\theta}_y \end{Bmatrix}_c +
$$

$$
\begin{bmatrix} 2k & 0 & 0 & 2k \\ 0 & 2k & -2k & 0 \\ 0 & -2k & 52k & 0 \\ 2k & 0 & 0 & 52k \end{bmatrix} \begin{Bmatrix} x \\ y \\ \theta_x \\ \theta_y \end{Bmatrix}_c = me\Omega^2 \begin{Bmatrix} \cos(\Omega t) \\ \sin(\Omega t) \\ 9\sin(\Omega t) \\ -9\cos(\Omega t) \end{Bmatrix}
$$

$$(4.2\text{-}19)$$

where the parameters: m, I_d, I_p, k, c, and me are given in the above text. Once the motion at the center of mass is known, the displacement at any other stations can be easily obtained from the rigid body characteristics of Eq. (4.2-1). For this system, the translational and rotational motions are coupled by the generalized stiffness and damping matrices, also two planes of motion are coupled by the gyroscopic matrix.

To validate the rigid rotor assumption, the complete rotor system is analyzed. The first three undamped synchronous forward critical speeds are found to be: 4,059 rpm, 7,009 rpm, and 145,729 rpm by using the Critical Speed Analysis. The associated mode shapes and potential energy distributions are plotted in Figure 4.2-3. For the first two modes, the entire shaft has less than 1 percent of potential energy while two bearings have more than 99 percent of potential energy. These two modes are classified as *rigid body modes*. Since the translational and rotational motions are coupled by the bearing stiffness and damping matrices, shown in Eq. (4.2-19), these two modes are not purely cylindrical or conical modes. However, the first mode is predominated by translational motion and the second mode is predominated by rotational motion. In the third mode, the associated frequency is much higher than the second frequency. The shaft exhibits significant deformation, and possesses more than 99 percent of potential energy. This mode is classified as a *bending mode*. For normal operation of the rotor system, the rotor speed is well below the third mode. Therefore this rotor system can be treated as a *rigid rotor*.

The Whirl Speed Map is plotted by using the Whirl Speed/Stability Analysis and shown in Figure 4.2-4. The critical speeds determined by the Whirl Speed Map are in agreement with the results from Critical Speed Analysis. It also shows that the gyroscopic effect has a minimal influence on the first mode since the motion is predominated by the translational motion. It does however, have a significant influence on the second mode since the motion is predominated by the rotational motion.

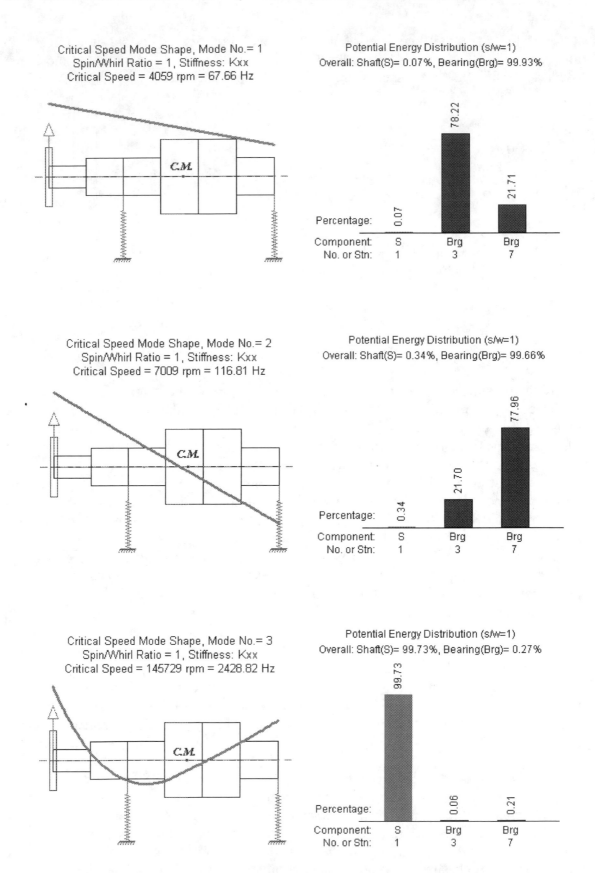

Figure 4.2-3 The first three critical speed mode shapes and associated potential energy

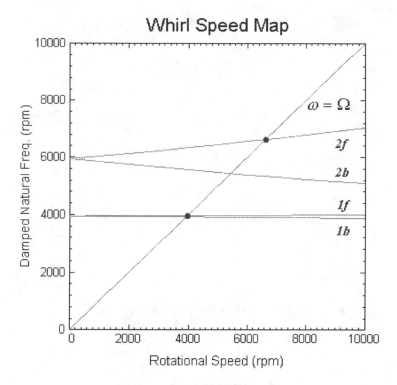

Figure 4.2-4 Whirl speed map

The steady-state unbalance response at the center of mass and the two bearing locations are plotted in Figure 4.2-5. The steady state shaft response at 4,000 rpm, 5,500 rpm, and 10,000 rpm is shown in Figure 4.2-6. Readers are encouraged to use the *Startup/Shutdown Animation* feature provided in **DyRoBeS** to visualize the rotor motion when it passes through the first two critical speeds.

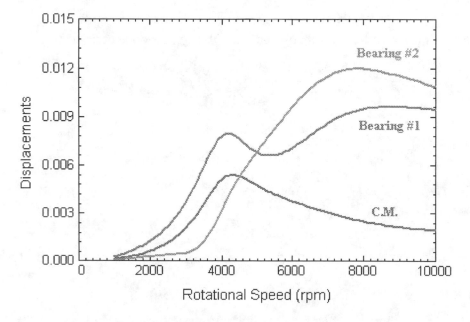

Figure 4.2-5 The steady state unbalance response

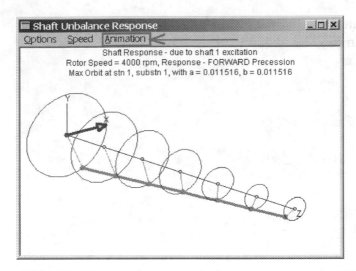

Figure 4.2-6 (a) The steady state unbalance response at 4000 rpm

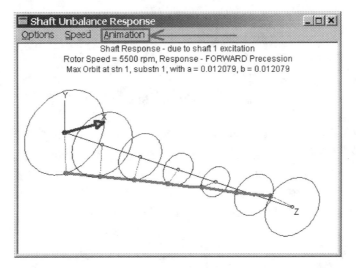

Figure 4.2-6 (b) The steady state unbalance response at 5500 rpm

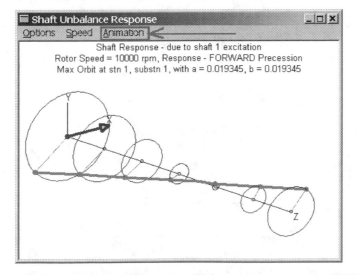

Figure 4.2-6 (c) The steady state unbalance response at 10000 rpm

Now, let us relocate bearing #1 to the left (away from CG location) by 2 inches as shown in Figure 4.2-7. In this case, the center of mass is at the center of the bearing span. The effective stiffness matrix becomes a diagonal matrix:

$$K = \begin{bmatrix} 2k & 0 & 0 & 0 \\ 0 & 2k & 0 & 0 \\ 0 & 0 & 72k & 0 \\ 0 & 0 & 0 & 72k \end{bmatrix}$$

The effective damping matrix has the same form. It shows that the translational and rotational motions are decoupled now, and can be solved independently; even the rotor is not symmetric.

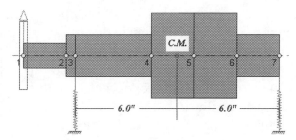

Figure 4.2-7 A rigid rotor with CM at the center of bearing span

The first two undamped synchronous forward critical speeds are calculated to be 4,180 rpm and 8,169 rpm by using the Critical Speed Analysis. Their associated mode shapes and potential energies are plotted in Figure 4.2-8. It shows that the first mode is a purely translational (cylindrical) mode, and the second mode is a purely rotational (conical) mode. It also shows that the potential energy is split evenly on both bearings for these two modes. Note that the small difference in values is due the shaft flexibility. The critical speeds can also be calculated analytically from the procedure discussed in Chapter 3 due to the decoupled translational and rotational motions.

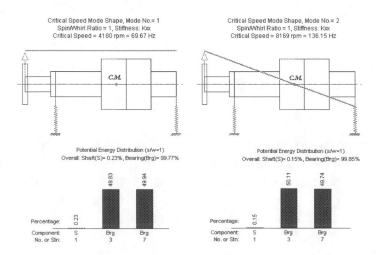

Figure 4.2-8 The first two critical speed mode shapes and associated potential energy

4.3 Rigid Disk on Flexible Rotor

In the previous section, the dynamics of a rigid rotor system on flexible bearings was discussed. In this section, a rigid disk on a flexible massless rotor, with rigid bearings is presented. Consider a single rigid disk mounted off-center on a flexible shaft with negligible mass and the shaft is supported by two rigid bearings, as illustrated in Figure 4.3-1. Before we discuss the finite element formulation of the rotating flexible shaft element, the flexibility influence coefficient method is presented here, to obtain the stiffness matrix for the entire flexible shaft. The importance of this technique is that in some applications, the shaft stiffness is very difficult to obtain analytically and must rely on the experiments. The influence coefficient a_{ij}, is defined as the displacement at the i location due to a unit force applied at the j location. For small deformations, the principle of superposition can be applied to determine the displacements in terms of the influence coefficients. Once the influence coefficient matrix is known, the inverse of the influence coefficient matrix is the stiffness matrix. For simply supported uniform flexible beam configurations, the deflections and rotations (slopes) due to forces and moments acting on the beam, are tabulated in most of the textbooks of strength of material. Since the shaft element is considered isotropic, only one plane is needed for the derivation of the stiffness coefficients, and the results can be expanded into two planes. For the disk mounted inside the bearing span, as illustrated in Figure 4.3-1, the deflection and rotation at the disk location due to force F_x and moment M_y action on the disk in the X-Z plane, are shown in Figure 4.3-2:

$$x = \left(\frac{a^2 b^2}{3EIL} \right) F_x + \left(\frac{ab(b-a)}{3EIL} \right) M_y \tag{4.3-1}$$

$$\theta_y = \left(\frac{ab(b-a)}{3EIL} \right) F_x + \left(\frac{a^3 + b^3}{3EIL^2} \right) M_y \tag{4.3-2}$$

The displacement-force relationship can be expressed in matrix form as:

$$\begin{Bmatrix} x \\ \theta_y \end{Bmatrix} = \begin{bmatrix} \left(\dfrac{a^2 b^2}{3EIL} \right) & \left(\dfrac{ab(b-a)}{3EIL} \right) \\ \left(\dfrac{ab(b-a)}{3EIL} \right) & \left(\dfrac{a^3 + b^3}{3EIL^2} \right) \end{bmatrix} \begin{Bmatrix} F_x \\ M_y \end{Bmatrix} \tag{4.3-3}$$

For the centrally mounted disk ($a = b$), the configuration becomes the classical Laval-Jeffcott rotor system, and the influence coefficient matrix is diagonal. That is the translational and rotational motions are decoupled and can be analyzed separately, as discussed in Chapter 3. However, for a non-centrally mounted disk ($a \neq b$), the deflection and slope are coupled together, and the translational and rotational motions have to be analyzed together.

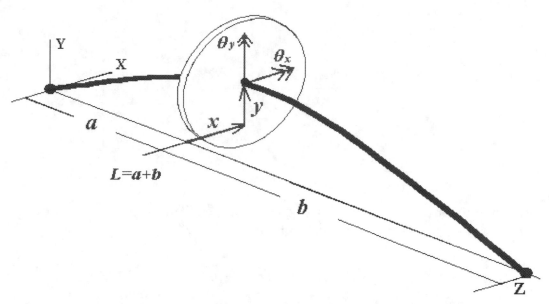

Figure 4.3-1 An off-centered rigid disk mounted on a flexible rotor

Deflection curves due to unit loading

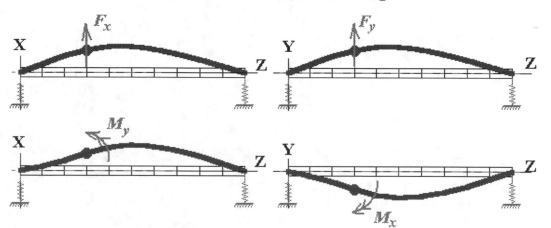

Figure 4.3-2 Deflection curves due to unit loading

The stiffness matrix is the inverse of the flexibility matrix.

$$\begin{Bmatrix} F_x \\ M_y \end{Bmatrix} = \begin{bmatrix} \left(3EI\dfrac{a^3+b^3}{a^3b^3}\right) & \left(3EI\dfrac{L(a-b)}{a^2b^2}\right) \\ \left(3EI\dfrac{L(a-b)}{a^2b^2}\right) & \left(3EI\dfrac{L}{ab}\right) \end{bmatrix} \begin{Bmatrix} x \\ \theta_y \end{Bmatrix} = \begin{bmatrix} k_{11} & k_{14} \\ k_{41} & k_{44} \end{bmatrix} \begin{Bmatrix} x \\ \theta_y \end{Bmatrix} \qquad (4.3\text{-}4)$$

Note that from the reciprocity theorem, the stiffness and flexibility matrices are symmetric about the diagonal ($k_{ij}=k_{ji}$). The complete stiffness matrix for configuration, as illustrated in Figure 4.3-1, is obtained by expanding the one plane into two planes motion:

$$\begin{Bmatrix} F_x \\ F_y \\ M_x \\ M_y \end{Bmatrix} = 3EI \begin{bmatrix} \dfrac{a^3+b^3}{a^3b^3} & 0 & 0 & \dfrac{L(a-b)}{a^2b^2} \\[2ex] 0 & \dfrac{a^3+b^3}{a^3b^3} & -\dfrac{L(a-b)}{a^2b^2} & 0 \\[2ex] 0 & -\dfrac{L(a-b)}{a^2b^2} & \dfrac{L}{ab} & 0 \\[2ex] \dfrac{L(a-b)}{a^2b^2} & 0 & 0 & \dfrac{L}{ab} \end{bmatrix} \begin{Bmatrix} x \\ y \\ \theta_x \\ \theta_y \end{Bmatrix} \qquad (4.3\text{-}5)$$

Following the same procedure described above, the stiffness matrix for an overhung rotor, as illustrated in Figure 4.3-3, can be obtained as:

$$\begin{Bmatrix} F_x \\ F_y \\ M_x \\ M_y \end{Bmatrix} = \begin{bmatrix} \dfrac{12EI}{a^3}\left(\dfrac{3a+L}{3a+4L}\right) & 0 & 0 & \dfrac{-6EI}{a^2}\left(\dfrac{3a+2L}{3a+4L}\right) \\[2ex] 0 & \dfrac{12EI}{a^3}\left(\dfrac{3a+L}{3a+4L}\right) & \dfrac{6EI}{a^2}\left(\dfrac{3a+2L}{3a+4L}\right) & 0 \\[2ex] 0 & \dfrac{6EI}{a^2}\left(\dfrac{3a+2L}{3a+4L}\right) & \dfrac{4EI}{a}\left(\dfrac{3a+3L}{3a+4L}\right) & 0 \\[2ex] \dfrac{-6EI}{a^2}\left(\dfrac{3a+2L}{3a+4L}\right) & 0 & 0 & \dfrac{4EI}{a}\left(\dfrac{3a+3L}{3a+4L}\right) \end{bmatrix} \begin{Bmatrix} x \\ y \\ \theta_x \\ \theta_y \end{Bmatrix}$$

$$(4.3\text{-}6)$$

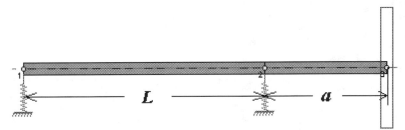

Figure 4.3-3 A rigid disk on an overhung rotor

Now the equations of motion for the rigid disk on a flexible shaft, as illustrated in Figure 4.3-1 or Figure 4.3-3, become:

$$(M_T^d + M_R^d)\ddot{q} + \Omega G^d \dot{q} + Kq = Q_{(4 \times 1)}^d \qquad (4.3\text{-}7)$$

Note that the gyroscopic matrix couples the two planes of motion by the rotational degrees-of-freedom, (θ_x, θ_y). For the non-centrally mounted disk $(a \neq b)$, the shaft stiffness matrix K, couples the translational and rotational displacements for each plane, (X-Z) and (Y-Z) planes, respectively. The shaft is considered to be isotropic; therefore two planes of deformation are not coupled by the stiffness matrix. For the centrally mounted disk $(a = b)$, the configuration becomes the classical Laval-Jeffcott rotor system and the stiffness matrix is diagonal.

Example 4.2: The Laval-Jeffcott Rotor System (4DOF)

To demonstrate the disk effects on a flexible rotor, a classical Laval-Jeffcott rotor system with the rigid disk mounted at the center of the shaft, as shown in Figure 4.3-4(a), is considered first. Then, the study extends to a more generalized Laval-Jeffcott rotor system with the rigid disk located off-center on the shaft, as shown in Figure 4.3-4(b). The rotor system is a 4-degrees-of-freedom model with two translational and two rotational displacements at the disk. The pertinent parameters for this generalized Laval-Jeffcott rotor are summarized in the figure. The shaft mass is negligible in this example. A small value of 1.0E-10 is entered in the material density, to provide a positive definite mass matrix and avoid numerical singularity. Since the shaft mass is neglected, a two-elements (3-stations) model is adequate for this 4 DOF system. The 3-stations model however, would not be able to "graphically" demonstrate the disk rotational DOF. Therefore, a 10-elements (11 stations) model is used in this example in order to graphically illustrate the disk rotational DOF. Each element has an equal length of 1 inch. The term "Station" is commonly used in rotordynamics, instead of the term "Node", which is generally used in finite element literature, because of the alternate meaning that Node has in the vibration mode shapes of rotordynamics.

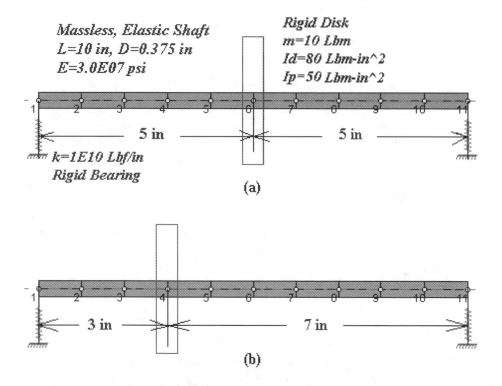

Figure 4.3-4 A rigid disk on a flexible rotor

Again, technically speaking, this is a 44 DOF model. However, since the shaft mass is extremely small compared to the disk mass, the additional degrees-of-freedom due to the distributed shaft mass possess very high natural frequencies compared to the 4 fundamental natural frequencies due to the rigid disk mass and moment of inertia, and they have no effect in these 4 fundamental modes. The equations of motion for the rigid disk from Eq. (4.3-7) are:

$$
\begin{bmatrix} m & 0 & 0 & 0 \\ 0 & m & 0 & 0 \\ 0 & 0 & I_d & 0 \\ 0 & 0 & 0 & I_d \end{bmatrix} \begin{Bmatrix} \ddot{x} \\ \ddot{y} \\ \ddot{\theta}_x \\ \ddot{\theta}_y \end{Bmatrix} + \Omega \begin{bmatrix} 0 & 0 & 0 & 0 \\ 0 & 0 & 0 & 0 \\ 0 & 0 & 0 & I_p \\ 0 & 0 & -I_p & 0 \end{bmatrix} \begin{Bmatrix} \dot{x} \\ \dot{y} \\ \dot{\theta}_x \\ \dot{\theta}_y \end{Bmatrix}
$$
$$
+ \begin{bmatrix} k_{11} & 0 & 0 & k_{14} \\ 0 & k_{22} & k_{23} & 0 \\ 0 & k_{32} & k_{33} & 0 \\ k_{41} & 0 & 0 & k_{44} \end{bmatrix} \begin{Bmatrix} x \\ y \\ \theta_x \\ \theta_y \end{Bmatrix} = \begin{Bmatrix} F_x \\ F_y \\ M_x \\ M_y \end{Bmatrix}
$$

(4.3-8)

From previous discussion, we learned that the shaft stiffness matrix is symmetric and the coefficients for this configuration are:

$$
k_{11} = k_{22} = 3EI \frac{a^3 + b^3}{a^3 b^3}
$$
$$
k_{44} = k_{33} = 3EI \frac{L}{ab}
$$
$$
k_{14} = k_{41} = -k_{23} = -k_{32} = 3EI \frac{L(a-b)}{a^2 b^2}
$$

(4.3-9)

The whirl speed and stability analysis is performed from 0 rpm to 10000 rpm with an increment of 1000 rpm. This analysis solves the complex eigenvalue problem of the complete two planes of motion. Since this rotor system is conservative (undamped), the imaginary parts of eigenvalues are also the undamped natural frequencies. The planar, forward synchronous, and backward synchronous critical speeds are calculated using the Critical Speed Analysis. The critical speeds determined by the whirl speed/stability analysis and by the critical speed analysis, are in agreement. Two cases are discussed in details with the centrally mounted disk first.

Case 1: Centrally mounted disk (*a=b=5*)

The first case is the symmetric rotor with the disk located at the center of the shaft. In this case (*a=b=5*), the translational and rotational motions are decoupled ($k_{14} = k_{41} = k_{23} = k_{32} = 0$) as expected and the dynamics of this decoupled system can be analytically determined using the discussion in Chapter 3. The four fundamental natural frequencies, calculated by using *DyRoBeS*, are plotted in the Whirl Speed Map, as shown in Figure 4.3-5. The Whirl Speed Map is a plot of natural frequencies vs. rotational speed. The Whirl Speed and Stability analysis solves the complex eigenvalue problems for each given speed.

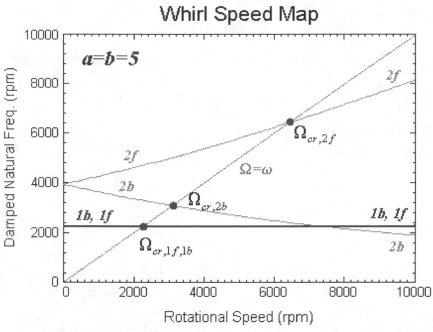

Figure 4.3-5 Whirl speed map for the centrally mounted disk

The natural frequencies and associated normal modes are numbered as 1b, 1f, 2b, and 2f. The "b" denotes the *backward whirl* and "f" denotes the *forward whirl*. A forward precessional mode is defined as a whirling motion in the same direction as the spin speed, while a backward precessional mode is in the opposite direction. It will be demonstrated from the analytical results and the calculated mode shapes, that the first two modes (1b, 1f) are purely translational motions at the disk, and have zero rotational motions at the disk location in this centrally mounted configuration. Further, the translational motion in both X and Y directions are not coupled, the first two natural frequencies are the same value due to the repeated eigenvalues and one for each plane of motion. The first two natural frequencies remain the same value as the speed increases due to the purely translational motion without the gyroscopic effect. However, the third and fourth modes (2b, 2f) are purely rotational motions at the disk without translational displacements. These two modes split into two directions with the rotor speed, and two planes of motion are coupled due to the gyroscopic effect. As the speed increases, the backward whirl frequency (2b) decreases and the forward whirl frequency (2f) increases. This is known as the gyroscopic stiffening effect on forward modes and softening effect on backward modes. At above 7500 rpm, the frequency of mode (2b) is even lower than the frequencies of (1b) and (1f) due to the gyroscopic effects. The synchronous excitation line is also overlapped in the map, and the intersections between the natural frequency (whirl speed) curves and the excitation line are referred to as *synchronous critical speeds*.

For a non-rotating system with zero rpm the natural frequencies occur in pairs: one for each plane of motion. At zero rpm, the forward and backward modes are not really applicable, but it is understandable and convenient to have these labels. The four fundamental modes of vibration at zero rpm are plotted in Figure 4.3-6. It is evident from Figure 4.3-6 that at the zero speed, the first two modes (repeated eigenvalues) are purely translational modes and one for each plane of motion, and the third and fourth modes (repeated eigenvalues) are purely rotational modes and again one for each plane of motion.

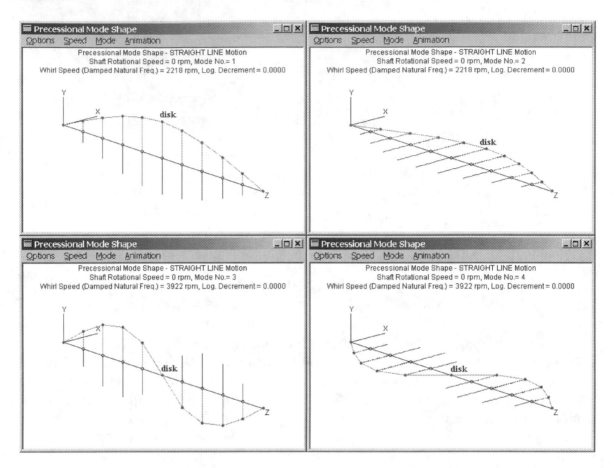

Figure 4.3-6 The first four mode shapes at zero rpm

The mode shapes at 10,000 rpm are plotted in Figure 4.3-7, in the order of the frequency values. Since there is no gyroscopic effect on modes 1b and 1f, and the two planes of motions are not coupled, the natural modes vibrate in their own plane, as illustrated in the mode shapes. The straight-line motion may be considered as being made up of two circular orbits of equal amplitudes, where one has a forward motion and the other has a backward motion. The gyroscopic effect splits modes 2b and 2f into two directions. The backward precessional mode has a lower natural frequency and the forward precessional mode has a higher natural frequency. Since this is an isotropic system and the gyroscopic effect couples two planes of motion, the whirling orbits are circular for modes 2b and 2f, instead of straight-line motions as for modes 1b and 1f. Also, this is an undamped system, so there is no phase lag in the responses. Therefore, the phase angles for all the stations are the same. These observations are in agreement with the discussions in Chapter 3, for the decoupled translational and rotational motions of a classical Laval-Jeffcott rotor system. Readers are strongly encouraged to use the animation option provided by the program to see the whirling motion. A picture is worth a thousand words.

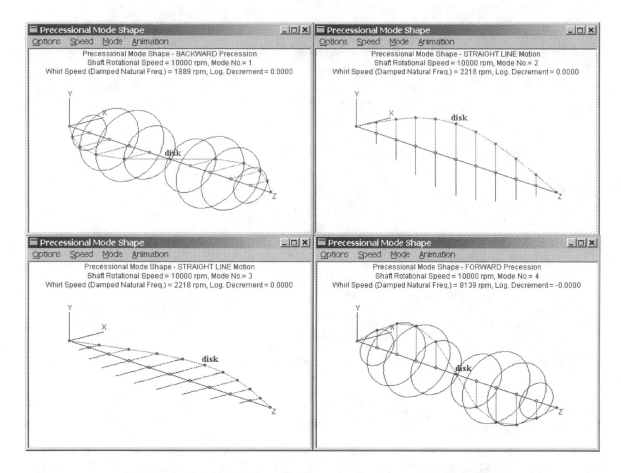

Figure 4.3-7 The first four mode shapes at 10,000 rpm

The critical speeds and any excitation resonance speeds are determined graphically by noting the coincidence of the shaft speed with the system natural frequencies for a given excitation line in the Whirl Speed Map. A synchronous excitation line is shown in the Whirl Speed Map, both the backward and forward synchronous critical speeds can be determined as indicated in the map. In this case, we have two synchronous critical speeds (forward and backward) for the translational motion and the two synchronous critical speeds (forward and backward) for the rotational motion. The forward synchronous critical speed for the rotational motion exists due to $I_d > I_p$, as explained in Chapter 3. For this symmetric Laval-Jeffcott rotor system, the critical speeds can be analytically determined as previously discussed. The whirl speeds at zero speed, and the synchronous critical speeds are determined as follows, for comparison purposes:

$$k_T = k_{11} = k_{22} = \frac{48EI}{L^3} = 1397.84 \text{ Lbf/in}$$

$$k_R = k_{33} = k_{44} = \frac{12EI}{L} = 34946 \text{ Lbf - in/rad}$$

$$k_{14} = k_{41} = k_{23} = k_{32} = 0$$

For translational motion, the natural frequencies are the critical speeds since the natural frequencies are independent on the speed in this case:

$$\Omega_{cr,1b,1f} = \omega_x = \omega_y = \sqrt{\frac{k_T}{m}} = \sqrt{\frac{1397.84}{10/386.088}} = 232.3 \text{ rad/sec} = 2218 \text{ rpm}$$

For rotational motion, the natural frequencies at zero speed are:

$$\omega_{\theta x} = \omega_{\theta y} = \sqrt{\frac{k_R}{I_d}} = \sqrt{\frac{34946}{80/386.088}} = 410.7 \text{ rad/sec} = 3922 \text{ rpm}$$

The forward synchronous critical speed due to purely rotational motion is:

$$\Omega_{cr,2f} = \sqrt{\frac{k_R}{I_d - I_p}} = \sqrt{\frac{34946}{(80-50)/386.088}} = 670.6 \text{ rad/sec} = 6404 \text{ rpm}$$

The backward synchronous critical speed due to purely rotational motion is:

$$\Omega_{cr,2b} = \sqrt{\frac{k_R}{I_d + I_p}} = \sqrt{\frac{34946}{(80+50)/386.088}} = 322.2 \text{ rad/sec} = 3076 \text{ rpm}$$

For an isotropic and undamped system, the critical speeds and other types of resonance speeds can be determined by using Critical Speed Analysis, without generating a complete whirl speed map. The Critical Speed Analysis calculates the undamped critical speeds and modes directly by solving a reduced eigenvalue problem associated with the system equations expressed in a rotating reference frame. These normal modes are circular due to isotropic characteristics and every finite element station has the same phase angle for each mode due to the undamped characteristics. Therefore, it is convenient to analyze the isotropic system in a rotating reference frame. The ratio of rotor spinning speed (Ω) to the whirling (rotating) speed (ω) of the rotating reference frame, determines the types of critical speeds to be solved. It is commonly called Spin/Whirl Ratio, and it is the inverse of the slope of the excitation line in the whirl speed map. The typical values for the *Spin/Whirl Ratio* (Ω/ω) are:

  1:    Forward synchronous critical speeds
 -1:   Backward synchronous critical speeds
  0:    Planar natural frequencies for non-rotating systems (zero speed)
  2:    Half frequency whirl (sub-synchronous criticals)
 1/n: Supsynchronous critical speeds excited by high order harmonic excitations

The forward synchronous critical speeds (Spin/Whirl Ratio = 1) are by far the most common, because they are the ones excited by unbalance, shaft residual bow, and disk skew. The planar natural frequencies and synchronous critical speeds determined by using the Critical Speed Analysis are tabulated below:

		Spin/Whirl Ratio	
Mode	0	1	-1
1	2218	2218	2218
2	3922	6404	3076

The results calculated by Critical Speed Analysis are identical to those calculated by Whirl Speed and Stability Analysis, and they are in agreement with the analytical results.

Now, let us examine the critical speed mode shapes and their associated energies. Figure 4.3-8 shows the first two forward synchronous critical speed mode shapes and their associated kinetic energies. For the first mode, the disk has a maximum deflection (translational displacement) and zero slope (rotational displacement). Therefore, the first mode is a purely translational mode where all the kinetic energy is due to the disk translational motion. For the second mode, the disk has a maximum slope (rotational displacement) and zero deflection (translational displacement). Therefore, the second mode is a purely rotational mode where all the kinetic energy is due to the disk rotatory inertia and gyroscopic moments. It should be noted that since this is a forward synchronous mode, the kinetic energy due to the gyroscopic effect is negative.

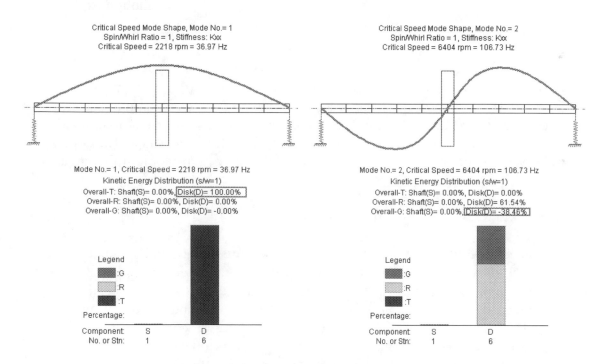

Figure 4.3-8 The first two forward synchronous critical speed mode shapes and their associated kinetic energies.

Figure 4.3-9 shows the first two backward synchronous critical speed mode shapes and their associated kinetic energies. The mode shapes are identical to the forward synchronous critical speed mode shapes in this case. However, it should be noted that since this is a backward synchronous mode, the kinetic energy due to the gyroscopic effect is positive.

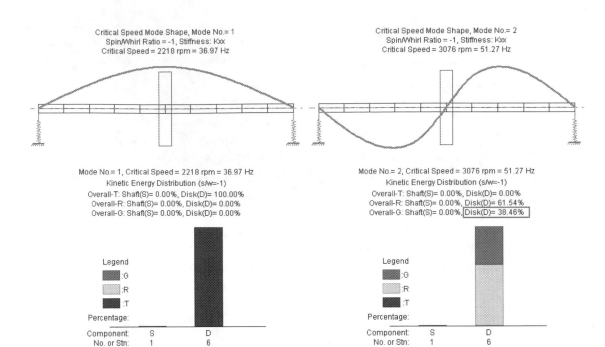

Figure 4.3-9 The first two backward synchronous critical speed mode shapes and their associated kinetic energies.

For the forward precessional mode, the gyroscopic effect contributes negative kinetic energy (much like inertia decreased effect), and tends to raise the corresponding forward whirl frequency (gyroscopic stiffening effect). For the backward precessional mode, the gyroscopic effect contributes positive kinetic energy (much like inertia added effect), and tends to lower the corresponding backward whirl frequency (gyroscopic softening effect).

Case 2: Off-Centrally mounted disk ($a=3$, $b=7$)

Now, let us consider the second case, that is, the rigid disk is located off-center. Let us move the disk location to station 4 as shown in Figure 4.3-4(b). That is, $a=3$ and $b=7$. The four fundamental natural frequencies are plotted against the rotor speed in Figure 4.3-10. Since the disk is located off-center, the translational and rotational motions are coupled. Also, two planes of motion are coupled by the gyroscopic effect. As a result, all the natural frequencies are affected by the disk gyroscopic moments. The natural frequencies associated with forward whirl increase as the speed increases, and the natural frequencies associated with backward whirl decrease as the speed increases. This gyroscopic effect is illustrated in Figure 4.3-10.

The four fundamental mode shapes at zero speed (0 rpm) are plotted in Figures 4.3-11. Again, at zero speed, the modes are planar in each direction. However, since the disk is off-center, the translational and rotational motions of the disk are coupled as illustrated in the mode shapes.

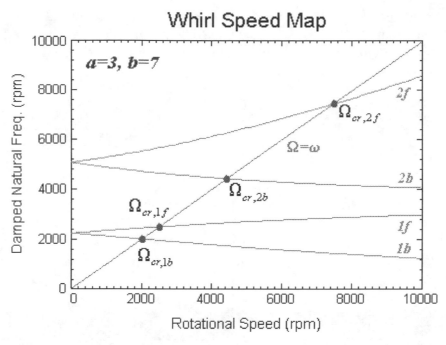

Figure 4.3-10 Whirl speed map for off-centered mounted disk

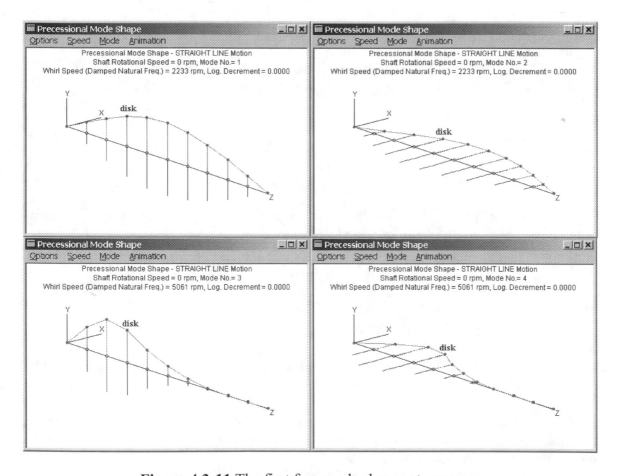

Figure 4.3-11 The first four mode shapes at zero rpm

The four fundamental mode shapes at 10,000 rpm are plotted in 4.3-12. Since the translational and rotational motions are coupled by the off-centered disk and the two planes of motion are coupled by the gyroscopic effect, straight-line motions no longer exist; so all the modes are circular.

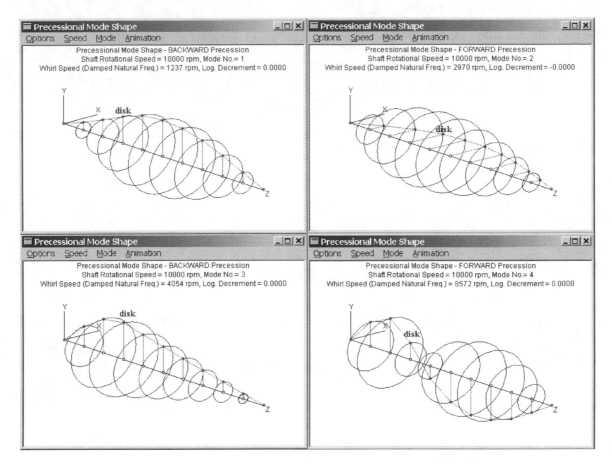

Figure 4.3-12 The first four mode shapes at 10,000 rpm

The planar natural frequencies and synchronous critical speeds determined by using the Critical Speed Analysis are tabulated below:

	Spin/Whirl Ratio		
Mode	0	1	-1
1	2233	2490	2006
2	5061	7410	4420

The results calculated by the Critical Speed Analysis are identical to those determined by the Whirl Speed and Stability Analysis.

It should be noted that due to the couplings among the motions, the mode shape sometimes changes with speed. Figure 4.3-13 shows the mode shapes for the second forward mode (2f) as the speed increases.

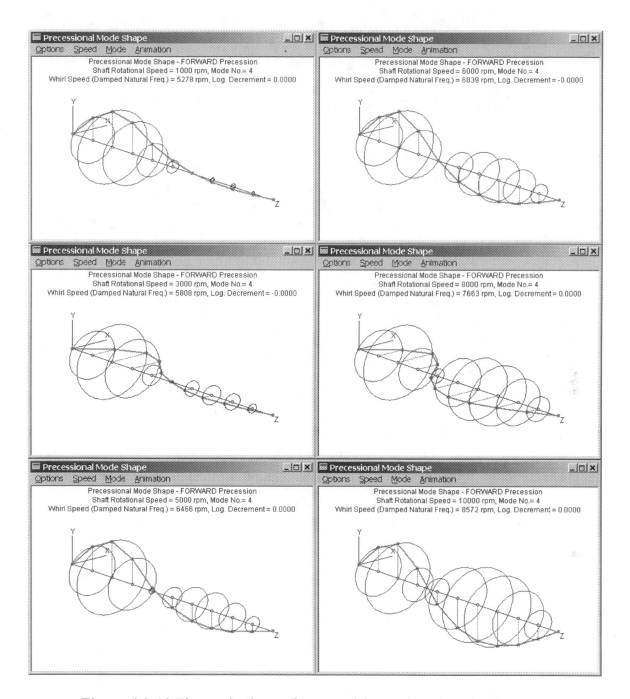

Figure 4.3-13 The mode shapes for second forward mode (2f) with speed

The above results can also be verified analytically by using the equations presented previously. For $a=3$ and $b=7$, the stiffness coefficients are:

$$k_{11} = k_{22} = 3EI \frac{a^3 + b^3}{a^3 b^3} = 3490.4 \text{ Lbf/in}$$

$$k_{44} = k_{33} = 3EI \frac{L}{ab} = 41602 \text{ Lbf - in/rad}$$

$$k_{14} = k_{41} = -k_{23} = -k_{32} = 3EI\frac{L(a-b)}{a^2b^2} = -7924.3\,\text{Lbf/rad}$$

The natural frequencies (eigenvalues) can be obtained from the homogeneous form of the above equations of motion, Eq. (4.3-8), by assuming the eigensolution $q = Ze^{\lambda t}$.

$$(M\lambda^2 + \Omega G\lambda + K\lambda)Z = 0_{(4x1)}$$

The characteristic equation is an eighth order polynomial equation if the above approach is adopted. For an isotropic and undamped system, the normal modes are circular due to isotropic characteristics and the phase angle is the same for each mode at all the finite element stations due to the undamped characteristics. The critical speeds and other types of resonance speeds can be determined by solving a reduced eigenvalue problem associated with the system equations expressed in a rotating reference frame. The eigensolution (normal mode) is planar and constant relative to the rotating reference frame. Since the two planes of motion are 90 degrees out-of-phase and decoupled in the rotating reference frame, it is possible to simplify the case of a two-planes rotordynamics problem into a planar one. Let us consider the X-Z plane of equation (4.1-9), the eigenvalue problem in the rotating reference frame becomes:

$$\left\{\begin{bmatrix} k_{11} & k_{14} \\ k_{41} & k_{44} \end{bmatrix} - \omega^2\left(\begin{bmatrix} m & 0 \\ 0 & I_d \end{bmatrix} - \frac{\Omega}{\omega}\begin{bmatrix} 0 & 0 \\ 0 & I_p \end{bmatrix}\right)\right\}\begin{Bmatrix} X \\ Z \end{Bmatrix} = \begin{Bmatrix} 0 \\ 0 \end{Bmatrix}$$

The characteristic equation for the above equation is:

$$\begin{vmatrix} k_{11} - \omega^2 m & k_{14} \\ k_{41} & k_{44} - \omega^2 I_{eq} \end{vmatrix} = 0$$

or

$$mI_{eq}\omega^4 - \left(k_{11}I_{eq} + k_{44}m\right)\omega^2 + \left(k_{11}k_{44} - k_{14}k_{41}\right) = 0$$

where

$$I_{eq} = I_d - \frac{\Omega}{\omega}I_p$$

For the planar (zero speed) natural frequencies:

$$\frac{\Omega}{\omega} = 0,\ I_{eq} = 80/386.088,$$

$$\omega_1 = 233.83\,rad/\sec = 2233\,rpm$$

$$\omega_2 = 529.96\,rad/\sec = 5061\,rpm$$

For the forward synchronous critical speeds:

$$\frac{\Omega}{\omega} = 1, \; I_{eq} = (80 - 50)/386.088,$$

$$\omega_1 = 260.78 \, rad/\sec = 2490 \, rpm$$

$$\omega_2 = 775.99 \, rad/\sec = 7410 \, rpm$$

For the backward synchronous critical speeds:

$$\frac{\Omega}{\omega} = -1, \; I_{eq} = (80 + 50)/386.088,$$

$$\omega_1 = 210.04 \, rad/\sec = 2006 \, rpm$$

$$\omega_2 = 462.81 \, rad/\sec = 4420 \, rpm$$

These results are identical to the computational results from **DyRoBeS**.

Figure 4.3-14 shows the first two forward synchronous critical speed mode shapes and their associated kinetic energy. Again, the kinetic energy due to gyroscopic effect for the forward modes is negative. For the off-center disk, the translational and rotational motions are coupled. Therefore, the gyroscopic effect exists in both modes.

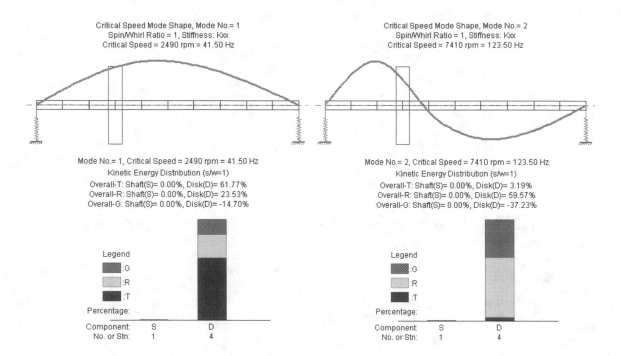

Figure 4.3-14 The first two forward synchronous critical speed mode shapes and their associated kinetic energies.

4.4 Flexible Disk Equation

If the attachment of the disk to the shaft is not stiff, or the natural frequency of the first diametral mode of a disk is close to the operating speed range of the rotor system, the disk is considered to be flexible. For a rigid disk, there are four degrees-of-freedom (2 translational and 2 rotational displacements) to describe the motion of the disk as described in previous sessions. These DOF are the same as the DOF of the shaft finite element station, to which the disk is attached. However, for a flexible disk, two additional rotational DOF are introduced. That is, there is a total of 6 DOF for each flexible disk. The flexible disk option can be important for large overhung rotors, such as large gas turbines where disk flexibility must be taken into consideration. The diametral and polar moments of inertia for the inner and outer disks are typically carefully adjusted to match the first disk diametral resonant frequency. Since the outer disk has only 2 rotational DOF and no translational DOF, it only possesses the moments of inertia and the entire mass of the disk is lumped into the inner disk. Figure 4.4-1 shows the mathematical models for the rigid (4 DOF) and flexible disks (6 DOF).

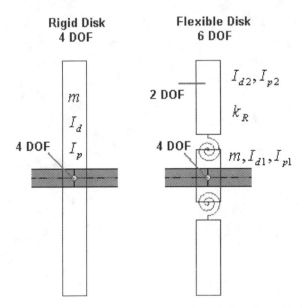

Figure 4.4-1 Models for rigid and flexible disks

The equations of motion for the flexible disk (6 DOF) in the fixed reference frame is the extension of the rigid disk equations:

$$M^d \ddot{q} + \Omega G^d \dot{q} + K^d q = Q^d_{(6x1)} \tag{4.4-1}$$

where

$$q = \left(x, y, \theta_x, \theta_y, \theta_{x2}, \theta_{y2}\right)^T$$

$\left(\theta_{x2}, \theta_{y2}\right)$ are the additional DOF for the outer disk.

$$\boldsymbol{M}^d = \begin{bmatrix} m & 0 & 0 & 0 & 0 & 0 \\ 0 & m & 0 & 0 & 0 & 0 \\ 0 & 0 & I_{d1} & 0 & 0 & 0 \\ 0 & 0 & 0 & I_{d1} & 0 & 0 \\ 0 & 0 & 0 & 0 & I_{d2} & 0 \\ 0 & 0 & 0 & 0 & 0 & I_{d2} \end{bmatrix} \tag{4.4-2}$$

$$\boldsymbol{G}^d = \begin{bmatrix} 0 & 0 & 0 & 0 & 0 & 0 \\ 0 & 0 & 0 & 0 & 0 & 0 \\ 0 & 0 & 0 & I_{p1} & 0 & 0 \\ 0 & 0 & -I_{p1} & 0 & 0 & 0 \\ 0 & 0 & 0 & 0 & 0 & I_{p2} \\ 0 & 0 & 0 & 0 & -I_{p2} & 0 \end{bmatrix} \tag{4.4-3}$$

$$\boldsymbol{K}^d = \begin{bmatrix} 0 & 0 & 0 & 0 & 0 & 0 \\ 0 & 0 & 0 & 0 & 0 & 0 \\ 0 & 0 & k_R & 0 & -k_R & 0 \\ 0 & 0 & 0 & k_R & 0 & -k_R \\ 0 & 0 & -k_R & 0 & k_R & 0 \\ 0 & 0 & 0 & -k_R & 0 & k_R \end{bmatrix} \tag{4.4-4}$$

This component equation of motion is ready for assembly into the system equations of motion.

4.5 Offset Disk Equation

Very often, the disk is attached to the rotor by an offset as shown in Figure 4.5-1.

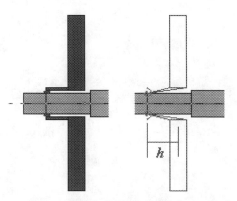

Figure 4.5-1 Offset disk model

An offset disk is cantilevered from a rotor station by an axial distance h.   The displacement vector at the rotor station is given as:

$$\mathbf{q}^T = \left(x, y, \theta_x, \theta_y\right) \tag{4.5-1}$$

And the displacement vector for an offset disk at the center of mass is given as:

$$\mathbf{q}_d^T = \left(x_d, y_d, \theta_{xd}, \theta_{yd}\right) \tag{4.5-2}$$

The attachment can be either rigid or flexible.   The component equations of motion for both cases are discussed below.

4.5.1 Rigid Offset Disk

If the attached disk is considered to be rigid, then the rigid disk displacement can be related to the rotor displacement by the following relationship:

$$\begin{aligned}
x_d &= x + h\theta_y \\
y_d &= y - h\theta_x \\
\theta_{xd} &= \theta_x \\
\theta_{yd} &= \theta_y
\end{aligned} \tag{4.5-3}$$

The kinetic energy for a rigid offset disk is given:

$$T = \frac{1}{2}m_d\left(\dot{x}_d^2 + \dot{y}_d^2\right) + \frac{1}{2}I_d\left(\dot{\theta}_{xd}^2 + \dot{\theta}_{yd}^2\right) + \frac{1}{2}\Omega I_p\left(\dot{\theta}_{xd}\theta_{yd} - \theta_{xd}\dot{\theta}_{yd}\right) + \frac{1}{2}\Omega^2 I_p \tag{4.5-4}$$

Substituting the disk displacements by the rotor displacements, and applying the Lagrange's equation, the translational mass and rotatory inertia matrices are:

$$\mathbf{M}_T^d = \begin{bmatrix} m_d & 0 & 0 & m_d h \\ 0 & m_d & -m_d h & 0 \\ 0 & -m_d h & m_d h^2 & 0 \\ m_d h & 0 & 0 & m_d h^2 \end{bmatrix}, \qquad \mathbf{M}_R^d = \begin{bmatrix} 0 & 0 & 0 & 0 \\ 0 & 0 & 0 & 0 \\ 0 & 0 & I_d & 0 \\ 0 & 0 & 0 & I_d \end{bmatrix} \tag{4.5-5}$$

The gyroscopic matrix $\Omega\mathbf{G}^d$ is a skew-symmetric matrix derived from the intermediate skew-symmetric matrix $\Omega\mathbf{g}^d$:

$$\boldsymbol{G}^d = \left(\boldsymbol{g}^d\right)^T - \boldsymbol{g}^d = -2\boldsymbol{g}^d = \begin{bmatrix} 0 & 0 & 0 & 0 \\ 0 & 0 & 0 & 0 \\ 0 & 0 & 0 & I_p \\ 0 & 0 & -I_p & 0 \end{bmatrix} \tag{4.5-6}$$

If gravitational force is included in the system, the gravitational forces at the rotor station are:

$$\begin{Bmatrix} F_x \\ F_y \\ M_x \\ M_y \end{Bmatrix}_g = \begin{Bmatrix} m_d g_x \\ m_d g_y \\ -m_d h g_y \\ m_d h g_x \end{Bmatrix} \tag{4.5-7}$$

In this case, no additional degrees-of-freedom is required and the component equation of motion is expressed in terms of the rotor displacements. Once the rotor displacements are known, the disk displacements can be obtained from Eq. (4.5-3).

4.5.2 Flexible Offset Disk

If the attached disk is considered to be flexible and the rotational flexibility is assumed to be at the attachment, then the flexible disk displacement can be related to the rotor displacement by the following relationship:

$$\begin{aligned} x_d &= x + h\theta_{yd} \\ y_d &= y - h\theta_{xd} \\ \theta_{xd} &\neq \theta_x \\ \theta_{yd} &\neq \theta_y \end{aligned} \tag{4.5-8}$$

Note that the rotational displacements of the disk are not equal to the rotational displacements of the rotor station for the flexible disk. They are connected by a rotational spring. Therefore, two additional rotational DOF are introduced for the flexible disk and the generalized displacement vector becomes:

$$\boldsymbol{q}^T = \left(x, y, \theta_x, \theta_y, \theta_{xd}, \theta_{yd}\right) \tag{4.5-9}$$

From the energy expressions and Lagrange's equation, the translational mass matrix, rotatory inertia matrix, gyroscopic matrix, and the additional stiffness matrix due to the disk flexibility are:

$$\boldsymbol{M}_T^d = \begin{bmatrix} m_d & 0 & 0 & 0 & 0 & m_d h \\ 0 & m_d & 0 & 0 & -m_d h & 0 \\ 0 & 0 & 0 & 0 & 0 & 0 \\ 0 & 0 & 0 & 0 & 0 & 0 \\ 0 & -m_d h & 0 & 0 & m_d h^2 & 0 \\ m_d h & 0 & 0 & 0 & 0 & m_d h^2 \end{bmatrix} \tag{4.5-10}$$

$$\boldsymbol{M}_R^d = \begin{bmatrix} 0 & 0 & 0 & 0 & 0 & 0 \\ 0 & 0 & 0 & 0 & 0 & 0 \\ 0 & 0 & 0 & 0 & 0 & 0 \\ 0 & 0 & 0 & 0 & 0 & 0 \\ 0 & 0 & 0 & 0 & I_d & 0 \\ 0 & 0 & 0 & 0 & 0 & I_d \end{bmatrix} \tag{4.5-11}$$

$$\boldsymbol{G}^d = \begin{bmatrix} 0 & 0 & 0 & 0 & 0 & 0 \\ 0 & 0 & 0 & 0 & 0 & 0 \\ 0 & 0 & 0 & 0 & 0 & 0 \\ 0 & 0 & 0 & 0 & 0 & 0 \\ 0 & 0 & 0 & 0 & 0 & I_p \\ 0 & 0 & 0 & 0 & -I_p & 0 \end{bmatrix} \tag{4.5-12}$$

$$\boldsymbol{K}^d = \begin{bmatrix} 0 & 0 & 0 & 0 & 0 & 0 \\ 0 & 0 & 0 & 0 & 0 & 0 \\ 0 & 0 & k_R & 0 & -k_R & 0 \\ 0 & 0 & 0 & k_R & 0 & -k_R \\ 0 & 0 & -k_R & 0 & k_R & 0 \\ 0 & 0 & 0 & -k_R & 0 & k_R \end{bmatrix} \tag{4.5-13}$$

If gravitational force is included in the system, the gravitational forces are:

$$\begin{Bmatrix} F_x \\ F_y \\ M_x \\ M_y \\ M_{xd} \\ M_{yd} \end{Bmatrix}_g = \begin{Bmatrix} m_d g_x \\ m_d g_y \\ 0 \\ 0 \\ -m_d h g_y \\ m_d h g_x \end{Bmatrix} \tag{4.5-14}$$

Note that the above expressions are for the flexible offset disk where the rotational flexibility is assumed at the attachment. If the flexibility occurs at the middle of the disk, then the displacement relationship becomes:

$$x_d = x + h\theta_y$$
$$y_d = y - h\theta_x$$
$$\theta_{xd} \neq \theta_x \qquad\qquad\qquad (4.5\text{-}15)$$
$$\theta_{yd} \neq \theta_y$$

In this case, the flexible offset disk can be modeled as the combination of a rigid offset disk and a flexible disk where ($m_d = I_{d1} = I_{p1} = 0$, and $k_r, I_{d2}, I_{p2} \neq 0$,) which are located at the same rotor station. **DyRoBeS-Rotor** allows multiple disks at the same rotor station to accommodate this situation. The offset disk is frequently used in the model of an aircraft engine as shown in Figure 4.5-2.

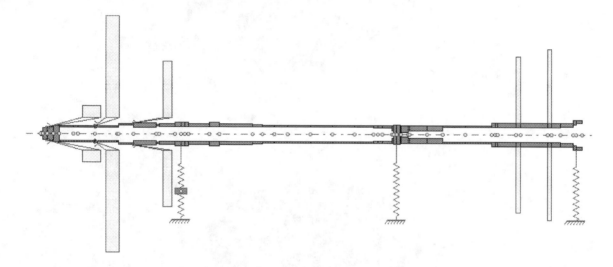

Figure 4.5-2 Offset disks in an aircraft engine model

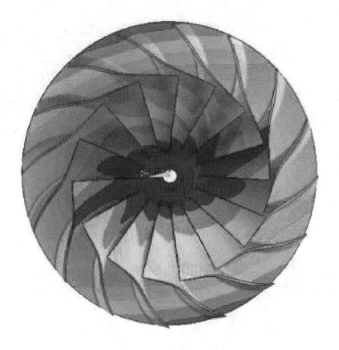

First Diametral Disk Mode

Second Diametral Disk Mode

Shaft Finite Element Equations **5**

5.1 General Considerations

Typical rotor systems, as illustrated in Figure 5.1-1 for a dual-shaft system and in Figure 5.1-2 for a single shaft system, consist of three primary component groups: rotating assemblies, non-rotating flexible supports, and massless (non-structural) interconnection components, such as bearings/dampers, seals, and aerodynamics cross-couplings, etc., that provide physical links or interacting forces between structural components. The rotating assembly is comprised of numerous shaft segments with various cross-sections and disks. The rotating assembly considered here is axisymmetric. The asymmetric rotating assembly introduces destabilizing effect and should be avoided. The governing equations and effects of the spinning disk have already been addressed in the previous chapter. In this chapter, the shaft element equations and effects will be discussed.

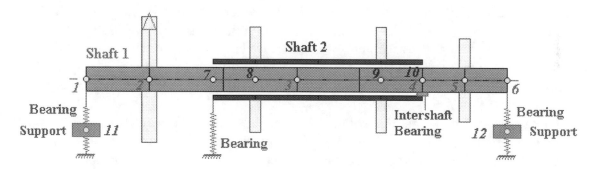

Figure 5.1-1 A dual-shaft system

The complete equations of motion for the shaft are obtained by assembling the equations of motion of each finite element segment. This is accomplished by relating the element coordinates to the chosen set of system coordinates through statements of displacement compatibility that insure the connectivity of the shaft. In the finite element approach, the shafts of the system are numbered consecutively from 1 to Ns (number of shafts) and the finite element stations of the model are numbered consecutively, starting with 1 at the left end of shaft 1 and continuing to the last station at the right end of shaft Ns. The shafts are made up of *Elements* with the numbering for each shaft starting at the left end. *Stations* are located at the ends of the Elements as illustrated in Figure 5.1-2.

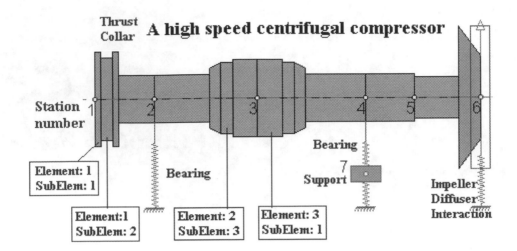

Figure 5.1-2 A typical single shaft system

Element i is located immediately to the right of *station i*. Each element may possess several sub-elements (starting with 1 at the left of the element) thereby allowing for reasonable flexibility in modeling shafts with several geometric discontinuities. The use of sub-elements is strongly encouraged when modeling large complicated rotor systems. This will save tremendous computational time with a minimal loss of accuracy in the results. However, when using sub-elements, it should be kept in mind that Disks and Bearings can only be placed at the finite element stations. Thus, stations should always be specified at major mass stations and bearings. In the finite element method, the degrees-of-freedom (DOF) at the finite element stations are the master DOF, which are kept in the assembled equations of motion, while the degrees-of-freedom at the internal sub-elements are considered to be slave DOF, which are condensed out before the assembly process. Once the displacements of the finite element stations (master DOF) have been found, the displacements of the sub-elements (slave DOF) can then be obtained by utilizing the condensation matrix. Sometimes the sub-element and element are referenced loosely here, since a sub-element is an element in the component equation level.

Three types of elements are commonly used in rotordynamics study and they are: *Cylindrical Element, Conical (Tapered) Element*, and *User's Supplied Element*. For other types of non-uniform cross-section elements, they can always be approximately modeled with these three basic element types. The user's supplied element is mainly used when the segment is too complicated to model and the flexibility influence coefficients matrix can be obtained by experiments. In such cases, the element stiffness can be obtained by inversing the flexibility matrix. The equations of motion of the cylindrical and conical element types can be obtained analytically. For the non-uniform cross-section elements, the cross-section area and area moment of inertia are dependent upon the spatial coordinate. Therefore, the coefficients in the elemental matrices must be numerically integrated and the closed-form expression is not readily available. To avoid the expense of numerical integration, the non-uniform cross-section element is commonly modeled with several cylindrical or conical elements, where the closed form expressions are available.

Very often different diameters are used for the kinetic energy and potential energy calculations, as illustrated in Figure 5.1-3 for an industrial motor. The wirings of the motor contribute to the kinetic energy (mass), but not to the potential energy (stiffness); Therefore, the diameter used in the kinetic energy calculation is greater than the diameter used in the potential energy calculation. The choice of the diameters depends on the experimental data, engineering judgment, and practical experience.

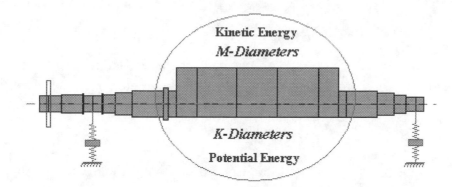

Figure 5.1-3 Different Diameters for the kinetic and potential energies

5.2 Cylindrical Element

A typical finite rotating shaft element (or sub-element) is illustrated in Figure 5.2-1. The element is considered to be homogeneous with distributed mass and elasticity. The lateral motion of a typical cross section of the element, located at a distance of s from the left end point, is described by two translational and two rotational displacements. Since the internal displacements $(x, y, \theta_x, \theta_y)$ are functions of spatial coordinate (s) and time (t), the finite element method is utilized to separate the variables. By utilizing the finite element formulation, the governing partial differential equations of motion become ordinary differential equations and the continuous system with infinite DOF become a discrete finite DOF system.

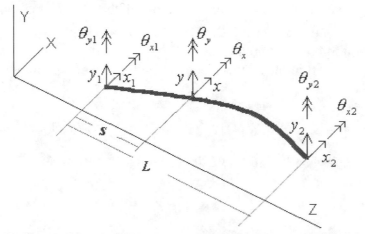

Figure 5.2-1 Coordinates for a typical finite shaft element

The internal displacements of a typical element can be approximately expressed in terms of the eight end-point displacements by specifying spatial shape function as:

$$
\begin{Bmatrix} x(s,t) \\ y(s,t) \\ \theta_x(s,t) \\ \theta_y(s,t) \end{Bmatrix}_{4\times 1} = \begin{bmatrix} \Psi_T(s) \\ \Psi_R(s) \end{bmatrix}_{4\times 8} q^e(t)_{8\times 1}
\tag{5.2-1}
$$

The displacement vector $q^e(t)$ is the time dependent end-point displacements (two translations and two rotations at each end) of the finite shaft element.

$$
q^e = (q_1 \mid q_2)^T = (x_1, y_1, \theta_{x1}, \theta_{y1} \mid x_2, y_2, \theta_{x2}, \theta_{y2})^T
\tag{5.2-2}
$$

The shape function matrix, $\Psi(s)$, is established by utilizing the beam elasticity theory, which can include the transverse shear deformation effect.   The individual shape functions represent the static displacement modes associated with a unit displacement of one of the end-point coordinates with all other coordinates constrained to zero. Since the shaft element is assumed to be isotropic and axisymmetric about the axis of rotation, only one plane of deformation is needed for the derivation of the shape functions. Subsequently, they can be expanded into two planes of deformation.

5.2.1 Bernoulli-Euler Beam Theory

For Bernoulli-Euler Beams without shear deformation effect, the translations and rotations, as shown in Figure 5.2-2, are related by:

$$
\theta_x = -\frac{\partial y}{\partial s}, \qquad \theta_y = \frac{\partial x}{\partial s}
\tag{5.2-3}
$$

Note that the negative sign is added in the slope of the y deformation for proper θ_x direction as illustrated in Figure 5.2-2. To derive the shape functions, consider the element deformation in the (X-Z) plane only.

The displacement within an element is expressed in terms of the end displacements and shape functions:

$$
x(s,t) = N_1(s)x_1(t) + N_2(s)\theta_{y1}(t) + N_3(s)x_2(t) + N_4(s)\theta_{x1}(t)
\tag{5.2-4}
$$

The shape functions must satisfy the four end (boundary) conditions of an element:

$$
\begin{aligned}
x(0,t) &= x_1(t) & \theta_y(0,t) &= \frac{\partial x}{\partial s} = \theta_{y1}(t) \\[2mm]
x(L,t) &= x_2(t) & \theta_y(L,t) &= \frac{\partial x}{\partial s} = \theta_{y2}(t)
\end{aligned}
\tag{5.2-5}
$$

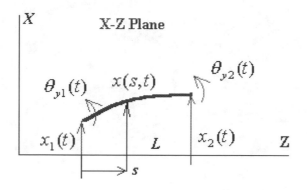

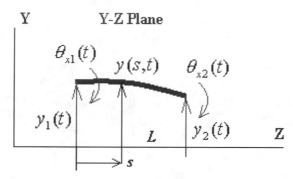

Figure 5.2-2 Displacement and slope relationships

Since there are a total of four end conditions in an element, a cubic polynomial with four parameters is used to describe the displacement:

$$x(s,t) = c_0 + c_1 s + c_2 s^2 + c_3 s^3 \qquad (5.2\text{-}6)$$

Solving the four parameters by using the four end conditions, and substituting the result, we obtain the following shape functions and their derivatives with respect to the s coordinate:

$$N_1 = 1 - 3\xi^2 + 2\xi^3 \qquad\qquad N_1' = \frac{1}{L}\left(-6\xi + 6\xi^2\right)$$

$$N_2 = L\left(\xi - 2\xi^2 + \xi^3\right) \qquad\qquad N_2' = 1 - 4\xi + 3\xi^2$$

$$N_3 = 3\xi^2 - 2\xi^3 \qquad\qquad N_3' = \frac{1}{L}\left(6\xi - 6\xi^2\right) \qquad (5.2\text{-}7)$$

$$N_4 = L\left(-\xi^2 + \xi^3\right) \qquad\qquad N_4' = -2\xi + 3\xi^2$$

Where the non-dimensional parameter: $\xi = \dfrac{s}{L}$ $\qquad\qquad$ (5.2-8)

These shape functions are known as *Hermitian* or *cubic interpolation functions* and are pictured in Figure 5.2-3.

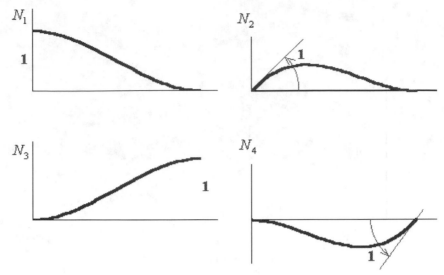

Figure 5.2-3 Hermitian shape functions

Once the shape functions are known for one plane of motion, they can be expanded into two planes of motion to form the complete shape function matrix. The shape function matrix is:

$$\mathbf{\Psi} = \begin{bmatrix} \Psi_T(s) \\ \Psi_R(s) \end{bmatrix} = \begin{bmatrix} N_1 & 0 & 0 & N_2 & N_3 & 0 & 0 & N_4 \\ 0 & N_1 & -N_2 & 0 & 0 & N_3 & -N_4 & 0 \\ 0 & -N_1' & N_2' & 0 & 0 & -N_3' & N_4' & 0 \\ N_1' & 0 & 0 & N_2' & N_3' & 0 & 0 & N_4' \end{bmatrix} \quad (5.2\text{-}9)$$

5.2.2 Timoshenko Beam Theory

With the inclusion of shear deformation, which is often called Timoshenko Beam Theory, the translation of the cross-section centerline is described by two displacements (x, y). It consists of two parts; one caused by bending (x_b, y_b), and the other one by shear deformation (x_s, y_s), as illustrated in Figure 5.2-4 for the X-Z plane. The rotation of the cross-section is described by two rotational angles (θ_x, θ_y), which are associated with the bending deflection of the element. Because of the shear alone, the element undergoes distortion but no rotation. The slope of the total deflection curve at position s is:

$$\frac{\partial x(s,t)}{\partial s} = \frac{\partial x_b}{\partial s} + \frac{\partial x_s}{\partial s} = \theta_y + \beta_y \quad (5.2\text{-}10)$$

$$\frac{\partial y(s,t)}{\partial s} = \frac{\partial y_b}{\partial s} + \frac{\partial y_s}{\partial s} = -\theta_x - \beta_x \quad (5.2\text{-}11)$$

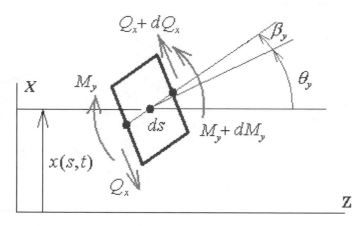

Figure 5.2-4 Element with shear deformation

where β_x, β_y are the angles of distortion due to shear. The relation between the internal bending moment and the bending deformation is:

$$M_x = -EI\frac{\partial^2 y_b}{\partial s^2} = EI\frac{\partial \theta_x}{\partial s}, \qquad M_y = EI\frac{\partial^2 x_b}{\partial s^2} = EI\frac{\partial \theta_y}{\partial s} \qquad (5.2\text{-}12)$$

The relation between the internal shearing force and shear deformation is:

$$Q_x = kAG\frac{\partial x_s}{\partial s} = kAG\left(\frac{\partial x}{\partial s} - \theta_y\right), \qquad Q_y = kAG\frac{\partial y_s}{\partial s} = kAG\left(\frac{\partial y}{\partial s} + \theta_x\right) \qquad (5.2\text{-}13)$$

where k is the shape factor depending on the shape of the cross-section and the material Poisson ratio.

There are two approaches used to derive shape functions with the inclusion of shear deformation. One approach uses four degrees-of-freedom $(x, y, \theta_x, \theta_y)$ at a typical internal cross-section of the element as presented before. The shear deformation effect is included in the shape functions. The derivations of shape functions by using four DOF are quite complicated. However, this approach has been presented and the coefficients for the elemental matrices are well documented in previous literature (Nelson, 1980), therefore, it is used here also for the cylindrical element formulation. The other approach includes shear deformations as additional two degrees-of-freedom. There are six degrees-of-freedom $(x, y, \theta_x, \theta_y, \beta_x, \beta_y)$ at a typical internal cross-section of the element in this approach. Since shear deformation coordinates are explicit in the displacement vector, the shape functions presented for the Bernoulli-Euler Beams can be directly employed for the translational and rotational shape functions, and linear interpolation functions are used for the shear deformation shape functions. However, this approach requires the use of Guyan reduction to condense out the shear deformation coordinates in order to obtain the conventional 8 DOF elements (4 at each end). This approach will be used for the conical element formulation since the closed-form expressions for the coefficients are not readily available for the conical element. The derivation of the shape functions for the first approach is well documented and is not repeated here. The results are included below for reference purposes. The shape function matrix is:

$$
\Psi = \begin{bmatrix} \Psi_T(s) \\ \Psi_R(s) \end{bmatrix} = \begin{bmatrix} N_1 & 0 & 0 & N_2 & N_3 & 0 & 0 & N_4 \\ 0 & N_1 & -N_2 & 0 & 0 & N_3 & -N_4 & 0 \\ 0 & -Nr_1 & Nr_2 & 0 & 0 & -Nr_3 & Nr_4 & 0 \\ Nr_1 & 0 & 0 & Nr_2 & Nr_3 & 0 & 0 & Nr_4 \end{bmatrix}
\tag{5.2-14}
$$

where the translational interpolation functions:

$$
N_i = \frac{1}{1+\Phi}(\alpha_i + \Phi\beta_i) \quad , \ i=1,2,3,4
\tag{5.2-15}
$$

$$
\begin{aligned}
\alpha_1 &= 1 - 3\xi^2 + 2\xi^3 & \beta_1 &= 1 - \xi \\
\alpha_2 &= L\left(\xi - 2\xi^2 + \xi^3\right) & \beta_2 &= \frac{L}{2}\left(\xi - \xi^2\right) \\
\alpha_3 &= 3\xi^2 - 2\xi^3 & \beta_3 &= \xi \\
\alpha_4 &= L\left(-\xi^2 + \xi^3\right) & \beta_4 &= \frac{L}{2}\left(-\xi + \xi^2\right)
\end{aligned}
$$

and the rotational interpolation functions:

$$
Nr_i = \frac{1}{1+\Phi}(\alpha_i + \Phi\beta_i) \quad , \ i=1,2,3,4
\tag{5.2-16}
$$

$$
\begin{aligned}
\alpha_1 &= \frac{1}{L}\left(-6\xi + 6\xi^2\right) & \beta_1 &= 0 \\
\alpha_2 &= 1 - 4\xi + 3\xi^2 & \beta_2 &= 1 - \xi \\
\alpha_3 &= \frac{1}{L}\left(6\xi - 6\xi^2\right) & \beta_3 &= 0 \\
\alpha_4 &= -2\xi + 3\xi^2 & \beta_4 &= \xi
\end{aligned}
$$

and

$$
\xi = \frac{s}{L} \ , \qquad \Phi = \frac{12EI}{kAGL^2}
\tag{5.2-17}
$$

If shear deformation effects are ignored, the parameter Φ is zero and the above shape functions are reduced to the Hermitian functions, as described in the Bernoulli-Euler Beam theory.

5.2.3 Energy Equations

In the previous chapter, the flexible shaft was considered with negligible mass distribution. However in practice, the shaft mass cannot and should not be neglected. That is, the kinetic energy due to the distributed shaft mass should be included in the formulation of equations of motion. The kinetic energy of an infinitesimal rotor element has the same form as the kinetic energy of a rigid disk. The total kinetic energy of a finite element segment is obtained by integrating the differential energy for an infinitesimal rotor element over the length of the element:

$$T = \frac{1}{2}\int_0^L \left\{ m^e\left(\dot{x}^2 + \dot{y}^2\right) + I_d^e\left(\dot{\theta}_x^2 + \dot{\theta}_y^2\right) + \Omega I_p^e\left(\dot{\theta}_x\theta_y - \theta_x\dot{\theta}_y\right) \right\} ds$$

$$+ \frac{1}{2}\Omega^2 \int_0^L I_p^e\, ds \tag{5.2-18}$$

where m^e, I_d^e, I_p^e are the mass, diametral, and polar moment of inertia per unit length. For circular cross-section, we have $I_p^e = 2I_d^e$. When substituting the shape functions into the element kinetic energy expression and neglecting the last term (spinning energy), which does not depend upon the vibrational displacements, the kinetic energy has the form:

$$T = \frac{1}{2}\dot{q}^T\left(M_T^e + M_R^e\right)\dot{q} + q^T\,\Omega g^e\,\dot{q} \tag{5.2-19}$$

where the element translational mass matrix, rotational inertia matrix, and gyroscopic matrix are:

$$M_T^e = \int_0^L m^e\boldsymbol{\Psi}_T^T\boldsymbol{\Psi}_T ds = \int_0^L \rho A\,\boldsymbol{\Psi}_T^T\boldsymbol{\Psi}_T ds \tag{5.2-20}$$

$$M_R^e = \int_0^L I_d^e\boldsymbol{\Psi}_R^T\boldsymbol{\Psi}_R ds = \int_0^L \rho I\,\boldsymbol{\Psi}_R^T\boldsymbol{\Psi}_R ds \tag{5.2-21}$$

$$G^e = \left(g^e\right)^T - g^e = -2g^e = \int_0^L 2\rho I\boldsymbol{\Psi}_R^T\begin{bmatrix} 0 & 1 \\ -1 & 0 \end{bmatrix}\boldsymbol{\Psi}_R ds \tag{5.2-22}$$

Note that the element translational mass matrix and rotational inertia matrix are symmetric matrices and the element gyroscopic matrix is a skew-symmetric matrix.

The potential (strain) energy of the rotating shaft element consists of elastic bending energy due to the bending moments, shear energy due to the shear forces, and work done due to the constant axial load. The translational displacements of a typical point internal to the element consist of the deformations due to the bending moment and shear force, as illustrated in Figure 5.2-4. However, rotational displacements are only related to the bending deformation. An element under shear force alone will only possess distortion but no rotation. The potential energy of a rotating shaft element under a constant axial load is:

$$V = \frac{1}{2} \int_0^L EI \left[\left(\frac{\partial \theta_x}{\partial z} \right)^2 + \left(\frac{\partial \theta_y}{\partial z} \right)^2 \right] ds$$

$$+ \frac{1}{2} \int_0^L kGA \left[\left(\frac{\partial x}{\partial z} - \theta_y \right)^2 + \left(\frac{\partial y}{\partial z} + \theta_x \right)^2 \right] ds \qquad (5.2\text{-}23)$$

$$+ \frac{1}{2} \int_0^L P \left[\left(\frac{\partial x}{\partial z} \right)^2 + \left(\frac{\partial y}{\partial z} \right)^2 \right] ds$$

where *EI* is the bending modulus, *kGA* is the effective shear modulus, and *P* is the axial load. Tension is defined as a positive axial force. Substituting the shape functions relationship, potential energy can be written in matrix form as:

$$V = \frac{1}{2} q^T \left(K_b + K_\beta + K_a \right) q \qquad (5.2\text{-}24)$$

where the bending stiffness is:

$$K_b = \int_0^L EI \left(\Psi_R' \right)^T \left(\Psi_R' \right) ds \qquad (5.2\text{-}25)$$

The shear stiffness is:

$$K_\beta = \int_0^L kGA \left(\Psi_T' + \begin{bmatrix} 0 & -1 \\ 1 & 0 \end{bmatrix} \Psi_R \right)^T \left(\Psi_T' + \begin{bmatrix} 0 & -1 \\ 1 & 0 \end{bmatrix} \Psi_R \right) ds \qquad (5.2\text{-}26)$$

The geometric stiffness matrix due to axial force is:

$$K_a = \int_0^L P \left(\Psi_T' \right)^T \left(\Psi_T' \right) ds \qquad (5.2\text{-}27)$$

The shaft element stiffness matrices derived from the potential (strain) energy are all symmetric and conservative in nature. The effect of shear deformation is included in the derivation of shape functions of the element. Therefore, it is taken into account not only in the potential energy calculation but also in the kinetic energy calculation. The total effect of the shear deformation is to lower the natural frequencies (Timoshenko et. al., 1974). The shear deformation effect can be important when analyzing a short stubby rotor system. For a long flexible rotor system, the shear deformation effect tends to be minimal.

Since the constant rotor rotational speed is assumed, the effect of axial torque on the lateral dynamics of a rotor system is usually neglected. The virtual work by the axial torque is non-conservative. Therefore, potential energy cannot be established and the effect due to the axial torque must be determined by virtual work. The virtual work done by the axial torque is given as:

$$\delta W = \int_0^L T_{orque} \begin{Bmatrix} x' \\ y' \end{Bmatrix}^T \begin{bmatrix} 0 & -1 \\ 1 & 0 \end{bmatrix} \begin{Bmatrix} \delta x'' \\ \delta y'' \end{Bmatrix} ds \tag{5.2-28}$$

where the incremental stiffness matrix is non-symmetric (circulatory) due to its non-conservative nature.

$$\boldsymbol{K}_{Tor} = \int_0^L T_{orque} \left(\boldsymbol{\Psi}_T'\right)^T \begin{bmatrix} 0 & -1 \\ 1 & 0 \end{bmatrix} \left(\boldsymbol{\Psi}_T''\right) ds \tag{5.2-29}$$

The effects of internal viscous and hysteretic damping can be mathematically incorporated into the equations of motion. The internal viscous damping contributes to the dissipative matrix and circulatory matrix, while the hysteretic damping contributes only to the circulatory matrix. The internal viscous damping destabilizes the rotor system when the rotor speed is above the critical speed, where the circulatory force is greater than the dissipative force, as illustrated in Chapter 3. It provides the stabilizing effect below the critical speed where the dissipative force dominates. As opposed to the internal viscous damping, the internal hysteretic damping destabilizes the rotor system at all the rotor speeds since it only contributes to the circulatory force. However, it is still questionable whether either type of internal damping accurately simulates the real physics of dynamic behavior. For metal materials, these effects are small and can be neglected.

For the cylindrical shaft elements, the cross-section area A and area moment of inertia I are constant, therefore, the coefficients in the elemental matrices can be mathematically integrated to obtain the closed-form expression. The elemental matrices for a cylindrical element are well documented in references (Nelson, 1980) and are included in the Appendix for reference. The equations of motion for a typical eight degrees-of-freedom rotating element are:

$$\left(\boldsymbol{M}_T^e + \boldsymbol{M}_R^e\right)\ddot{\boldsymbol{q}}^e + \Omega \boldsymbol{G}^e \dot{\boldsymbol{q}}^e + \boldsymbol{K}^e \boldsymbol{q}^e = \boldsymbol{Q}_{(8x1)}^e \tag{5.2-30}$$

where

$$q^e = (q_1 \mid q_2)^T = (x_1, y_1, \theta_{x1}, \theta_{y1} \mid x_2, y_2, \theta_{x2}, \theta_{y2})^T \qquad (5.2\text{-}31)$$

5.3 Conical Element

For the linearly tapered conical element, the closed-form expressions can also be obtained by considering the shear deformations as additional coordinates at each element end. This results in twelve degrees-of-freedom per element, i.e. 2 translations, 2 rotations, and 2 shear deformations per element end. The additional shear deformation coordinates can then be condensed out of the elemental matrices by using Guyan reduction. This results in the conventional eight degrees-of-freedom elements for the consistence with the cylindrical element. The internal displacements of a typical conical element can be approximately expressed in terms of the 12 end-point displacements by specifying spatial shape function as:

$$\begin{Bmatrix} x(s,t) \\ y(s,t) \\ \theta_x(s,t) \\ \theta_y(s,t) \\ \beta_x(s,t) \\ \beta_y(s,t) \end{Bmatrix}_{(6\times1)} = \begin{bmatrix} \Psi_T(s) \\ \Psi_R(s) \\ \Psi_\beta(s) \end{bmatrix}_{(6\times12)} q(t)_{(12\times1)} \qquad (5.3\text{-}1)$$

where

$$q_{(12\times1)} = (x_1, y_1, \theta_{x1}, \theta_{y1} \mid x_2, y_2, \theta_{x2}, \theta_{y2} \mid \beta_{x1}, \beta_{y1}, \beta_{x2}, \beta_{y2})^T \qquad (5.3\text{-}2)$$

The above order of the displacements was chosen for easy implementation of the Guyan reduction to condense out the shear deformation coordinates. Since the shear deformation coordinates are explicit in the displacement vector, the Hermitian interpolation functions presented in the Bernoulli-Euler Beam theory are used for the translational and rotational shape functions and linear interpolation functions are used for the shear deformation shape functions. The energy expressions are the same as discussed before with the exception the stiffness matrix due to the shear deformation:

$$K_\beta = \int_0^L kGA\Psi_\beta^T \Psi_\beta \, ds \qquad (5.3\text{-}3)$$

Following the same procedure, the equations of motion for a twelve-degrees-of-freedom conical element are of the form:

$$(M_T^e + M_R^e)\ddot{q}^e + \Omega G^e \dot{q}^e + K^e q^e = Q_{(12\times1)}^e \qquad (5.3\text{-}4)$$

The (12x12) elemental matrices are documented in reference (Greenhill, et. al. 1985) and are not repeated here. However, these (12x12) matrices are reduced to the equivalent (8x8) matrices prior to assembly into the global system matrices. The condensation technique is based on the Guyan reduction. The procedure is presented below for reference. Partitioning the stiffness matrix and displacement vector of the static homogeneous equation, we have:

$$\begin{bmatrix} K_{mm} & K_{ms} \\ K_{sm} & K_{ss} \end{bmatrix} \begin{Bmatrix} q_m \\ q_s \end{Bmatrix} = 0 \tag{5.3-5}$$

where

$$q_m = (x_1, y_1, \theta_{x1}, \theta_{y1} \mid x_2, y_2, \theta_{x2}, \theta_{y2})^T_{(8x1)} \text{ are the master displacements to remain,}$$

and

$$q_s = (\beta_{x1}, \beta_{y1}, \beta_{x2}, \beta_{y2})^T_{(4x1)} \text{ are the slave shear deformation displacements to be reduced.}$$

From the second row of the equation, the shear deformation displacements are:

$$q_s = -K_{ss}^{-1} K_{sm} q_m \tag{5.3-6}$$

Therefore, the condensation matrix is of the form:

$$\begin{Bmatrix} q_m \\ q_s \end{Bmatrix}_{(12x1)} = \begin{bmatrix} I \\ -K_{ss}^{-1} K_{sm} \end{bmatrix}_{(12x8)} q_{m(8x1)} = T_{(12x8)} q_m \tag{5.3-7}$$

Then the condensed matrices used in the assembly process are:

$$\hat{A}_{(8x8)} = T^T_{(8x12)} A_{(12x12)} T_{(12x8)} \tag{5.3-8}$$

where A are the original (12x12) mass, gyroscopic, and stiffness matrices and $\hat{A}$ are the reduced (8x8) mass, gyroscopic, and stiffness matrices, respectively. The reduced force vector has the form:

$$\hat{Q}_{(8x1)} = T^T_{(8x12)} Q_{(12x1)} \tag{5.3-9}$$

5.4 User's Supplied Element

The user's supplied element is mainly used when the segment is too complicated to model and the flexibility influence coefficients matrix can be obtained by experiments. In such cases, the element stiffness can be obtained by inversing the flexibility matrix. A rotor element is assumed to be isotropic, and the two planes of (X-Z) and (Y-Z) have identical dynamic properties. Only the stiffness matrix in the X-Z plane is required. The full two planes of motion can be easily expanded from this single plane of motion as demonstrated in Appendix B.

The element total mass and diametral moment of inertia are used to establish the mass matrix. The mass matrix in the (X-Z) plane will be:

$$
\begin{bmatrix}
m/2 & 0 & 0 & 0 \\
0 & I_d/2 & 0 & 0 \\
0 & 0 & m/2 & 0 \\
0 & 0 & 0 & I_d/2
\end{bmatrix}
\quad \text{for} \quad
\begin{Bmatrix}
\ddot{x}_1 \\
\ddot{\theta}_{y1} \\
\ddot{x}_2 \\
\ddot{\theta}_{y2}
\end{Bmatrix}
\tag{5.4-1}
$$

The user supplied (4x4) stiffness matrix in the (X-Z) plane will be:

$$
\begin{bmatrix}
K_{11} & & Sym. & \\
K_{21} & K_{22} & & \\
K_{31} & K_{32} & K_{33} & \\
K_{41} & K_{42} & K_{43} & K_{44}
\end{bmatrix}
\quad \text{for} \quad
\begin{Bmatrix}
x_1 \\
\theta_{y1} \\
x_2 \\
\theta_{y2}
\end{Bmatrix}
\tag{5.4-2}
$$

The structural stiffness matrix is symmetric. The Material Number in the Shaft Elements data form should be set to **0** for User's Supplied Elements.

5.5 Sub-Element Condensation

The concept of the use of subelements is common in the Finite Element Method. The degrees-of-freedom (DOF) at the finite element "station" are called the Master DOF which are kept in the assembled equations of motion, and the DOF at the subelements are considered to be the Slave DOF which are condensed out prior to the assembly process. Once the displacements of the Master DOF have been found, the displacements of the Slave DOF can then be obtained by reversing the condensation process. Let us use the element number 2 of Figure 5.1-2 as an example to demonstrate this condensation process. This element contains four (4) sub-elements (5 substations), as illustrated in Figure 5.5-1: three cylindrical subelements and one conical subelement. The elemental matrices and force of the conical subelement are reduced to the conventional eight degrees-of-freedom, as outlined in the previous section, prior to the assembly process.

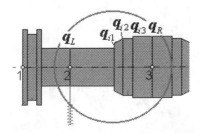

Figure 5.5-1 Sub-Elements

The assembled equations of motion for element number 2 are of the form:

$$\left(M_T + M_R\right)\ddot{q} + \Omega G\,\dot{q} + K\,q = Q_{(20x1)} \tag{5.5-1}$$

where

$$q = \left(q_L, q_{i1}, q_{i2}, q_{i3}, q_R\right)^T = \left(q_L, q_i, q_R\right)^T \tag{5.5-2}$$

The displacement vector q_L contains the four displacements at station 2, which is the left end of element 2 and also the right end of element 1. The displacement vector q_R contains the four displacements at station 3, which is the right end of element 2 and also the left end of element 3. The displacement vector q_i contains the 12 internal displacements, which is to be condensed out. The mass, gyroscopic, and stiffness matrices are obtained from the assemblage of the subelement matrices shown in the schematic form:

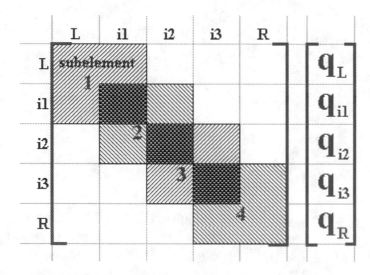

Figure 5.5-2 Elemental matrix

The single shaded areas (8x8) represent the element (sub-element) entries. The double shaded areas (4x4) represent entries that are the sum of the overlapping entries to ensure the continuity requirements. Note that the matrices are banded. Computational efficiency can be obtained by using the banded matrix manipulation. The procedure to condense out the internal displacements is described below. Consider the static, homogeneous equations:

$$\begin{bmatrix} K_{LL} & K_{LI} & K_{LR} \\ K_{IL} & K_{II} & K_{IR} \\ K_{RL} & K_{RI} & K_{RR} \end{bmatrix} \begin{Bmatrix} q_L \\ q_I \\ q_R \end{Bmatrix} = 0 \tag{5.5-3}$$

From the second row, the internal displacements can be expressed by the end displacements:

$$q_I = -K_{II}^{-1} K_{IL} q_L - K_{II}^{-1} K_{IR} q_R \tag{5.5-4}$$

Then the transformation of coordinates may be written as:

$$\begin{Bmatrix} q_L \\ q_I \\ q_R \end{Bmatrix} = \begin{bmatrix} I & 0 \\ -K_{II}^{-1} K_{IL} & -K_{II}^{-1} K_{IR} \\ 0 & I \end{bmatrix} \begin{Bmatrix} q_L \\ q_R \end{Bmatrix} = T \, q^e_{(8x1)} \tag{5.5-5}$$

By using the above transformation, the original 20 DOF equations are reduced to be 8 DOF equations. The condensed matrices used in the assembly process are:

$$\hat{A}_{(8x8)} = T^T A T \tag{5.5-6}$$

where A are the original (20x20) mass, gyroscopic, and stiffness matrices and $\hat{A}$ are the reduced (8x8) mass, gyroscopic, and stiffness matrices, respectively. The reduced force vector has the form:

$$\hat{Q}_{(8x1)} = T^T Q \tag{5.5-7}$$

5.6 Couplings

In many applications, there are several rotors connected by couplings and spacers as illustrated in Figure 5.6-1. Very often coupling is modeled as a user's supplied element (an elastic component) with isotropic translational stiffness of K_T and rotational stiffness of K_R between station i and station j as shown in Figure 5.6-2. The mass properties (mass and inertia) of the coupling element can be lumped into these two connecting stations.

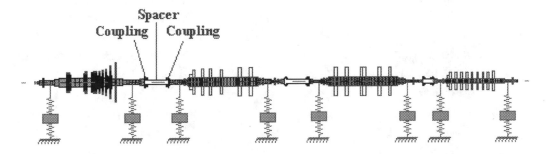

Figure 5.6-1 Rotors connected by couplings (courtesy of Malcolm Leader)

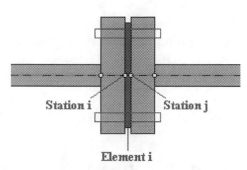

Figure 5.6-2 Coupling model

Then the stiffness matrix of this flexible coupling is of the form:

$$
\begin{bmatrix}
K_T & 0 & 0 & 0 & -K_T & 0 & 0 & 0 \\
0 & K_T & 0 & 0 & 0 & -K_T & 0 & 0 \\
0 & 0 & K_R & 0 & 0 & 0 & -k_R & 0 \\
0 & 0 & 0 & K_R & 0 & 0 & 0 & -K_R \\
-K_T & 0 & 0 & 0 & K_T & 0 & 0 & 0 \\
0 & -K_T & 0 & 0 & 0 & K_T & 0 & 0 \\
0 & 0 & -K_R & 0 & 0 & 0 & K_R & 0 \\
0 & 0 & 0 & -K_R & 0 & 0 & 0 & K_R
\end{bmatrix}
\begin{Bmatrix}
x_i \\
y_i \\
\theta_{xi} \\
\theta_{yi} \\
x_j \\
y_j \\
\theta_{xj} \\
\theta_{yj}
\end{Bmatrix}
\tag{5.6-1}
$$

There are several ways to model coupling in **DyRoBeS** and two simple approaches are explained below. The first approach is to create a pseudo linear bearing with principle translational stiffness and rotation stiffness, as shown in Figure 5.6-3. The mass and inertia of the coupling will be modeled as disks at both connecting stations. A dummy element with zero mass and stiffness connecting these two stations is needed for graphical presentation as illustrated in Figure 5.6-2.

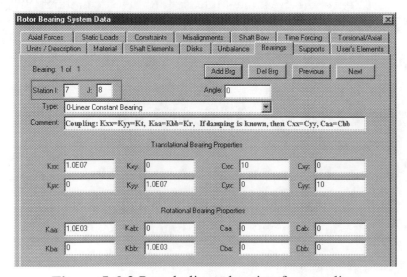

Figure 5.6-3 Pseudo linear bearing for coupling

The second approach is to utilize the User's Supplied Element as illustrated in Figure 5.6-4. In this case, the mass and inertia can be modeled into this element. Note that only (X-Z) plane is needed for the isotropic element.

Note that if $K_T = \infty$, and $K_R = 0$, the coupling model becomes a hinge connecting two stations with equal translational displacements and slope discontinuity. In this case, the coupling is degenerated to a single station with the Moment Release properties. The data input for this case is illustrated in Figure 5.6-5.

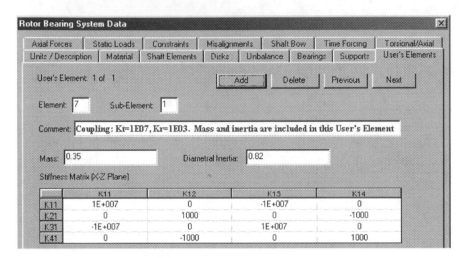

Figure 5.6-4 User's element for coupling

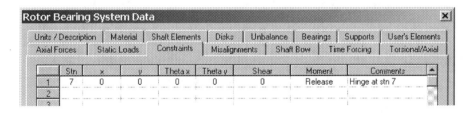

Figure 5.6-5 Moment release for hinged coupling

5.7 Examples

Three examples are used here to demonstrate the finite element approach to the rotating flexible shaft. Various boundary conditions are utilized to validate the computational results with the analytical results. The effects of rotatory inertia, shear deformation, gyroscopic moments, and axial loads on the system natural frequencies are studied. The effect of the number of finite elements is also discussed.

Example 5.1: A Uniform Beam with Various Boundary Conditions

The natural frequencies and natural modes of a uniform beam, as shown in Figure 5.7-1, with various boundary conditions are examined in this example. The closed-form solutions are readily available in almost every vibration textbook for comparison purposes (Thomson, 1981, Timoshenko, et al. 1974). The effects of the shaft rotatory inertia, shear deformation, and external axial load are also studied in this example. The beam under study is a uniform shaft with a length of 50 inches and a diameter of 4 inches. The weight density of the shaft is 0.283 Lb_m/in^3 (7.33E-04 $Lb_f\text{-}s^2/in^4$) and the Young's modulus is 3.0E07 Lb_f/in^2 (psi). The uniform beam is modeled with 10 finite elements (11 finite element stations). For comparison purposes, the shaft is modeled initially as a Bernoulli-Euler beam neglecting the rotatory inertia, shear deformation, and gyroscopic effects. These effects will be added later to examine their influence on the natural frequencies. The program is capable of modeling/analyzing various combinations of boundary conditions, however, only four sets of boundary conditions are presented here:

1. Simply supported beam (Pinned-Pinned)
2. Cantilever beam (Clamped-Free)
3. Beam with free ends (Free-Free)
4. Clamped-Clamped beam

A uniform beam with various boundary conditions
L=50 in, D=4 in, rho=0.283 Lbm/in^3, E=3E07 Lbf/in^2

Figure 5.7-1 A uniform beam with various boundary conditions

Since this is an isotropic, undamped and non-rotating system, two planes of motions, (X-Z) and (Y-Z), are decoupled and only one plane of motion is needed to determine the natural frequencies and natural modes of planar motion. The other plane has the same frequencies. The planar natural frequencies and modes can easily be determined by using the Critical Speed Analysis with a zero Spin/Whirl Ratio. Although it is named Critical Speed Analysis in rotordynamics and designed to analyze the rotating systems, it can calculate the natural frequencies and modes for a non-rotating system with a specified zero Spin/Whirl Ratio. The material properties of the shaft and the finite elements data are entered in the *Material* and *Shaft Elements* tabs as shown in Figure 5.7-2. For the Free-Free boundary conditions, no additional data is needed. For other boundary conditions, the constraints are entered in the *Constraints* tab as shown in Figure 5.7-3. The constraint is represented by a triangle in the configuration display. The run time parameters for the Critical Speed Analysis are given in Figure 5.7-4. Note that the shaft rotatory inertia, shear deformation, and gyroscopic effect are neglected in this case, and only the planar modes (spin/whirl ratio = 0) are calculated.

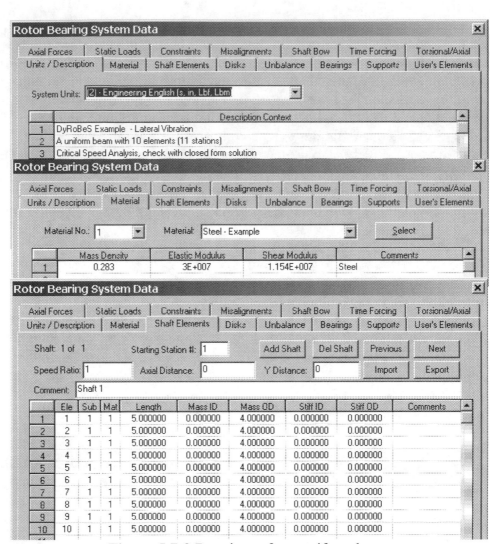

Figure 5.7-2 Data input for a uniform beam

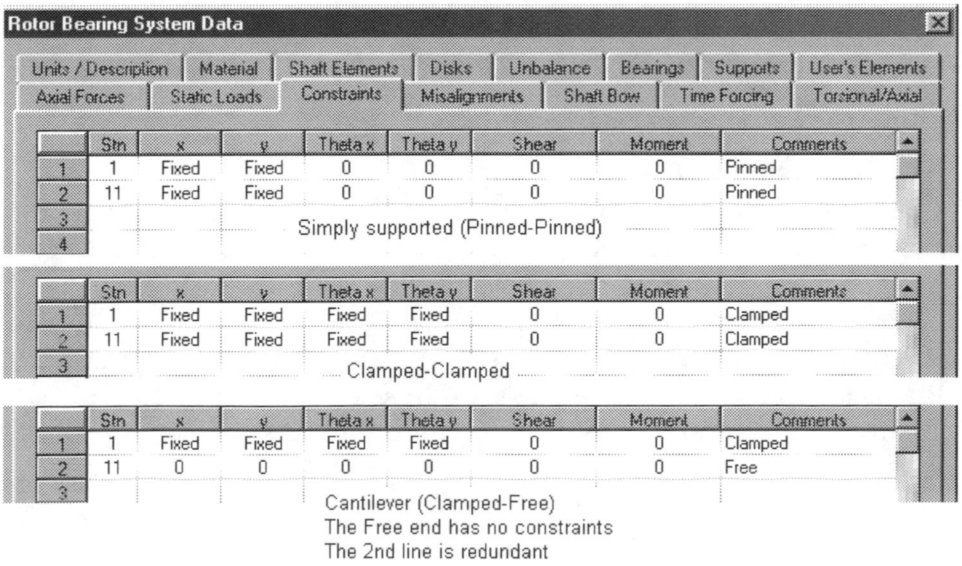

Figure 5.7-3 Various boundary conditions

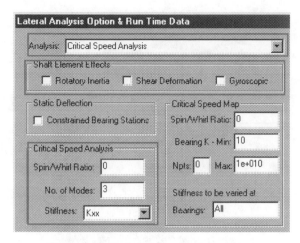

Figure 5.7-4 Analysis data input

Since this example is a uniform beam, the natural frequencies for various boundary conditions can be obtained analytically. The natural frequencies for the above boundary conditions are tabulated in many textbooks in the following form:

$$\omega_i = C_i \sqrt{\frac{EI}{\rho A L^4}} \quad \text{rad/sec} \qquad i=1,2,\ldots\ldots ndof \qquad (5.7\text{-}1)$$

where C_i are the constants depending on the boundary conditions, and they are listed below for reference.

Boundary condition	First Mode	Second Mode	Third Mode
Simply supported	$9.87\,(\pi^2)$	$39.5\,(4\pi^2)$	$88.9\,(9\pi^2)$
Cantilever	3.52	22.0	61.7
Free-Free	0	0	22.4
Clamped-Clamped	22.4	61.7	121

The calculated natural frequencies by using **DyRoBeS** and the analytical solutions for the above four boundary conditions are tabulated below for comparison purposes:

1. Simply supported (Pinned-Pinned) beam

The first three natural frequencies for a pinned-pinned beam in Hz are:

Mode No.	Analytical	**DyRoBeS**
1	127.1	127.1
2	508.5	508.5
3	1144.0	1144.6

The associated natural mode shapes are presented in the same plot in Figure 5.7-5. Readers are encouraged to use the Animation option provided in **DyRoBeS** to visualize the mode movement, as illustrated in Figure 5.7-6.

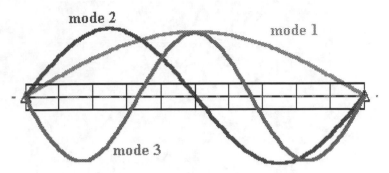

Figure 5.7-5 The first three mode shapes for a simply supported beam

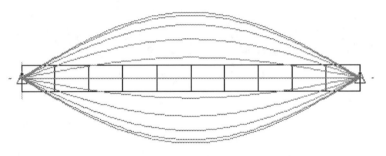

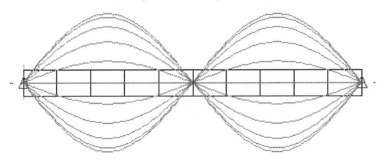

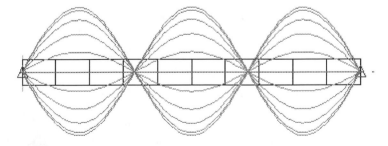

Figure 5.7-6 Mode animation

2. Cantilever (Clamped-Free) beam

The first three natural frequencies for a clamped-free beam in Hz are:

Mode No.	Analytical	*DyRoBeS*
1	45.3	45.3
2	283.8	283.8
3	794.6	794.8

The associated natural mode shapes are sketched in the same plot in Figure 5.7-7.

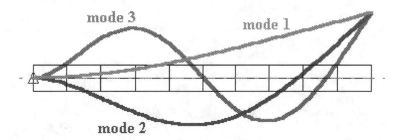

Figure 5.7-7 The first three mode shapes for a clamped-free beam

3. Free-Free beam

The first three natural frequencies for a free-free beam in Hz are:

Mode No.	Analytical	*DyRoBeS*
1	0	0 (rigid body mode)
2	0	0 (rigid body mode)
3	288.1	288.2

In this case, the shaft is unrestrained and it can move like a rigid body in (X-Z) plane. Therefore, there are two *rigid body modes* with zero frequencies. For the rigid body mode, the potential energy is zero and only kinetic energy exists. The associated natural mode shapes are sketched in the same plot in Figure 5.7-8.

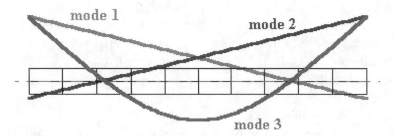

Figure 5.7-8 The first three mode shapes for a free-free beam

4. Clamped-Clamped beam

The first three natural frequencies for a clamped-clamped beam in Hz are:

Mode No.	Analytical	*DyRoBeS*
1	288.1	288.2
2	794.3	794.5
3	1557.1	1558.7

For the clamped-clamped condition, the first natural frequency is the same as the third natural frequency of the free-free condition. The associated natural mode shapes are sketched in the same plot in Figure 5.7-9.

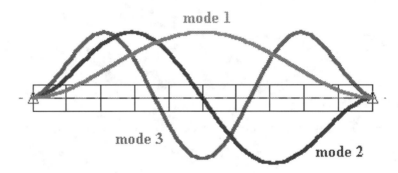

Figure 5.7-9 The first three mode shapes for a clamped-clamped beam

The above results calculated by *DyRoBeS*, using the finite element method, are in very good agreements with the analytical solutions. One may note that for the higher modes, the natural frequencies calculated using the finite element method are slightly higher than those obtained from the closed form solutions. The discrepancy is due to the discretization utilized in the finite element formulation. The analytical solutions are obtained from a continuous system with infinite degrees-of-freedom. However, the finite element solutions are obtained from a discrete system with finite degrees-of-freedom. The basic finite element principle is to discretize the continuous system into many small finite elements and try to simulate the original continuous system with a discrete model. It is like some artificial constraints (stiffness) introduced into the system, therefore, the calculated natural frequencies are slightly higher than those obtained from the analytical solutions. One can always increase the number of elements to obtain more accurate results between the finite element solutions and analytical solutions. The effect of number of finite elements on the natural frequencies will be presented later.

The Whirl Speed and Stability analysis is not strictly applicable to this example because the beam is isotropic, undamped and non-rotating. However, it can be interesting to compare the results calculated by the Whirl Speed Analysis with those obtained from the Critical Speed Analysis. The Whirl Speed/Stability Analysis calculates the complex eigenvalue problems for both planes of motion. The damping coefficients (real parts of the eigenvalues) should be zero since the system is undamped. The damped natural frequencies (imaginary parts of the eigenvalues) should be the same as the undamped natural frequencies obtained from the Critical Speed Analysis. Since the

Whirl Speed Analysis calculates the damped natural frequencies for both planes and the system is isotropic, there will be two identical damped natural frequencies with two different mode shapes for each undamped natural frequency. The whirl Speeds at zero rpm have been calculated for the simply supported beam. The results are listed below:

```
******************** Whirl Speed and Stability Analysis ********************

Shaft  1      Speed=         .00 rpm  =        .00 R/S =         .00 Hz
              40 Precessional Modes,      0 Pure Real Modes
*************************** Precessional Modes ***************************
           *********** Frequency ************   Damping      Log.   Damping
  Mode     rpm           R/S         Hz        Coefficient Decrement Factor

    1    7626.84       798.681      127.11       .0000      .000      .000
    2    7626.84       798.681      127.11       .0000      .000      .000
    3    30510.4       3195.04      508.51       .0000      .000      .000
    4    30510.4       3195.04      508.51       .0000      .000      .000
    5    68677.8       7191.92      1144.6       .0000      .000      .000
    6    68677.8       7191.92      1144.6       .0000      .000      .000
```

These results are in agreement with the results from the Critical Speed Analysis, thus this example helps to verify the whirl speed calculation as well.

Effects of Rotatory Inertia and Shear Deformation

The effects of rotatory inertia, shear deformation, and gyroscopic effect can be included in or neglected from the analysis option as shown below. The gyroscopic effect has been studied before, therefore, only the rotatory inertia and shear deformation effects are studied in this simple example.

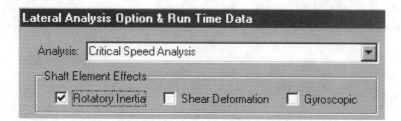

Figure 5.7-10 Shaft element effects

Let us bring back the simply supported beam that was analyzed before. The effects of rotatory inertia and shear deformation on the natural frequencies are listed in the table below:

Effects included	Mode 1 (Hz)	Mode 2 (Hz)	Mode 3 (Hz)
None	127.1	508.5	1144.6
Rotatory inertia	126.9 (-0.157%)	504.5 (-0.787%)	1124.8 (-1.730%)
Shear Deformation	126.4 (-0.551%)	497.5 (-2.163%)	1092.9 (-4.517%)
Rotatory Inertia and Shear Deformation	126.1	493.9	1077.1

It shows that rotatory inertia and shear deformation both have the effect of lowering the natural frequencies. The correction due to shear deformation is about 3 times as important as the correction due to rotatory inertia. Similar results were documented by Timoshenko et al. (1981) and Craig (1981) for a rectangular beam.

Effect of Axial Load

The effect of axial force (tension or compression) on the natural frequencies is examined here. The axial force is entered under the Axial Forces Tab as shown below. Tension is defined to be positive in magnitude, and compression is defined to be negative in magnitude. The natural frequencies for the simply supported beam with 100,000 Lb_f tension and compression are calculated and compared with the analytical solutions (Timoshenko et al., 1981). The natural frequencies from the analytical solution are:

$$\omega_i = i^2 \pi^2 \sqrt{\frac{EI}{\rho A L^4}} \sqrt{1 + \frac{SL^2}{i^2 \pi^2 EI}} \qquad \text{where } S \text{ is the axial load.} \qquad (5.7\text{-}2)$$

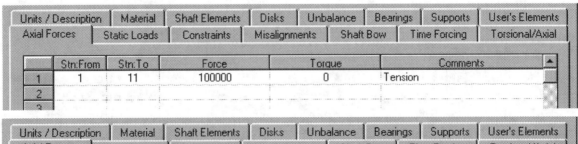

Figure 5.7-11 Axial load input

Mode No.	No Axial Force Analytical/*DyRoBeS*	Tension Analytical/*DyRoBeS*	Compression Analytical/*DyRoBeS*
1	127.1/127.1	131.3/131.3	122.8/122.8
2	508.5/508.5	512.7/512.8	504.2/504.2
3	1144.0/1144.6	1148.3/1148.9	1139.7/1140.4

Again, the results from *DyRoBeS* are in good agreement with the analytical solutions. The frequencies are higher when the shaft is under tension, and lower when the shaft is under compression. As the axial compression force increases, the frequencies decrease. When the first natural frequency becomes zero, the axial force is called the *Euler Buckling Load*. It is the smallest load at which a state of neutral equilibrium is possible. A larger load will cause the shaft to buckle. One should not confuse this buckling phenomenon with the zero frequency of rigid body mode obtained from the free-free boundary conditions.

Effect of Number of Finite Elements

The following table shows how the number of elements affects the first four natural frequencies (Hz) for the simply supported beam, which was previously discussed.

	Mode 1	Mode 2	Mode 3	Mode 4
Analytical	127.1	508.5	1144	2034
2 – elements	127.6	564.3	1419	2586
4 – elements	127.1	510.5	1165	2257
5 – elements	127.1	509.3	1153	2081
8 – elements	127.1	508.6	1145	2042
10 – elements	127.1	508.5	1145	2037

The percentage errors are plotted in the figure below. It shows that the percentage error decreases as the number of elements increases. It also shows that the % of error will be less than 5% for the first four modes when the number of elements is greater than and equal to 5. The choice of the number of elements depends on the number of frequency modes, which are of interest. If only the first two modes are relevant in the analysis, then the 4-elements model will produce very good results in this example.

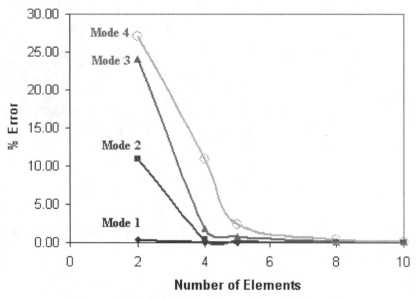

Figure 5.7-12 Effect of number of finite elements

Example 5.2: A Uniform Beam Supported by Elastic Springs

In reality the geometric constraints as described in the previous example are not practical. The geometric constraints are usually replaced by elastic bearings. A beam supported by elastic bearings with translational restraints at both ends may be considered as a condition between the free-free and pinned-pinned conditions. For comparison purposes, the beam is supported at both ends by two identical bearings with a linear bearing stiffness of 60,000 Lb$_f$/in. To build this model, one can simply uses the previous example, delete the constraints, and add the linear bearings under the bearings tab as shown in Figure 5.7-13.

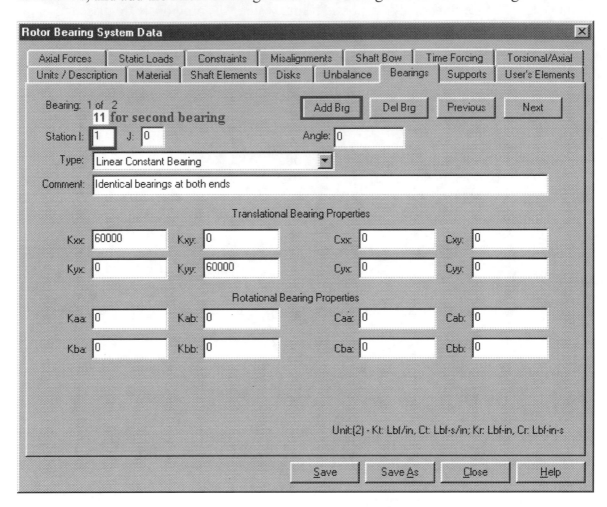

Figure 5.7-13 Bearing input

Now, the constraints are replaced by the bearings, which are represented by coils connected to the ground as shown in Figure 5.7-14.

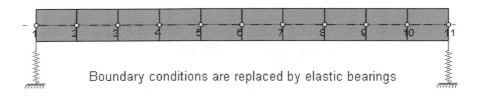

Figure 5.7-14 A uniform beam supported by two elastic springs

The natural frequencies calculated by *DyRoBeS* are in agreement with published data (Lund, 1974). One purpose of this example is to demonstrate the Critical Speed Map Analysis provided by *DyRoBeS*. The Critical Speed Map Analysis is an outgrowth of the Critical Speed Analysis. This analysis calculates the undamped critical speeds for a range of bearing stiffnesses. This map is used as a check of the design of the shaft and the bearings from a stiffness point of view. Again, since this is a non-rotating beam, the natural frequencies can be calculated with a zero Spin/Whirl Ratio in the analysis. Spin/Whirl Ratio = 1, is most commonly used in the Critical Speed Analysis for rotating machines since unbalance excitation is a synchronous rotating force. The first three natural frequencies for bearing stiffnesses from 100 to 1.0E09 Lbf/in are plotted in 5.7-15.

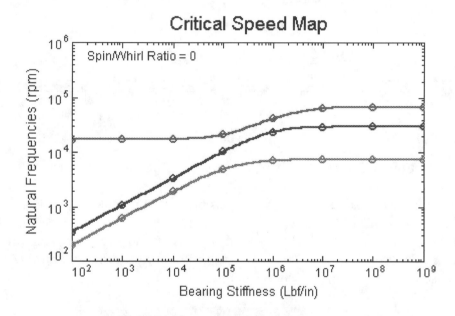

Figure 5.7-15 The first three natural frequencies vs. bearing stiffness

The map can be approximately divided into three zones according to the bearing stiffness value. The first zone is for bearing stiffness less than 1.0E04 Lbf/in, the second zone is from 1.0E04 to 1.0E07 Lbf/in, and the third zone is for bearing stiffness above 1.0E07 Lbf/in. The first zone shows that as the bearing stiffness decreases, the first two modes of frequency quickly approach zero and the third mode of frequency approaches to a constant. It becomes the free-free boundary condition when the bearing stiffness approaches zero. In the first zone, the first two modes can be characterized as *rigid rotor modes* where the potential energy is mainly in the bearings, and the shaft possesses much less potential energy. The third mode can be characterized as a *flexible rotor mode* where the shaft possesses most of the potential energy, and the bearings have little contribution. In the third zone where the bearing stiffness is above 1.0E07 Lbf/in, the bearing stiffness has little effect on the natural frequencies and it becomes the pinned-pinned boundary conditions. The first two modes now are characterized as *flexible rotor modes*. There are no rigid rotor modes for bearings with very high stiffness. The mode shapes for various bearing stiffnesses are plotted in Figure 5.7-16.

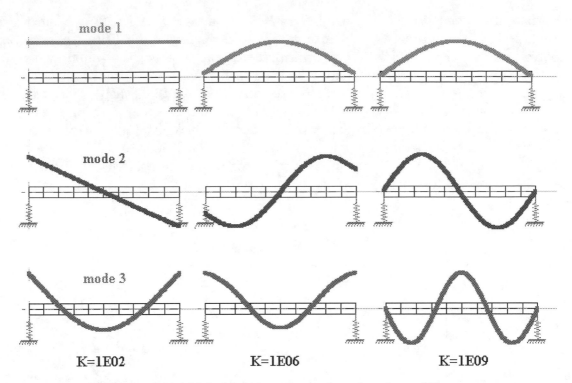

K=1E02   K=1E06   K=1E09

Figure 5.7-16 Mode shapes for various bearing stiffness

The gyroscopic effect due to disk inertia was presented in Chapter 4. The gyroscopic effect of the shaft is discussed here. Since the gyroscopic effect is dependent on shaft spin speed, Whirl Speed and Stability is used for this study. Let us change the bearing stiffness back to 60,000 Lbf/in in this study. The Whirl Speed and Stability Analysis is performed based on the parameters shown below:

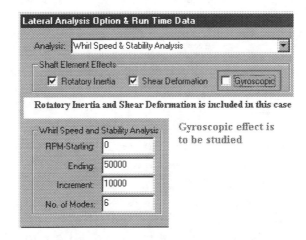

Figure 5.7-17 Analysis input for the whirl speed/stability analysis

The first six natural frequencies with and without gyroscopic effect are plotted in Figure 5.7-18. Since this is an undamped isotropic system, without gyroscopic effect the two planes of motion are decoupled. Therefore, repeated roots (natural frequencies) are calculated from the complex eigenvalue problem. For each natural frequency there is an associated mode in each plane of motion. Although the modes are labeled as 1b, 1f, 2b, 2f, and 3b, 3f, technically speaking these modes are straight-line motions as shown in Figure 5.7-19 for spin speed at 10,000 rpm.

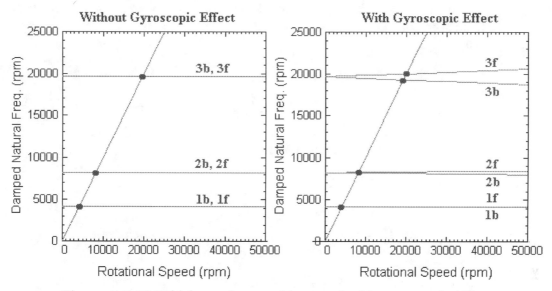

Figure 5.7-18 Whirl speed map without and with gyroscopic effect

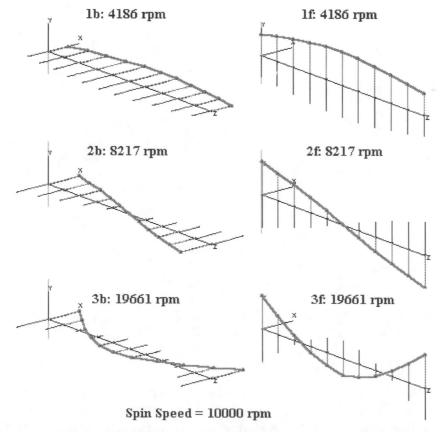

Figure 5.7-19 Mode shapes without gyroscopic effect

With the gyroscopic effect, two planes of motion are coupled and the forward precessional modes are stiffening and backward precessional modes are softening due to the gyroscopic moments. Since the gyroscopic effect is proportional to the spin speed, whirl frequency, slopes (rotational displacements), and moment of inertia, the gyroscopic effect has more influence on the higher modes than on the lower modes. The first six mode shapes at a spin speed of 10000 rpm are plotted in Figure 5.7-20.

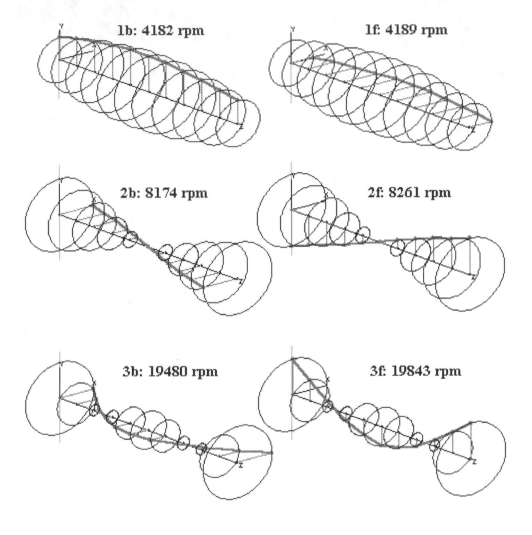

Spin Speed = 10000 rpm

Figure 5.7-20 Mode shapes with gyroscopic effect

Again, since this is an undamped system, the phase angles are the same for all the finite element stations.

Example 5.3: Two Rotors Connected by a Spline Coupling

The rotor assembly presented in this example consists of two shafts connected by a spline coupling as shown in Figure 5.7-21. This system was analyzed by using the transfer matrix method (Rao, 1983). The first rotor (station 1 to 7) is a cantilever rotor with zero translational and rotational displacements at station 1 (clamped left end). The second rotor (station 8 to 17) is a simply supported rotor with zero translational displacements at stations 11 and 17 (pinned support). Two rotors are connected by a spline coupling from station 7 of shaft 1 to station 8 of shaft 2. The translational displacements at coupling (stations 7 and 8) are equal and continuous. The disk at station 4 has mass and moment of inertia as shown in the sketch. The coupling and pulley have masses and negligible moments of inertia. There are several ways to model this system in *DyRoBeS*. Some of them are described below for reference. The results obtained by using different modeling techniques are identical and readers are encouraged to go through these modeling techniques. Four modeling techniques are described below:

1. Use a two-shaft system as shown in Figure 5.7-21, divide and enter the coupling mass into stations 7 and 8 by using *Disks* tab. Create a linear bearing connecting stations 7 and 8 with infinite translational stiffness (1.0E10 Lbf/in is high enough in this case) and zero rotational stiffness.
2. Use one shaft system. Model the coupling as a *User's Element* (element 7). Both the mass and stiffness of the coupling can be entered into this artificial element.
3. Use one shaft system. Model the coupling as a linear bearing connecting stations 7 and 8 and two disks at stations 7 and 8.
4. Use one shaft system. However, two rotors are connecting by a coupling at a single station 7 with Moment Release as a Constraint.

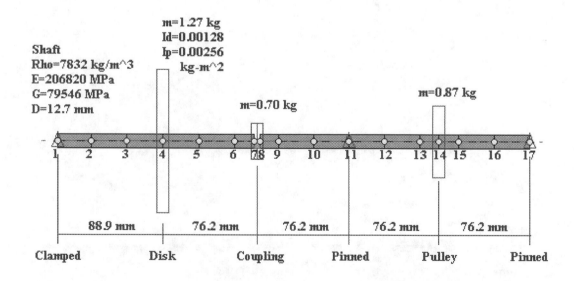

Figure 5.7-21 Two shafts connected by a coupling

The first seven forward and backward synchronous critical speeds are plotted in Figure 5.7-22.

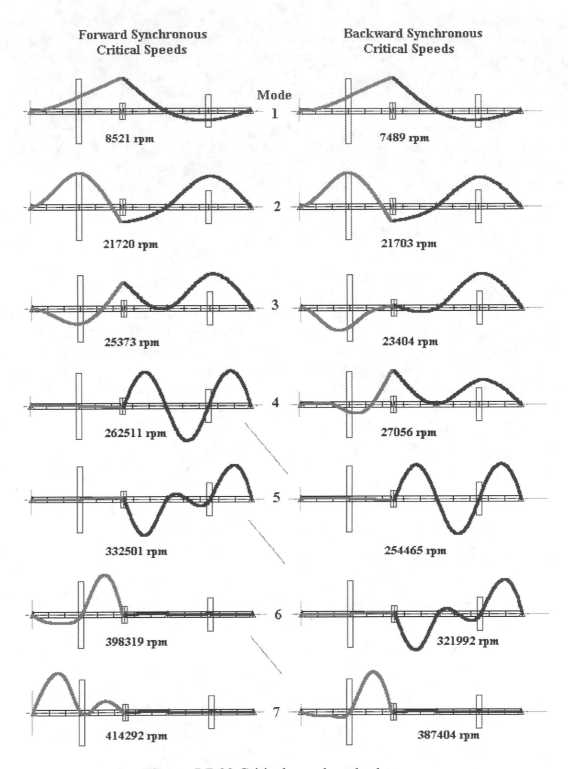

Figure 5.7-22 Critical speed mode shapes

By comparing the forward and backward mode shapes, it shows that the mode shapes for the first and second modes are essentially the same for both forward and backward modes. The gyroscopic effect has very little influence on the second mode since the slopes are very small (nearly zero) at the disk and pulley stations. Therefore, their critical speeds and natural frequencies are very close. However, the forward and backward mode shapes for the third mode are quite different. This is because the third backward mode is very close to the second forward mode at high speed and they are influenced by each other. The mode shapes of backward modes 4, 5, 6, and 7 are similar to the forward modes 3, 4, 5, and 6. This is caused by the gyroscopic effect, which stiffens the forward modes and softens the backward modes. This phenomenon can be better explained in the Whirl Speed Map shown in Figure 5.7-23. Again, an excellent agreement is obtained between the Critical Speed Analysis and the Whirl Speed/Stability Analysis.

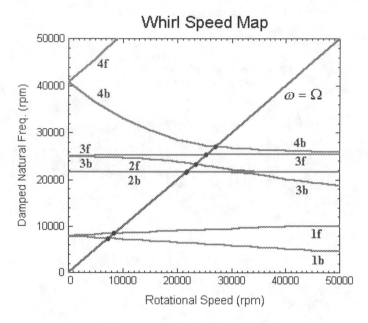

Figure 5.7-23 Whirl speed map

The moment (rotational) stiffness for a coupling is often known. The bending moment is not zero at coupling connection and moment release is not suitable. In this case, a pseudo linear bearing and user's element as explained before are adequate for this type of coupling. If the translational and rotational stiffness at the coupling are very large, then two rotors are essentially rigidly connected and the results will be the same as without coupling.   If the translational and rotational stiffness are zero, the two rotors are essentially decoupled and the results are identical as they are analyzed separately. Readers are encouraged to simulate these two extreme cases by using *DyRoBeS*.

Critical Speed Mode Shape, Mode No.= 1
Spin/Whirl Ratio = 1, Stiffness: Kxx
Critical Speed = 637 rpm = 10.62 Hz

Critical Speed Mode Shape, Mode No.= 2
Spin/Whirl Ratio = 1, Stiffness: Kxx
Critical Speed = 706 rpm = 11.77 Hz

Critical Speed Mode Shape, Mode No.= 3
Spin/Whirl Ratio = 1, Stiffness: Kxx
Critical Speed = 734 rpm = 12.24 Hz

Critical Speed Mode Shape, Mode No.= 4
Spin/Whirl Ratio = 1, Stiffness: Kxx
Critical Speed = 757 rpm = 12.61 Hz

The first four critical speed mode shapes for an 1150 MW Turbine Generator

Bearings and other Interconnection Components **6**

6.1 General Consideration

A typical rotor system consists of three primary component groups: rotating assembly (shafts and disks), non-rotating support structure (housing, casing, stator, etc), and various non-structural interconnection components, such as bearings/dampers, seals, and fluid induced aerodynamics cross-couplings, etc., that provide physical links or interacting forces between structures. The governing equations of motion of the rotating shafts and disks have been discussed in the previous chapters. In this chapter, the interconnection components will be presented. All these interconnection components can be treated as "bearings", some are real and some are pseudo. Many different types of bearings will be discussed in this chapter. In the applications, these bearings can be combined together either in series or in parallel, such as the squeeze film damper is used in series with the rolling element or fluid film bearings and/or in parallel with centering spring.

In general, the forces and moments acting on, or interacting between, the structures (rotating assemblies and/or the support structures) from these interconnection components are of the form:

$$F_x = F_x(x, y, \dot{x}, \dot{y}, t, B)$$

$$F_y = F_y(x, y, \dot{x}, \dot{y}, t, B)$$

$$M_x = M_x(\theta_x, \theta_y, \dot{\theta}_x, \dot{\theta}_y, t, D)$$ (6.1-1)

$$M_y = M_y(\theta_x, \theta_y, \dot{\theta}_x, \dot{\theta}_y, t, D)$$

where B and D contain the operating parameters which change with the machinery operating conditions (such as speed, load, temperature, etc.) and the geometric parameters which are fixed once the parts are made (such as clearance, length, preload, etc.). Almost all these interconnecting forces are nonlinear in nature and nonlinear transient analysis is required to predict the dynamic behavior of the nonlinear system. The nonlinear transient analysis can be very time consuming and not practical in the preliminary design stage. Many of these interconnecting forces, such as fluid film bearings, seals, etc. can be linearized around the static equilibrium position. Some of them may not be linearized and nonlinear theory must be applied, such as squeeze film dampers. Two primary considerations need to be addressed in the design of bearings. One is the bearing static performance such as minimum film thickness, maximum film pressure, temperature rise, power loss, and flow requirement. The other one is the dynamic characteristics of the rotor bearing systems. Therefore, it is advantageous to decouple the rotor equations and the lubrication equations in the preliminary design stage.

Linearized Bearing Dynamic Coefficients

For small vibrations $(x, y, \dot{x}, \dot{y})$ in the vicinity of the static equilibrium position, the bearing forces acting on the rotor can be expressed by the Taylor's expansion:

$$
\begin{Bmatrix} F_x \\ F_y \end{Bmatrix} = \begin{Bmatrix} F_{x0} \\ F_{y0} \end{Bmatrix} + \begin{bmatrix} \dfrac{\partial F_x}{\partial x} & \dfrac{\partial F_x}{\partial y} \\ \dfrac{\partial F_y}{\partial x} & \dfrac{\partial F_y}{\partial y} \end{bmatrix}_0 \begin{Bmatrix} x \\ y \end{Bmatrix} + \begin{bmatrix} \dfrac{\partial F_x}{\partial \dot{x}} & \dfrac{\partial F_x}{\partial \dot{y}} \\ \dfrac{\partial F_y}{\partial \dot{x}} & \dfrac{\partial F_y}{\partial \dot{y}} \end{bmatrix}_0 \begin{Bmatrix} \dot{x} \\ \dot{y} \end{Bmatrix} + \text{high order terms}
$$

$$
= \begin{Bmatrix} F_{x0} \\ F_{y0} \end{Bmatrix} - \begin{bmatrix} k_{xx} & k_{xy} \\ k_{yx} & k_{yy} \end{bmatrix} \begin{Bmatrix} x \\ y \end{Bmatrix} - \begin{bmatrix} c_{xx} & c_{xy} \\ c_{yx} & c_{yy} \end{bmatrix} \begin{Bmatrix} \dot{x} \\ \dot{y} \end{Bmatrix} + \text{high order terms} \tag{6.1-2}
$$

At static equilibrium, the static forces are equal to the external loads in the opposite directions.

$$
\begin{Bmatrix} F_{x0} \\ F_{y0} \end{Bmatrix} = \begin{Bmatrix} -W_x \\ -W_y \end{Bmatrix} \tag{6.1-3}
$$

The eight linearized bearing stiffness and damping coefficients are evaluated in the static equilibrium position and can be expressed in more compact forms:

$$
k_{ij} = -\left(\frac{\partial F_i}{\partial x_j} \right)_0 \quad \text{and} \quad c_{ij} = -\left(\frac{\partial F_i}{\partial \dot{x}_j} \right)_0 \quad \text{where } i = x, y \text{ and } j = x, y \tag{6.1-4}
$$

with i represents the direction of force, and j represents the direction of displacement or velocity. The subscript "o" denotes that the derivatives are evaluated at the static equilibrium position.

Therefore, the bearing static performance and the eight dynamic coefficients can be obtained by solving the lubrication equation without knowing the details of the rotor. Once the bearing static performance is satisfied, the dynamic coefficients can be used in the linear rotor dynamic analysis. For example when determining the critical speeds, rotor stability in the linear sense, and steady state unbalance response. It is not common, although it is possible to have the moment coefficients in the linearized bearing model.

When including the moment coefficients, the linearized bearing dynamic forces are of the form:

$$\begin{Bmatrix} dF_x \\ dF_y \\ dM_x \\ dM_y \end{Bmatrix} = -\begin{bmatrix} k_{xx} & k_{xy} & 0 & 0 \\ k_{yx} & k_{yy} & 0 & 0 \\ 0 & 0 & k_{aa} & k_{ab} \\ 0 & 0 & k_{ba} & k_{bb} \end{bmatrix} \begin{Bmatrix} x \\ y \\ \theta_x \\ \theta_y \end{Bmatrix} - \begin{bmatrix} c_{xx} & c_{xy} & 0 & 0 \\ c_{yx} & c_{yy} & 0 & 0 \\ 0 & 0 & c_{aa} & c_{ab} \\ 0 & 0 & c_{ba} & c_{bb} \end{bmatrix} \begin{Bmatrix} \dot{x} \\ \dot{y} \\ \dot{\theta}_x \\ \dot{\theta}_y \end{Bmatrix} \qquad (6.1\text{-}5a)$$

or in matrix form:

$$dQ = -K_b q - C_b \dot{q} \qquad (6.1\text{-}5b)$$

The above equations are for the bearings connected to the rigid ground. For bearings connecting two finite element stations (stations i and j), the bearing model becomes:

$$\begin{Bmatrix} dQ_i \\ dQ_j \end{Bmatrix} = -\begin{bmatrix} K_b & -K_b \\ -K_b & K_b \end{bmatrix} \begin{Bmatrix} q_i \\ q_j \end{Bmatrix} - \begin{bmatrix} C_b & -C_b \\ -C_b & C_b \end{bmatrix} \begin{Bmatrix} \dot{q}_i \\ \dot{q}_j \end{Bmatrix} \qquad (6.1\text{-}6)$$

where

$$q_i = \left(x, y, \theta_x, \theta_y\right)^T_i \quad \text{and} \quad q_j = \left(x, y, \theta_x, \theta_y\right)^T_j \qquad (6.1\text{-}7)$$

6.2 Bearing Coordinate Systems

Two coordinate systems commonly used by bearing analysts and rotordynamicists for the bearing analysis are shown in Figure 6.2-1:

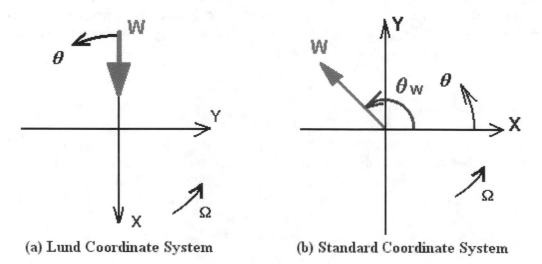

(a) Lund Coordinate System   **(b) Standard Coordinate System**

Figure 6.2-1 Bearing coordinate systems

6.2.1 Lund Convention

The first Cartesian coordinate system (X,Y,Z), used to describe bearing orientation and geometry, is commonly used by the bearing analysts (Lund and Thomsen, 1978). The X axis is chosen to be collinear with the bearing load vector (W). Note that the direction of the X axis does not have to be pointing vertically downward, as shown in Figure 6.2-1a. It can be in any direction as long as the X axis is aligned with the bearing load vector (W), i.e., the load vector can be in any direction. The Y axis is perpendicular with the X axis in the direction of shaft rotation. θ is the circumferential angular coordinate, measured from the negative load vector (negative X axis) in the direction of shaft rotation.

6.2.2 Standard Convention

The second Cartesian coordinate system (X,Y,Z), used to describe the bearing geometry and orientation, is a standard convention as shown in Figure 6.2-1b. That is, X axis is to the right and Y axis is to the top. The circumferential angular coordinate θ is measured from the positive X axis in the direction of shaft rotation. The load vector (W) can be in any direction with respect to the X axis, by a specified angle (θ_w).

Each coordinate system has its advantages. The first coordinate system, commonly referred to as Lund's convention, describes the bearing geometry and load vector orientation by aligning the X axis with the load vector. This coordinate system is convenient for the bearing analysis. The disadvantage is that the bearing geometric data are dependent on the loading direction. For a gear driven rotor, load vector can be in any direction, due to the power level (loading condition) of the rotor. In this case, when the same bearing is analyzed using a different loading direction, the bearing geometric data (leading and trailing edges of the lobe) must be re-entered. For the second coordinate system, commonly referred to as Standard convention, the bearing geometric data are independent on the load vector. An additional parameter is required to specify the load vector. The loading direction is specified by an angle (θ_w). However, it is desirable to know the stiffness and damping coefficients in the loading direction and it's perpendicular axis. Therefore, a coordinate transformation may be needed to transform the bearing coefficients to the loading direction.

6.2.3 Coordinate Transformation

The linearized bearing coefficients are obtained from the bearing analysis or from the tabulated data books. These dynamic coefficients are usually expressed in their local coordinate systems and need to be transformed to the global coordinate system, which is used to describe the motion of the rotor system. This is especially true for the gear driven systems, since the resultant bearing load directions are normally different for the bearings. Consider the two coordinate systems shown in Figure 6.2-2.

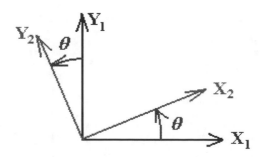

Figure 6.2-2 Coordinate transformation

The coordinate transformations are:

$$\begin{Bmatrix} X \\ Y \end{Bmatrix}_2 = \begin{bmatrix} \cos\theta & \sin\theta \\ -\sin\theta & \cos\theta \end{bmatrix} \begin{Bmatrix} X \\ Y \end{Bmatrix}_1 \qquad \text{or in matrix form} \qquad X_2 = T\,X_1 \qquad (6.2\text{-}1)$$

$$\begin{Bmatrix} X \\ Y \end{Bmatrix}_1 = \begin{bmatrix} \cos\theta & -\sin\theta \\ \sin\theta & \cos\theta \end{bmatrix} \begin{Bmatrix} X \\ Y \end{Bmatrix}_2 \qquad \text{or in matrix form} \qquad X_1 = T^T X_2 \qquad (6.2\text{-}2)$$

The linearized bearing forces in the (X_1, Y_1) coordinates are:

$$\begin{Bmatrix} F_x \\ F_y \end{Bmatrix}_1 = -\begin{bmatrix} k_{xx} & k_{xy} \\ k_{yx} & k_{yy} \end{bmatrix}_1 \begin{Bmatrix} x \\ y \end{Bmatrix}_1 - \begin{bmatrix} c_{xx} & c_{xy} \\ c_{yx} & c_{yy} \end{bmatrix}_1 \begin{Bmatrix} \dot{x} \\ \dot{y} \end{Bmatrix}_1 \qquad \text{or} \qquad F_1 = -K_1 X_1 - C_1 \dot{X}_1 \qquad (6.2\text{-}3)$$

and in the (X_2, Y_2) coordinates are:

$$\begin{Bmatrix} F_x \\ F_y \end{Bmatrix}_2 = -\begin{bmatrix} k_{xx} & k_{xy} \\ k_{yx} & k_{yy} \end{bmatrix}_2 \begin{Bmatrix} x \\ y \end{Bmatrix}_2 - \begin{bmatrix} c_{xx} & c_{xy} \\ c_{yx} & c_{yy} \end{bmatrix}_2 \begin{Bmatrix} \dot{x} \\ \dot{y} \end{Bmatrix}_2 \qquad \text{or} \qquad F_2 = -K_2 X_2 - C_2 \dot{X}_2 \qquad (6.2\text{-}4)$$

By utilizing the coordinate transformations, we have the following relationship:

$$F_1 = T^T F_2, \quad K_1 = T^T K_2\,T, \qquad C_1 = T^T C_2\,T \qquad (6.2\text{-}5)$$

or

$$F_2 = T\,F_1, \quad K_2 = T\,K_1\,T^T, \qquad C_2 = T\,C_1\,T^T \qquad (6.2\text{-}6)$$

Two commonly used coordinate systems in describing the linearized bearing coefficients are shown in Figure 6.2-3. The transformation matrix for these two coordinate systems is:

$$T = \begin{bmatrix} 0 & 1 \\ -1 & 0 \end{bmatrix} \qquad \text{for} \quad \theta = 90^0 \qquad (6.2\text{-}7)$$

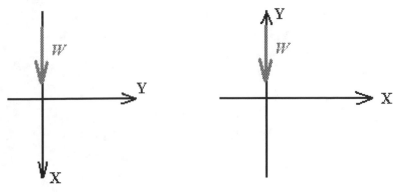

Coordinate System: 1   Coordinate System: 2

Figure 6.2-3 Two commonly used coordinate systems

The relationships for bearing coefficients are:

$$\begin{bmatrix} k_{xx} & k_{xy} \\ k_{yx} & k_{yy} \end{bmatrix}_1 = \begin{bmatrix} 0 & -1 \\ 1 & 0 \end{bmatrix}\begin{bmatrix} k_{xx} & k_{xy} \\ k_{yx} & k_{yy} \end{bmatrix}_2 \begin{bmatrix} 0 & 1 \\ -1 & 0 \end{bmatrix} = \begin{bmatrix} k_{yy2} & -k_{yx2} \\ -k_{xy2} & k_{xx2} \end{bmatrix} \qquad (6.2\text{-}8)$$

and

$$\begin{bmatrix} k_{xx} & k_{xy} \\ k_{yx} & k_{yy} \end{bmatrix}_2 = \begin{bmatrix} 0 & 1 \\ -1 & 0 \end{bmatrix}\begin{bmatrix} k_{xx} & k_{xy} \\ k_{yx} & k_{yy} \end{bmatrix}_1 \begin{bmatrix} 0 & -1 \\ 1 & 0 \end{bmatrix} = \begin{bmatrix} k_{yy1} & -k_{yx1} \\ -k_{xy1} & k_{xx1} \end{bmatrix} \qquad (6.2\text{-}9)$$

Similarity for the damping coefficients.

6.3 Fundamentals of Hydrodynamic Journal Bearings

Hydrodynamic bearings are commonly used in rotating machinery. The fluid film bearings are often the major source of the damping, which attenuates the synchronous vibrations. However, their cross-coupled stiffness properties (or tangential forces) can introduce a major destabilizing effect, which may create large sub-synchronous vibration when rotor speed exceeds the instability threshold. The simplest journal bearing is a plain journal bearing, as sketched in Figure 6.3-1. Clearances are exaggerated in the figure for illustrative purposes. The clearance is generally very small (on the order of one-thousandth) compared to the journal diameter. For illustrative purposes, the bearing is presented with the load vector being is a downward position.

The fundamental purpose of a journal bearing is to provide radial support (load carrying capability) to a rotating shaft. Three basic factors are required to develop the hydrodynamic pressure film. They are: converging geometry, relative motion, and viscous lubricant.

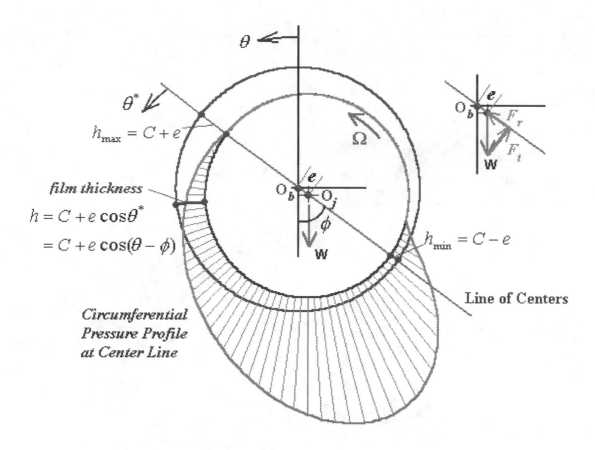

Figure 6.3-1 Plain cylindrical journal bearing

At equilibrium, the force generated by the hydrodynamic film pressure is equal to the static load acting on the bearing. Various lubricants are used in rotating machinery, ranging from very high viscous powders to low viscous gases. However, most industrial applications utilize mineral or synthetic lubricants with an ISO VG viscosity grade numbers ranging from 5 to 150. Details of the lubricant properties are discussed in the Appendix C and therefore not presented here. In this session, only fluid film bearings with incompressible flow are discussed. Bearings with compressible flow will be discussed later.

The relative motion is provided by the shaft rotational speed. Under load, the centers of journal and bearing are not coincident; the journal center is displaced from the bearing center by a distance of e. This eccentric arrangement provides a converging geometry that combines with the relative motion of the journal and bearing, and with the viscous effect of the lubricant, so that hydrodynamic pressure is developed, which produces the load carry-capability as shown in Figure 6.3-1. The rotating shaft drags the oil into the converging region ($0 \le \theta^* \le \pi$). As the clearance decreases, the pressure and pressure gradient must be generated within the converging oil film, to establish a basis for flow, in order to maintain the flow continuity. In the diverging region ($\pi \le \theta^* \le 2\pi$), the pressure ends with zero pressure and zero gradients; the oil then ruptures and becomes streamers in the cavitated zone. The pressure profile for Figure 6.3-1 is shown with different views, along with the film thickness in Figure 6.3-2. It shows that the pressure film begins slightly ahead of the maximum film thickness (at the point where the film starts to converge), and ends very subtly at some point beyond the minimum film thickness, with zero pressure and gradients.

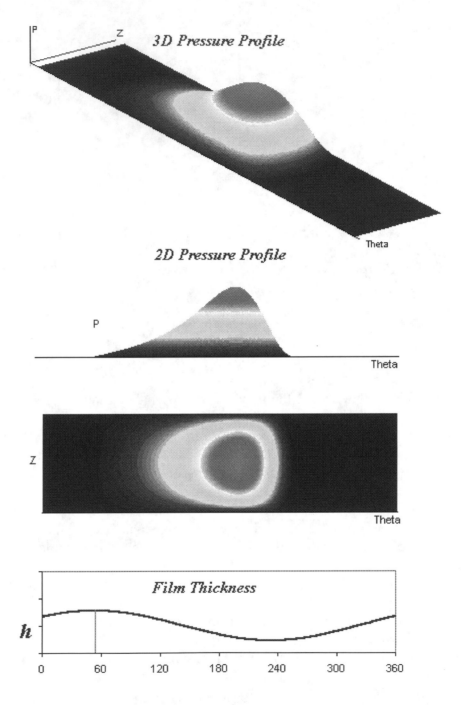

Figure 6.3-2 Pressure profiles and film thickness

At steady state condition, the journal static equilibrium position is defined by journal eccentricity (e) and attitude angle (ϕ). The attitude angle is the angle between the load direction and the line of centers (journal center and bearing center). The attitude angle in the direction of shaft rotation is a very unique characteristic of the fixed profile journal bearing, since it produces the positive tangential force, which is a destabilizing force, and is responsible for the unstable oil whirl/whip phenomena. Oil whirl/whip is the sub-synchronous vibration, where the rotor exhibits a forward precessional motion. At *oil whirl*, the rotor vibrates at its natural frequency that is close to half the rotor's rotational speed (normally, 45%-50%), in the forward precessional rigid body mode. At whip, the rotor vibrates at the natural frequency of the first forward bending mode that typically does not vary with rotor speed.

From the rotor stability point of view, bearing, which produces zero attitude angle, is the most desirable configuration. However, there are many important factors besides rotor stability that affect bearing design and selection. In general, they fall into four categories: cost, manufacturability and installation, static performance (minimum film thickness, maximum film pressure, power loss, temperature rise, flow requirement), and dynamic characteristics (synchronous and sub-synchronous vibrations, critical speeds, system stability). There are many handbooks on bearing selection, lubrication, and design criteria available. They should only be used as general guidelines. Engineers (designers) have to use their own judgment and make decisions based on their experience and the design limits established by their companies.

The locus of the static equilibrium position for the journal center is shown in Figure 6.3-3. Note that the journal does not move in the loading direction other than with an attitude angle of ϕ. For the given operating conditions (Ω and W), the journal has an equilibrium position defined by the eccentricity e and attitude angle ϕ. At the static equilibrium position, the fluid film reaction force (F) and the static load (W) acting on the journal are balanced. For a constant static load, the journal rests at the bottom of the bearing ($\varepsilon=e/C=1$) at zero rotational speed. As the speed increases, the journal center rises up along the static equilibrium locus and then reaches the center of the bearing position ($\varepsilon=0$) when the speed becomes infinity. For a constant rotational speed, the journal center is at the center of the bearing when static load is zero. As the static load increases, the journal center moves down along the locus and then reaches the bottom of the bearing, when static load is infinity. However, this static equilibrium locus, which is shown in Figure 6.3-3, is based on the hydrodynamic lubrication theory. Figure 6.3-4 shows the journal eccentricity ratio and film thickness verses Sommerfeld number, which is a non-dimensional number, frequently used in the bearing design. For a very high eccentricity regime (very small Sommerfeld Number), the film thickness is very small, and the heat generation and friction are not dependent on the lubricant viscosity. Therefore in this regime, the hydrodynamic theory does not apply and is called *Boundary Lubrication*. The transition regime between the hydrodynamic lubrication and boundary lubrication is commonly called *mixed lubrication*. In the mixed and boundary lubrication regimes, the surface finish (roughness) of the mating surface plays an important role. In practice, it is very difficult to achieve this boundary lubrication operating condition without damaging the bearing surface or breaking down the lubricant. The film temperature and pressure limitations proposed in many handbooks are really used to prevent these types of operating conditions. However, for high speed and low load conditions, the attitude angle becomes large in the very low eccentricity ratio regime, and the tangential force generated by the fluid film is prone to destabilize the rotor system. The operating region with low eccentricity is not desirable from a rotor stability point of view.

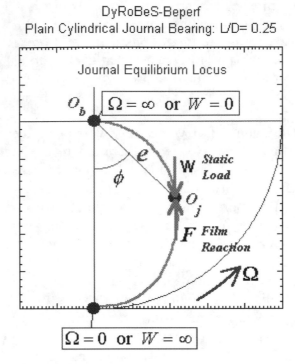

Figure 6.3-3 Journal equilibrium locus

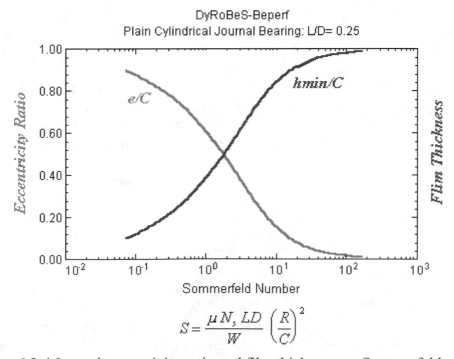

$$S = \frac{\mu N_s LD}{W} \left(\frac{R}{C}\right)^2$$

Figure 6.3-4 Journal eccentricity ratio and film thickness vs. Sommerfeld number

6.4 Fixed Profile Journal Bearings

There are many types of fixed profile journal bearings and some of them are shown in Figure 6.4-1. The configurations can be observed with very simple plain cylindrical journal bearing up to complicated multi-lobe bearings. In general, the fixed profile bearing is made up of a number of fixed circular arc segments called "lobes" or "pads". The lobes are separated by axial lubricant supply grooves.

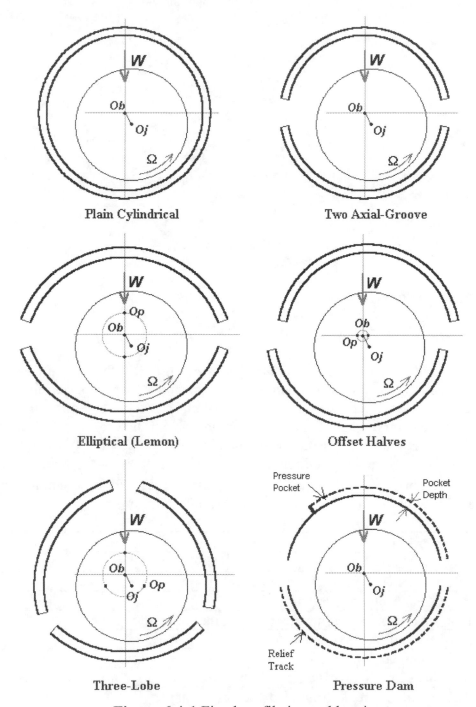

Figure 6.4-1 Fixed profile journal bearings

A three-lobe bearing is sketched in Figure 6.4-2 to illustrate the parameters used to describe the bearing geometry. Again, the clearances are exaggerated in the figure for illustrative purposes.

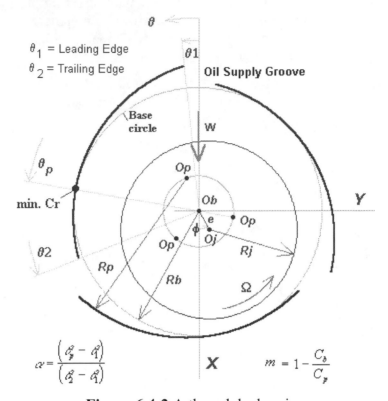

Figure 6.4-2 A three-lobe bearing

Two coordinate systems can be used in **DyRoBeS-BePerf**. They were described in the previous section. As described before, it is convenient, when performing bearing analysis, to use a local Cartesian coordinate system (X,Y,Z) to describe the bearing orientation and geometry, where the X axis is collinear with the bearing load vector (W). The journal static equilibrium position is defined by journal eccentricity (e) and attitude angle (ϕ). Under dynamic conditions, the journal oscillates with small amplitudes around this equilibrium position. The nomenclatures used to describe the bearing geometry are listed in the following table with a brief description:

Symbol	Description
O_b	Bearing center
O_j	Journal center
O_p	Pad (lobe) center of curvature
R_b, D_b	Bearing assembled radius (diameter) at minimum clearance
R_j, D_j	Journal shaft radius (diameter)
R_p, D_p	Pad machined radius (diameter)
e	Journal eccentricity, distance from bearing center to journal center
ϕ	Attitude angle, angle from load vector (X axis) to the line connecting bearing center and journal center

θ_1	Angle from the negative load vector (negative X axis) to leading edge of the first lobe
θ_2	Angle from the negative load vector (negative X axis) to trailing edge of the first lobe
θ_p	Angle from the negative load vector (negative X axis) to the line connecting the bearing center and the pad center of curvature
χ	Lobe or Pad arc length, $\chi = \theta_2 - \theta_1$
χ_p	Angle from leading edge to the minimum clearance point for a centered shaft for fixed lobe bearings, or, Angle from leading edge to pad pivot point for tilting pad bearings, $\chi_p = \theta_p - \theta_1$
Npad	Number of pads (lobes)
L	Bearing (babbitt) axial length
C_b	Bearing minimum assembled radial clearance, $C_b = R_b - R_j$
C_p	Pad machined radial clearance, $C_p = R_p - R_j$

The bearing radius at minimum clearance (R_b) for a centered shaft can be described as the radius of the largest shaft that could be inserted into the bearing. A circle drawn, based on R_b is referred to as a *bearing base circle*. For a positive preloaded bearing, pad radius (R_p) is greater than bearing radius (R_b) and the circular pads are moved inward towards the center of the bearing. Thus, when the journal is centered in the bearing, the pads are loaded by the geometry effect. The fraction of the distance between pad center of curvature and bearing center to pad radial clearance is called "*Preload*":

$$m = \frac{\left(C_p - C_b\right)}{C_p} = 1 - \frac{C_b}{C_p} \qquad (6.4\text{-}1)$$

When the preload is zero ($C_b = C_p$), the pad centers of curvature coincide with the bearing center and the bearing is cylindrical. When the preload has a value of 1 ($C_b = 0$), the shaft touches all the pads and the bearing minimum radial clearance is zero. Typical preload value for a fixed lobe bearing ranges from 0.4 to 0.75. Figure 6.4-3 illustrates four different preloads (0, 0.5, 1, -0.5) for a 3-lobe bearing. Note that negative preload is not desirable.

Another key parameter used to describe the preloaded bearing geometry is the fraction of the converging pad length to the full arc length. This parameter is called "*Offset*" or "*Tilt*" and is given by the following expression:

$$\alpha = \frac{\left(\theta_p - \theta_1\right)}{\left(\theta_2 - \theta_1\right)} = \frac{\chi_p}{\chi} \qquad (6.4\text{-}2)$$

The value of offset is meaningful only when the bearing is preloaded. At θ_p, the bearing has a minimum clearance for a centered shaft, and the lobe arcs intersect with the bearing base circle. A lobe which is symmetrically located with respect to the centered journal, i.e. offset = 0.5, is defined as having no "lobe tilt". In this case, the clearance space has equal convergent and divergent arcs. An offset of 0.5 is commonly used to accommodate the reversal rotation of the shaft and also to avoid the problem of the bearing being installed backwards. Typical offset ranges from 0.5 to 1.0. For an "offset halves" bearing, the offset could be larger than 1, depending on the position of pad center of curvature. An offset less than 0.5, increases the diverging film thickness and is not desirable. Figure 6.4-4 illustrates four different offsets (0.2, 0.5, 0.8, 1.0) for a 3-lobe bearing with 0.5 preload. Note that an offset less than 0.5 is not desirable.

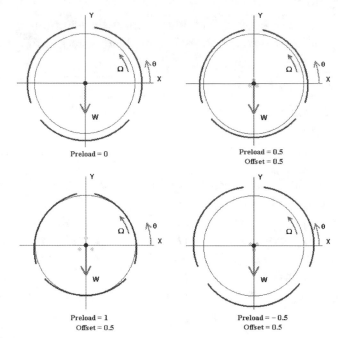

Figure 6.4-3 Bearing configuration for different preloads

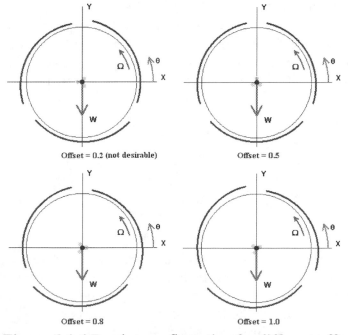

Figure 6.4-4 Bearing configuration for different offsets

Design Consideration

In addition to the bearing dynamic coefficients used in the rotor dynamics study, the bearing static performance is also an important factor in the design of the hydrodynamic journal bearings. The effects of the bearing dynamic coefficients on the rotor dynamics are discussed in details in other sessions. Design considerations on the bearing static performance are discussed here. Three basic operating characteristics need to be considered and limited in bearing design in order to ensure satisfactory operation. A general guideline is also provided for references. Please note that these are general guidelines and should not be strictly applied without any engineering judgment and operating experience.

1. **Minimum film thickness**. For most applications, 1.0 mils (0.001 inch) of the minimum film thickness is recommended. However, value of 0.5 mils has also been used for smaller shaft diameters. An empirical expression proposed by Gardner for the minimum allowable film thickness (h, mils) for hydrodynamic oil film bearings in terms of the bearing diameter (D, inches) and speed (N, rpm) is given below for reference purposes:

$$h\,(mils) = 6.85 \times 10^{-3}\,D^{0.71}\,N^{0.43}$$

(6.4-3)

2. **Maximum film pressure**. The babbitts are the most widely used bearing materials. They are either tin- or lead-base alloys having excellent embeddability and conformability characteristics. However, they also have relatively low load-carrying capability. Therefore, these alloys are metallurgical bonded to stronger backing materials such as steel, cast iron, or bronze. The maximum hydrodynamic film pressure occurs around (slightly ahead of) the minimum film thickness. The extent of plastic deformation occurring in the lining under the area of maximum film pressure is limited not only by the steel backing but by the lining alloy outside the area of maximum film pressure where the deformation is elastic. The alloy tensile strength decreased significantly as the temperature increases. The value of 1000-1500 psi is recommended for the maximum allowable film pressure under normal operation.

3. **Maximum bearing temperature**. The high temperature decreases the strength of the babbitts also affects the lubricant performance. At bearing metal temperature around 230°F-250°F (110°C-120°C), the journal and bearing can experience discoloration through oil lacquer formation. The golden brown or yellow discoloration does not cause any operational problems. However, at higher temperature, depending on the lubricants and their additives, the oil lacquer deposits become carbonized coke-like deposits around the minimum film thickness and trailing edges of the pad, which can decrease the bearing clearance and cause substantial operational problems. Materials other than babbitts and synthetic lubricants are utilized if higher operating temperature is required.

There are many proposed design guidelines on the bearing clearance. Figure 6.4-5 is recommended by Garner (1981) on the minimum diametral clearance.

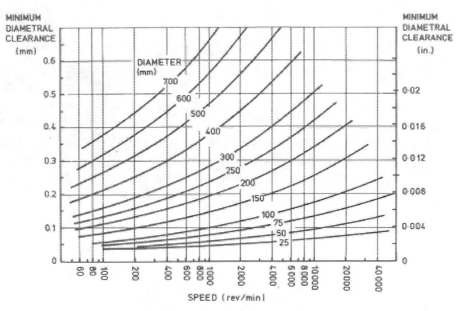

Figure 6.4-5 Recommended minimum Diametral Clearance by Garner (1981)

Author has performed many bearing tests on 3-lobe bearings and tilting pad bearings with various preloads and offsets. The mineral light turbine oil (ISO VG 32) was used during the tests. The bearing clearance effects on the rotor dynamics and bearing performance are summarized in Figure 6.4-6. The bearings with high clearance ratio in Figure 6.4-6 are mainly tilting pad bearings. There are two types of bearing failures, one is due to high temperature and bearing fretting/frosting, which results in high synchronous vibration, and the other is due to rotor instability, which results in high sub-synchronous vibration.

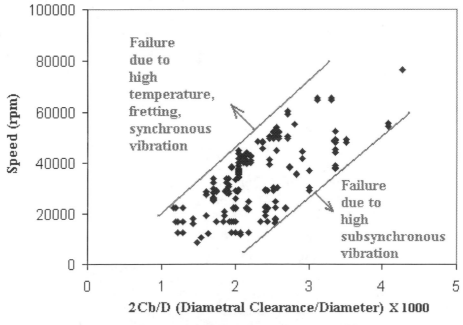

Figure 6.4-6 Bearing clearance effects

Figure 6.4-7 shows the temperature readings for a 3-lobe bearing during machine startup. Five temperature sensors were installed under the bearing babbitts as illustrated in the Figure 6.4-7. The machine was started with 110°F oil supply and inlet valve partially closed (bearing was not fully loaded) and it reached the full speed after 12 seconds. After 60 seconds, the machine was loaded and remained the same operating condition during test. The final inlet oil temperature was regulated by a mixing value and remained at 115°F. It shows that the trailing edge of the loaded pad has highest temperature reading. It also shows the temperature reached the equilibrium very quickly.

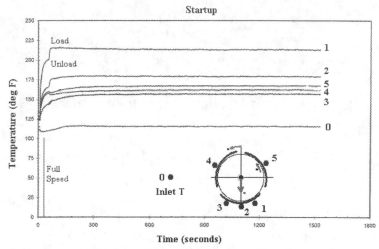

Figure 6.4-7 Temperature readings

Figure 6.4-8 shows the temperature readings for a bearing with smaller bearing clearance. It shows that all the temperature readings were above 350°F at one time right after the machine reached the full speed, and then it settled down with lower temperatures with inlet valve partially closed. However, these steady state temperatures were still higher than normal. The machine was disassembled and the bearing was found to have frosting as shown in Figure 6.4-9a. It only took one startup to damage the bearing. Continuous operation of this bearing resulted in heavy vanishing deposit as shown in Figure 6.4-9b due to the high temperature and low clearance.

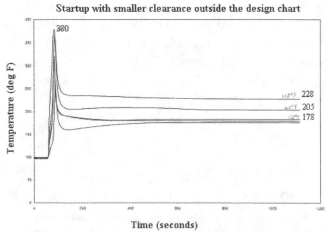

Figure 6.4-8 Temperature readings with smaller bearing clearance

Figure 6.4-9 Bearing failure due to tight clearance

6.5 Reynolds Equation

The governing equation for pressure distribution in a fluid film journal bearing is Reynolds equation, which is derived from the Navier-Stokes equation and the continuity equation. The nonlinear fluid film forces acting on the journal are determined by the application of boundary conditions and the integration of pressure distribution. The Reynolds equation for a journal bearing with incompressible turbulent flow has the form (Booser, 1983):

$$\frac{\partial}{\partial \bar{x}}\left(G_{\bar{x}}\frac{h^3}{\mu}\frac{\partial P}{\partial \bar{x}}\right) + \frac{\partial}{\partial \bar{y}}\left(G_{\bar{y}}\frac{h^3}{\mu}\frac{\partial P}{\partial \bar{y}}\right) = \frac{U}{2}\frac{\partial h}{\partial \bar{x}} + \frac{\partial h}{\partial t} \qquad (6.5\text{-}1)$$

where $\bar{x} = R\theta$ is the circumferential coordinate in the direction of rotation, and $\bar{y}$ is the axial coordinate along the rotor center. Following the right-hand rule, the $\bar{z}$ axis is the radial coordinate along the film thickness. The fluid film is very thin compared to its extent (length and diameter). Therefore, for a givn journal position, the pressure P is only a function of the circumferential and axial coordinates and does not vary across the film thickness. In general, the film thickness h is a function of the journal position and the circumferential coordinate. However, it can also be dependent upon the axial coordinate, such as the pressure dam or multiple-pocket bearings. For a given journal position, we have:

$$P = P(\bar{x}, \bar{y}) \qquad\qquad h = h(\bar{x}, \bar{y}) \qquad (6.5\text{-}2)$$

U is the surface velocity. For a conventional journal bearing, the surface velocity equals the journal surface velocity ($U = U_j = R\Omega$). For a floating ring bushing, the surface velocity equals the sum of the journal surface velocity and the ring surface velocity ($U = U_j + U_b$). Figure 6.5-1 shows an unwrapped journal-bearing configuration. The viscosity μ is constant throughout the film. For the hydrodynamic journal bearing, one can reasonably assume that the lubricant properties are only a function of temperature. Lubricant properties are discussed in the Appendix C. $G_{\bar{x}}$ and $G_{\bar{y}}$ are called the turbulent flow coefficients, which are the correctional terms of viscosity, caused by the turbulent diffusion. The following approximations are proposed by Constantinescu (1973):

$$\frac{1}{G_{\bar{x}}} = 12 + 0.0136\,\mathrm{Re}^{0.90} \qquad \text{in circumferential direction} \tag{6.5-3}$$

$$\frac{1}{G_{\bar{y}}} = 12 + 0.0043\,\mathrm{Re}^{0.96} \qquad \text{in axial direction} \tag{6.5-4}$$

where

$$\mathrm{Re} = \frac{\rho\,U h}{\mu} \qquad\qquad \text{is the local Reynolds number.} \tag{6.5-5}$$

For laminar flow, turbulent flow coefficients are $G_{\bar{x}} = G_{\bar{y}} = 1/12$.

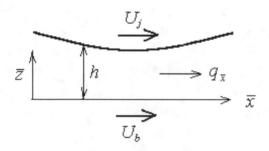

Figure 6.5-1 Unwrapped journal bearing

The volume flows in the direction of motion (circumferential direction) and in axial direction (perpendicular to the direction of motion) per unit width are:

$$q_{\bar{x}} = \frac{\left(U_j + U_b\right)h}{2} - \frac{G_{\bar{x}}h^3}{\mu}\frac{\partial P}{\partial \bar{x}} \tag{6.5-6}$$

$$q_{\bar{y}} = -\frac{G_{\bar{y}}h^3}{\mu}\frac{\partial P}{\partial \bar{y}} \tag{6.5-7}$$

The shear stress from Newton's viscosity law is:

$$\tau_{\bar{z}} = C_f\frac{\mu\left(U_j - U_b\right)}{h} - \frac{1}{2}\left(h - 2\bar{z}\right)\frac{\partial P}{\partial \bar{x}} \tag{6.5-8}$$

Where C_f is the turbulent Couette shear stress factor. For laminar flow, $C_f = 1$ and for turbulent flow (Constantinescu, 1973):

$$C_f = 1 + 0.0012 \, \text{Re}^{0.94} \tag{6.5-9}$$

Thus, the shear stress acting on the shaft is:

$$\tau_s = \tau_{\bar{z}=h} = C_f \frac{\mu(U_j - U_b)}{h} + \frac{h}{2}\frac{\partial P}{\partial \bar{x}} \tag{6.5-10}$$

The turbulence effect increases the effective fluid viscosity. Therefore it increases the load carry capability and power loss compared to those of the laminar flow. For high-speed applications, the turbulence effect can increase the power loss by a significant amount.

6.6 Bearing Static and Dynamic Characteristics

At static equilibrium, the position of the journal center is defined by coordinates (x_0, y_0) or (e_0, ϕ_0). The bearing is assumed to be concentric such that the journal axis is parallel to the bearing axis. Under dynamic conditions, the journal center oscillates around the static equilibrium position with small amplitudes of Δx and Δy, as shown in Figure 6.6-1. In addition to the translational displacements, the journal center has two rotational (angular) displacements, $(\Delta\theta_x, \Delta\theta_y)$. Readers should not be confused by the coordinate system (x, y, z), which is used to describe the rotor motion, and the coordinate system $(\bar{x}, \bar{y}, \bar{z})$, which is used in the Reynolds equation.

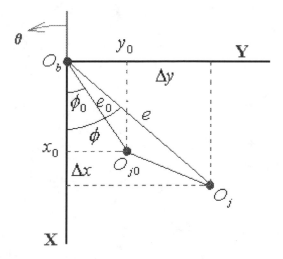

Figure 6.6-1 Journal static and dynamic positions

For a plain journal bearing, the film thickness h is:

$$h = C + e\cos(\theta - \phi) = C + (e\cos\phi)\cos\theta + (e\sin\phi)\sin\theta$$
$$= C + x\cos\theta + y\sin\theta \tag{6.6-1}$$

For small vibration ($\Delta x, \Delta y, \Delta\dot{x}, \Delta\dot{y}, \Delta\theta_x, \Delta\theta_y, \Delta\dot{\theta}_x, \Delta\dot{\theta}_y$), the film thickness becomes:

$$h = h_0 + \Delta h = h_0 + (\Delta x + z\Delta\theta_y)\cos\theta + (\Delta y - z\Delta\theta_x)\sin\theta \tag{6.6-2}$$

where the axial coordinate, z, is measured from the bearing centerplane. When substituting the film thickness into the Reynolds equation, z coordinate is replaced by $\bar{y} - \dfrac{L}{2}$, where $\bar{y}$ is the axial coordinate measured from the end of the bearing and L is the bearing axial length.

By assuming the vibration amplitudes to be small, a first order Taylor's expansion of the hydrodynamic pressure can be written as:

$$P = P_0 + \Delta P$$

$$= P_0 + \left(\frac{\partial P}{\partial x}\right)_0 \Delta x + \left(\frac{\partial P}{\partial y}\right)_0 \Delta y + \left(\frac{\partial P}{\partial \dot{x}}\right)_0 \Delta\dot{x} + \left(\frac{\partial P}{\partial \dot{y}}\right)_0 \Delta\dot{y}$$

$$+ \left(\frac{\partial P}{\partial \theta_x}\right)_0 \Delta\theta_x + \left(\frac{\partial P}{\partial \theta_y}\right)_0 \Delta\theta_y + \left(\frac{\partial P}{\partial \dot{\theta}_x}\right)_0 \Delta\dot{\theta}_x + \left(\frac{\partial P}{\partial \dot{\theta}_y}\right)_0 \Delta\dot{\theta}_y \tag{6.6-3}$$

$$= P_0 + P_x\Delta x + P_y\Delta y + P_{\dot{x}}\Delta\dot{x} + P_{\dot{y}}\Delta\dot{y} + P_{\theta_x}\Delta\theta_x + P_{\theta_y}\Delta\theta_y + P_{\dot{\theta}_x}\Delta\dot{\theta}_x + P_{\dot{\theta}_y}\Delta\dot{\theta}_y$$

Where h_0 and P_0 are the film thickness and pressure under static equilibrium condition. All the derivatives are evaluated at the static equilibrium condition.

The nonlinear fluid film forces and moments acting on the journal are determined by the application of boundary conditions and the integration of pressure distribution.

$$\begin{Bmatrix} F_x \\ F_y \end{Bmatrix} = \sum_{i=1}^{Npad} \int_0^L \int_{\theta_1}^{\theta_2} P \begin{Bmatrix} \cos\theta \\ \sin\theta \end{Bmatrix} d\bar{x}d\bar{y} \tag{6.6-4}$$

$$\begin{Bmatrix} M_x \\ M_y \end{Bmatrix} = \sum_{i=1}^{Npad} \int_0^L \int_{\theta_1}^{\theta_2} (\bar{y} - L/2) P \begin{Bmatrix} \cos\theta \\ \sin\theta \end{Bmatrix} d\bar{x}d\bar{y} \tag{6.6-5}$$

where θ_1, θ_2 are the leading and trailing edges of the pressure film for each pad under integration. $d\bar{x}$ is replaced by $R\,d\theta$ during the integration. For small amplitudes, the nonlinear fluid film forces and moments can be linearized about the static equilibrium position:

$$\begin{Bmatrix} F_x \\ F_y \end{Bmatrix} = \begin{Bmatrix} F_{x0} \\ F_{y0} \end{Bmatrix} + \begin{bmatrix} \dfrac{\partial F_x}{\partial x} & \dfrac{\partial F_x}{\partial y} \\ \dfrac{\partial F_y}{\partial x} & \dfrac{\partial F_y}{\partial y} \end{bmatrix}_0 \begin{Bmatrix} \Delta x \\ \Delta y \end{Bmatrix} + \begin{bmatrix} \dfrac{\partial F_x}{\partial \dot{x}} & \dfrac{\partial F_x}{\partial \dot{y}} \\ \dfrac{\partial F_y}{\partial \dot{x}} & \dfrac{\partial F_y}{\partial \dot{y}} \end{bmatrix}_0 \begin{Bmatrix} \Delta \dot{x} \\ \Delta \dot{y} \end{Bmatrix}$$ (6.6-6)

$$= \begin{Bmatrix} F_{x0} \\ F_{y0} \end{Bmatrix} - \begin{bmatrix} k_{xx} & k_{xy} \\ k_{yx} & k_{yy} \end{bmatrix} \begin{Bmatrix} \Delta x \\ \Delta y \end{Bmatrix} - \begin{bmatrix} c_{xx} & c_{xy} \\ c_{yx} & c_{yy} \end{bmatrix} \begin{Bmatrix} \Delta \dot{x} \\ \Delta \dot{y} \end{Bmatrix}$$

$$\begin{Bmatrix} M_x \\ M_y \end{Bmatrix} = \begin{Bmatrix} M_{x0} \\ M_{y0} \end{Bmatrix} + \begin{bmatrix} \dfrac{\partial M_x}{\partial \theta_x} & \dfrac{\partial M_x}{\partial \theta_y} \\ \dfrac{\partial M_y}{\partial \theta_x} & \dfrac{\partial M_y}{\partial \theta_y} \end{bmatrix}_0 \begin{Bmatrix} \Delta \theta_x \\ \Delta \theta_y \end{Bmatrix} + \begin{bmatrix} \dfrac{\partial M_x}{\partial \dot{\theta}_x} & \dfrac{\partial M_x}{\partial \dot{\theta}_y} \\ \dfrac{\partial M_y}{\partial \dot{\theta}_x} & \dfrac{\partial M_y}{\partial \dot{\theta}y} \end{bmatrix}_0 \begin{Bmatrix} \Delta \dot{\theta}_x \\ \Delta \dot{\theta}_y \end{Bmatrix}$$ (6.6-7)

$$= \begin{Bmatrix} M_{x0} \\ M_{y0} \end{Bmatrix} - \begin{bmatrix} k_{aa} & k_{ab} \\ k_{ba} & k_{bb} \end{bmatrix} \begin{Bmatrix} \Delta \theta_x \\ \Delta \theta_y \end{Bmatrix} - \begin{bmatrix} c_{aa} & c_{ab} \\ c_{ba} & c_{bb} \end{bmatrix} \begin{Bmatrix} \Delta \dot{\theta}_x \\ \Delta \dot{\theta}_y \end{Bmatrix}$$

Assuming the journal axis is aligned with the bearing axis at static equilibrium, the force and moment equations (translational and rotational motions) are decoupled. By substituting the perturbed pressure equation (6.6-3) into the fluid film forces and moments equations (6.6-4 and 6.6-5), and collecting the first order terms according to the perturbed variables, the bearing dynamic coefficients then are related to the pressure derivatives in the following forms:

The translational stiffness coefficients from the linearized forces equations:

$$\begin{Bmatrix} k_{xx} \\ k_{yx} \end{Bmatrix} = \sum_{i=1}^{Npad} \int_0^L \int_{\theta_1}^{\theta_2} -P_x \begin{Bmatrix} \cos\theta \\ \sin\theta \end{Bmatrix} d\bar{x} d\bar{y}$$ (6.6-8)

$$\begin{Bmatrix} k_{xy} \\ k_{yy} \end{Bmatrix} = \sum_{i=1}^{Npad} \int_0^L \int_{\theta_1}^{\theta_2} -P_y \begin{Bmatrix} \cos\theta \\ \sin\theta \end{Bmatrix} d\bar{x} d\bar{y}$$ (6.6-9)

The translational damping coefficients from the linearized forces equations:

$$\begin{Bmatrix} c_{xx} \\ c_{yx} \end{Bmatrix} = \sum_{i=1}^{Npad} \int_0^L \int_{\theta_1}^{\theta_2} -P_{\dot{x}} \begin{Bmatrix} \cos\theta \\ \sin\theta \end{Bmatrix} d\bar{x} d\bar{y}$$ (6.6-10)

$$\begin{Bmatrix} c_{xy} \\ c_{yy} \end{Bmatrix} = \sum_{i=1}^{Npad} \int_0^L \int_{\theta_1}^{\theta_2} -P_{\dot{y}} \begin{Bmatrix} \cos\theta \\ \sin\theta \end{Bmatrix} d\bar{x} d\bar{y}$$ (6.6-11)

The rotational stiffness coefficients from the linearized moments equations:

$$\begin{Bmatrix} k_{aa} \\ k_{ba} \end{Bmatrix} = \sum_{i=1}^{Npad} \int_0^L \int_{\theta_1}^{\theta_2} -(\bar{y}-L/2)P_{\theta_x} \begin{Bmatrix} \cos\theta \\ \sin\theta \end{Bmatrix} d\bar{x}d\bar{y} \qquad (6.6\text{-}12)$$

$$\begin{Bmatrix} k_{ab} \\ k_{bb} \end{Bmatrix} = \sum_{i=1}^{Npad} \int_0^L \int_{\theta_1}^{\theta_2} -(\bar{y}-L/2)P_{\theta_y} \begin{Bmatrix} \cos\theta \\ \sin\theta \end{Bmatrix} d\bar{x}d\bar{y} \qquad (6.6\text{-}13)$$

The rotational damping coefficients from the linearized moments equations:

$$\begin{Bmatrix} c_{aa} \\ c_{ba} \end{Bmatrix} = \sum_{i=1}^{Npad} \int_0^L \int_{\theta_1}^{\theta_2} -(\bar{y}-L/2)P_{\dot{\theta}_x} \begin{Bmatrix} \cos\theta \\ \sin\theta \end{Bmatrix} d\bar{x}d\bar{y} \qquad (6.6\text{-}14)$$

$$\begin{Bmatrix} c_{ab} \\ c_{bb} \end{Bmatrix} = \sum_{i=1}^{Npad} \int_0^L \int_{\theta_1}^{\theta_2} -(\bar{y}-L/2)P_{\dot{\theta}_y} \begin{Bmatrix} \cos\theta \\ \sin\theta \end{Bmatrix} d\bar{x}d\bar{y} \qquad (6.6\text{-}15)$$

At static equilibrium, the static forces are equal to the external loads in the opposite directions.

$$\begin{Bmatrix} F_{x0} \\ F_{y0} \end{Bmatrix} = \begin{Bmatrix} -W_x \\ -W_y \end{Bmatrix} = \sum_{i=1}^{Npad} \int_0^L \int_{\theta_1}^{\theta_2} P_0 \begin{Bmatrix} \cos\theta \\ \sin\theta \end{Bmatrix} d\bar{x}d\bar{y} \qquad (6.6\text{-}16)$$

The journal is assumed to be aligned with bearing in static equilibrium. The static pressure is symmetric with respect to the bearing centerplane. At equilibrium, the static moments are zero. Furthermore, the moment coefficients (k_{aa},...etc.) are small compared to the force coefficients (k_{xx},...etc.) and their contributions to the dynamics of rotor systems are negligible. Therefore, the perturbed rotational displacements, which produce the moment coefficients, are neglected in the following discussion.

By substituting the perturbed pressure and film thickness expressions into the Reynolds equation, omitting the high order terms and perturbed rotational displacements, retaining only the first order terms, and collecting terms according to the perturbation variables, we get the following five equations:

$$\frac{\partial}{\partial \bar{x}}\left(G_{\bar{x}}\frac{h_0^3}{\mu}\frac{\partial P_i}{\partial \bar{x}}+f_i\right)+\frac{\partial}{\partial \bar{y}}\left(G_{\bar{y}}\frac{h_0^3}{\mu}\frac{\partial P_i}{\partial \bar{y}}+g_i\right)=0 \qquad i=0,\Delta x,\Delta y,\Delta \dot{x},\Delta \dot{y} \qquad (6.6\text{-}17)$$

where

$$f_0 = -\frac{Uh_0}{2} \qquad\qquad g_0=0 \qquad\qquad (6.6\text{-}18)$$

$$f_{\Delta x} = \left(3\frac{G_{\bar{x}}}{\mu}h_0^2\frac{\partial P_0}{\partial x} - \frac{U}{2} \right)\cos\theta \qquad g_{\Delta x} = \left(3\frac{G_{\bar{y}}}{\mu}h_0^2\frac{\partial P_0}{\partial y} \right)\cos\theta \qquad (6.6\text{-}19)$$

$$f_{\Delta y} = \left(3\frac{G_{\bar{x}}}{\mu}h_0^2\frac{\partial P_0}{\partial x} - \frac{U}{2} \right)\sin\theta \qquad g_{\Delta y} = \left(3\frac{G_{\bar{y}}}{\mu}h_0^2\frac{\partial P_0}{\partial y} \right)\sin\theta \qquad (6.6\text{-}20)$$

$$f_{\Delta\dot{x}} = -R\sin\theta \qquad\qquad\qquad g_{\Delta\dot{x}} = 0 \qquad\qquad\qquad (6.6\text{-}21)$$

$$f_{\Delta\dot{y}} = R\cos\theta \qquad\qquad\qquad g_{\Delta\dot{y}} = 0 \qquad\qquad\qquad (6.6\text{-}22)$$

Now we have five second-order partial differential equations involving a scalar-valued function P_i with specified boundary conditions. The finite element formulation of solving the partial differential equations will be discussed later. If the moment coefficients are interested, then four additional equations are obtained by remaining the four perturbed rotational displacements.

It should be noted that the cross-coupling damping coefficients are derived from self-adjoint operators. Therefore, the cross-coupling damping coefficients are equal ($c_{xy} = c_{yx}$) and the damping matrix is symmetric (Lund, 1987, Shang and Dien, 1989). A symmetric damping matrix is referred to as a dissipative damping matrix and always stabilizes the rotor system as discussed in Chapter 3. However, in view of the equations for the cross-coupling stiffness coefficients, they are generally not equal ($k_{xy} \neq k_{yx}$). Thus, the stiffness matrix is a real general matrix and can be decomposed into a conservative (elastic) matrix and a non-conservative (circulatory) matrix. This circulatory matrix can supply the energy to the forward whirl motion and destabilize the rotor system (Adams and Padovan, 1981, Lund, 1987).

At static equilibrium, the journal center is defined by coordinates (x_0, y_0) or (e_0, ϕ_0). The static pressure distribution P_0 is determined by solving the above zero order partial differential equation with specified boundary conditions. The fluid film reaction forces are obtained by integrating the static film pressure distribution. At static equilibrium, the forces acting on the journal center in the X and Y directions must be balanced. For a giving bearing operating condition (load and speed), the determination of the static equilibrium position is an iterative process involving two nonlinear equations (forces) and two unknowns (journal position x_0 and y_0).

For multiple lobe bearings, the film thickness is a function of circumferential coordinate and the journal position:

$$\begin{aligned} h_0(\theta, x_0, y_0) &= C_p + e_p\cos(\theta - \phi_p) \\ &= C_p + (e_p\cos\phi_p)\cos\theta + (e_p\sin\phi_p)\sin\theta \end{aligned} \qquad (6.6\text{-}23)$$

where C_p is the lobe radial clearance, e_p (lobe eccentricity) is the distance between the journal center and lobe (pad) center, and ϕ_p (lobe attitude angle) is the angle between the loading direction ($+X$ axis) to the line connecting the journal center and lobe center as shown in Figure 6.6-2.

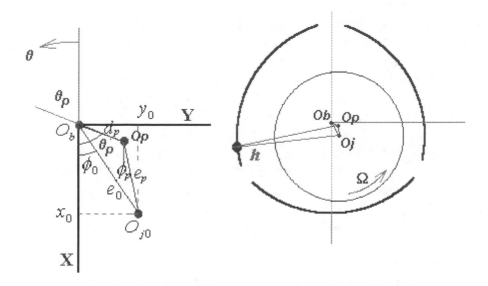

Figure 6.6-2 The relationship between the journal center, bearing center, and lobe center

The lobe eccentricity and attitude angle are related to the journal position by:

$$x_0 = e_0 \cos\phi_0 = d_p \cos\theta_p + e_p \cos\phi_p \qquad (6.6\text{-}24a)$$

$$y_0 = e_0 \sin\phi_0 = d_p \sin\theta_p + e_p \sin\phi_p \qquad (6.6\text{-}24b)$$

where $d_p = C_p\text{-}C_b$ is the distance between the bearing center and lobe center and θ_p is the angle from the loading direction ($+X$ axis) to the line connecting the bearing center and the pad center of curvature.

However, for pressure dam bearings and multiple pocket bearings, the film thickness can also be dependent on the axial coordinate. In general, the film thickness can be of the form:

$$h_0 = h_0(\overline{x}, \overline{y}, x_0, y_0) \qquad (6.6\text{-}25)$$

Again, readers should not be confused by the circumferential and axial coordinates ($\overline{x}, \overline{y}$) used in the Reynolds equation, and the journal coordinates (x_0, y_0) used to describe the journal equilibrium position.

Once the static equilibrium position is established by solving the steady state Reynolds equation in an iterative process, bearing static performance can easily be determined from the known static pressure distribution and journal static equilibrium position. If heat balance is required, then another outer loop iterative procedure is needed, since the lubricant properties are dependent on temperature. With the established film boundaries from the steady state solution, the four perturbed pressures can be calculated from the rest four partial differential equations without any iterative process. Then, the bearing dynamic coefficients are obtained by integrating the perturbed pressures over the established film domain.

6.7 Solutions for the Reynolds Equation

The steady-state pressure and four pressure perturbations are governed by five two dimensional second-order partial differential equations as previously described. These boundary value problems can be solved by the Finite Difference or Finite Element Methods. The use of the Finite Difference Method in solving these equations is well documented by Lund and Thomsen (1978) and is not repeated here. The Finite Element Method was presented by Allaire et al. (1977) and Nicholas (1977) in solving the steady state Reynolds equation, with linear interpolation functions and triangular elements. The bearing dynamic coefficients are obtained by repeatedly solving the steady state Reynolds equation with small amount of displacement and velocity perturbations about equilibrium. Due to the highly nonlinear nature of the fluid film forces, the numerical differentiation of the perturbed force with respect to displacement and velocity, to obtain the bearing dynamic coefficients, can be very complicated, inaccurate, and highly dependent on the perturbation size. The Finite Element Method with high order interpolation functions and rectangular elements is presented here. This procedure is also used to solve the perturbed pressures equations. The bearing dynamic coefficients can be directly obtained by integrating the perturbed pressures and when obtained by this procedure, are more consistent and accurate.

Both methods have been implemented in *DyRoBeS-BePerf*. It is the author's opinion that the computer program for the Finite Difference Method is more difficult to implement. Many details need to be considered, especially in the boundary iteration for the film cavitations. Once the computer code is implemented, the computational time is much less than the Finite Element Method. However, the Finite Difference Method is not suitable for bearings with clearance discontinuities, such as the pressure dam bearings, because the derivatives of the film thickness are required in the Finite Difference Procedure. The Finite Element Method on the other hand, is very easy to program, since it is very straightforward. It is suitable for complex bearing geometries, such as pressure dam bearings, because the derivatives of the film thickness are not required in Finite Element Method and only the film thickness and the derivatives of the interpolation function are needed. The turbulence effect can be easily included in the Finite Element Method.

In this section, the basic steps in the solutions of the Reynolds equation and the perturbed pressure equations by using Finite Element Formulation will be discussed in the following order:

1. Discretization and selection of element type
2. Determination of degrees-of-freedom and selection of interpolation functions
3. Elemental formulation
4. Assemblage equations for the entire domain
5. Boundary conditions
6. Solutions for the primary unknowns
7. Computation of derived quantities

6.7.1 Discretization and Selection of Element Type

For the journal bearing analysis, the discretization and selection of element type is quite simple. In general, the bearing can be unwrapped, and two global coordinates $(\bar{x}, \bar{y})$ are used to describe the pressure distribution, $P(\bar{x}, \bar{y})$, as shown in Figure 6.7-1.

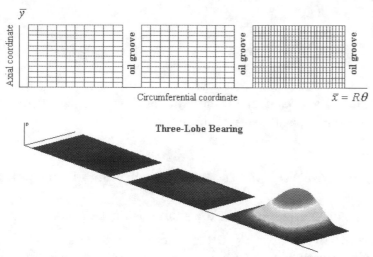

Figure 6.7-1 Finite element meshes for a three-lobe bearing

The circumferential coordinate $\bar{x} = R\theta$ ranges from 0 to 2π in the direction of shaft rotation. Also, a cyclic boundary condition is applied at 0 and 2π. The axial coordinate $\bar{y}$ ranges from 0 to L (bearing axial length). Very often only half the bearing axial length is required for the axial coordinate with bearings having symmetry around the centerplane.

In view of bearing geometry, the rectangular element is the simplest element type to use in journal bearing analysis. For a typical rectangular element, as shown in Figure 6.7-2, the element is defined by four nodes (vertices), which are numbered counterclockwise having the global coordinates $(\bar{x}_i, \bar{y}_i)$ $(i = 1, 2, 3, 4)$.

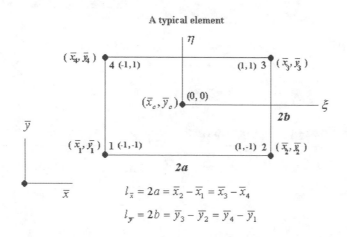

Figure 6.7-2 A typical rectangular element

A local non-dimensional normalized coordinate system (natural coordinates) with the origin at the center of the element is commonly used in the elemental formulation. The advantages of using the natural coordinates are: it is convenient in constructing the interpolation functions in the local coordinate system and the local coordinate can considerably facilitate the integrations and differentiations involved in the formulation of element matrices. The relationships between the two coordinates are:

$$\xi = \frac{\overline{x} - \overline{x}_c}{a} \quad \text{and} \quad \eta = \frac{\overline{y} - \overline{y}_c}{b} \tag{6.7-1}$$

For computational efficiency and accuracy, it is desirable to have finer meshes for the lobe where the load is, and coarser meshes for the rest of lobes where the pressure is small or lobes are unloaded, as illustrated in Figure 6.7-1.

6.7.2 Determination of Degrees-of-Freedom and Selection of Interpolation Functions

Static pressure gradients are required for the solutions of the perturbed pressures (P_x, P_y) and for the calculation of the bearing stiffness coefficients, as indicated in Eqs. (6.6-19) and (6.6-20). The solutions of these equations are extremely sensitive to the accuracy of the gradients. Also, the pressure gradients are needed for the determination of the film cavitation. Therefore, in addition to the pressure, the pressure gradients are also introduced as unknown variables (degrees-of-freedom). Hence, there are 3 degrees-of-freedom for each node and a total of 12 degrees-of-freedom for each finite element, as shown in Figure 6.7-3.

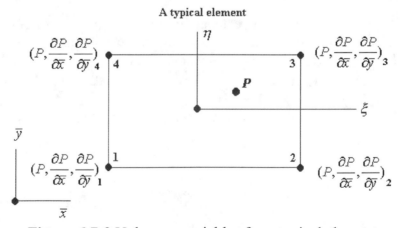

Figure 6.7-3 Unknown variables for a typical element

The pressure distribution in the interior of an element is approximated by the expression:

$$P(\overline{x}, \overline{y}) = P(\xi, \eta) = \sum_{i=1}^{4} \left(N_i P_i + N_{4+i} \frac{\partial P_i}{\partial \overline{x}} + N_{8+i} \frac{\partial P_i}{\partial \overline{y}} \right) \tag{6.7-2}$$

where the Hermite (cubic) interpolation functions for node i (i=1,2,3,4) are:

$$N_i = \frac{1}{8}(\xi_0 + 1)(\eta_0 + 1)(2 + \xi_0 + \eta_0 - \xi^2 - \eta^2)$$

$$N_{4+i} = \frac{a}{8}\xi_i(\xi_0 + 1)^2(\xi_0 - 1)(\eta_0 + 1) \qquad (6.7\text{-}3)$$

$$N_{8+i} = \frac{b}{8}\eta_i(\xi_0 + 1)(\eta_0 + 1)^2(\eta_0 - 1)$$

and

$$\xi = \frac{\overline{x} - \overline{x}_c}{a} \qquad \text{and} \qquad \eta = \frac{\overline{y} - \overline{y}_c}{b} \qquad (6.7\text{-}4)$$

$$\xi_0 = \xi\,\xi_i \qquad \text{and} \qquad \eta_0 = \eta\,\eta_i \qquad (6.7\text{-}5)$$

It can be written in the vector form:

$$P(\overline{x}, \overline{y}) = \boldsymbol{N}^T \boldsymbol{u}^e = \sum_{j=1}^{12} N_j u_j \qquad (6.7\text{-}6)$$

where

$$\boldsymbol{N}^T(\xi, \eta) = [N_1, N_2, N_3, N_4, N_5, N_6, N_7, N_8, N_9, N_{10}, N_{11}, N_{12}] \qquad (6.7\text{-}7)$$

and $\boldsymbol{u}^e$ is the unknown nodal variables (pressure and pressure gradients) which are to be determined:

$$\boldsymbol{u}^e = \left[P_1, P_2, P_3, P_4, \frac{\partial P_1}{\partial x}, \frac{\partial P_2}{\partial x}, \frac{\partial P_3}{\partial x}, \frac{\partial P_4}{\partial x}, \frac{\partial P_1}{\partial y}, \frac{\partial P_2}{\partial y}, \frac{\partial P_3}{\partial y}, \frac{\partial P_4}{\partial y}\right]^T \qquad (6.7\text{-}8)$$

6.7.3 Elemental Formulation

Two most commonly used procedures in the derivation of element equations in Finite Element Analysis are the energy methods and the variational formulations. Energy methods are widely used in the structural model where the energy expressions are readily available, such as the rotating disks and shaft elements, which were presented in the previous chapters. Variational formulations are widely used in the fluid model where the governing partial differential equations are available. The latter will be presented in this section. Two approaches in the variational formulations are available for obtaining the approximation and for formulating the element equations. The first one is the method of weighted residuals, also known as the Galerkin's method. It approaches the differential equations directly. The second one is the determination of variational functionals for which the stationary values are sought. The variational functional was presented by Allaire et al. (1977) for the steady state Reynolds equation. The bearing dynamic coefficients are obtained by using the numerical differentiations in their work. As presented in the previous sections, the bearing dynamic coefficients are more accurate if they are obtained from the perturbed pressure equations. Since we have five differential equations to be solved, Galerkin's method, which solves the five differential equations in the same way, is presented below. The five differential equations previously presented have the same form:

$$\frac{\partial}{\partial \bar{x}}\left(G_{\bar{x}} \frac{h_0^3}{\mu} \frac{\partial P_k}{\partial \bar{x}} + f_k \right) + \frac{\partial}{\partial \bar{y}}\left(G_{\bar{y}} \frac{h_0^3}{\mu} \frac{\partial P_k}{\partial \bar{y}} + g_k \right) = 0 \qquad k = 0, \Delta x, \Delta y, \Delta \dot{x}, \Delta \dot{y} \qquad (6.7\text{-}9)$$

Since the governing differential equation is valid over the entire domain, it is also valid over the individual element. The first step in the finite element formulation is to construct the variational form of the governing differential equation over the element. Let us multiply the differential equation by an arbitrary test function v and integrate over the element domain A^e:

$$\int_{A^e} v \left[\frac{\partial}{\partial \bar{x}}\left(G_{\bar{x}} \frac{h_0^3}{\mu} \frac{\partial P_k}{\partial \bar{x}} + f_k \right) + \frac{\partial}{\partial \bar{y}}\left(G_{\bar{y}} \frac{h_0^3}{\mu} \frac{\partial P_k}{\partial \bar{y}} + g_k \right) \right] dA^e = 0 \qquad (6.7\text{-}10)$$

From the integration by parts, also known as the gradient and divergence theorems, we obtain

$$\int_{A^e} -\left[\frac{\partial v}{\partial \bar{x}}\left(G_{\bar{x}} \frac{h_0^3}{\mu} \frac{\partial P_k}{\partial \bar{x}} + f_k \right) + \frac{\partial v}{\partial \bar{y}}\left(G_{\bar{y}} \frac{h_0^3}{\mu} \frac{\partial P_k}{\partial \bar{y}} + g_k \right) \right] dA^e +$$

$$\oint_{\Gamma^e} v \left\{ n_{\bar{x}}\left(G_{\bar{x}} \frac{h_0^3}{\mu} \frac{\partial P_k}{\partial \bar{x}} + f_k \right) + n_{\bar{y}}\left(G_{\bar{y}} \frac{h_0^3}{\mu} \frac{\partial P_k}{\partial \bar{y}} + g_k \right) \right\} ds = 0 \qquad (6.7\text{-}11)$$

Note that the second integration term is performed on the element boundary Γ^e where ds is the arc length of an infinitesimal element along the boundary, and $n_{\bar{x}}, n_{\bar{y}}$ are the directional cosines. It indicates that not only the integral of the differential equations must be satisfied, but also the boundary conditions must be satisfied. In view of the steady state Reynolds equation, the boundary conditions are simply the steady state flow continuity (balance) requirements. It is also called natural boundary condition that vanishes during the assembly process for the interior elements, except those at the bearing geometric boundaries. The boundary conditions are applied at the global level not in the element level, therefore, the natural boundary condition (second term) can be neglected. This variational form is a preliminary step towards the Galerkin's method for generating the element equations for the boundary value problems. For the Galerkin's method, the test functions v are chosen to be the interpolation functions N_i. Substituting the pressure distribution and test function into the variational form:

$$P_k(\overline{x},\overline{y}) = \sum_{j=1}^{12} N_j u_{kj} \qquad \frac{\partial P_k}{\partial \overline{x}} = \sum_{j=1}^{12} \frac{\partial N_j}{\partial \overline{x}} u_{kj} \qquad \frac{\partial P_k}{\partial \overline{y}} = \sum_{j=1}^{12} \frac{\partial N_j}{\partial \overline{y}} u_{kj} \qquad (6.7\text{-}12)$$

$$v = N_i \qquad (6.7\text{-}13)$$

We have:

$$\sum_{j=1}^{12} \left\{ \int_{A^e} \left[\frac{\partial N_i}{\partial \overline{x}} \left(G_{\overline{x}} \frac{h_0^3}{\mu} \frac{\partial N_j}{\partial \overline{x}} \right) + \frac{\partial N_i}{\partial \overline{y}} \left(G_{\overline{y}} \frac{h_0^3}{\mu} \frac{\partial N_j}{\partial \overline{y}} \right) \right] dA^e \right\} u_{kj}$$

$$= -\sum_{j=1}^{12} \int_{A^e} \left[\frac{\partial N_i}{\partial \overline{x}} (f_k) + \frac{\partial N_i}{\partial \overline{y}} (g_k) \right] dA^e \qquad i = 1,2,3\ldots12 \qquad (6.7\text{-}14)$$

or in matrix form:

$$\boldsymbol{K}^e \boldsymbol{u}_k^e = \boldsymbol{F}^e_{(12x1)} \qquad (6.7\text{-}15)$$

Note that the matrix $\boldsymbol{K}^e$ is symmetric ($K_{ij}=K_{ji}$). The above elemental equations are valid for all five differential equations with different f_k and g_k given in the previous sections. The numerical integration is performed in the local (natural) coordinates and Gauss-Legendre quadratures are used in the numerical integrations for the element matrix and load vector. The details of this integration are presented in most of Finite Element Textbook (Reddy, 1993) and are not repeated here.

6.7.4 Assemblage Equations for the Entire Domain

Since the element equation is derived for an arbitrarily typical element, it holds for any element from the finite element mesh. For journal bearing analysis, there is only one element type and its element equation is used repeatedly for all the elements and assembled to form the global equations. This global equation describes the behavior of the entire film domain. The assembly process is based on the law of compatibility or connectivity of the inter-element continuity conditions. The assemblage equations are of the form:

$$\boldsymbol{K}\boldsymbol{u} = \boldsymbol{F} \qquad (6.7\text{-}16)$$

The assembled stiffness matrix is a banded symmetric matrix. Skyline solver can be used to save computational time. Without imposing the essential boundary conditions, the stiffness matrix is singular and the equation cannot be solved.

6.7.5 Boundary Conditions

Consider a typical bearing pad (lobe) with an axial length of L and the pad arc starts from $\bar{x}_1 = R\theta_1$ to $\bar{x}_2 = R\theta_2$ as shown in Figure 6.7-4. There are three types of boundary conditions to be addressed. The first type of boundaries are along the bearing edges in the axial direction, the second type of boundaries are due to the oil grooves and pressure cavitations in the circumferential direction, the third type of boundaries are occurred in the interior of the film domain due to the geometric discontinuity. The first two types of boundaries are shown in Figure 6.7-4.

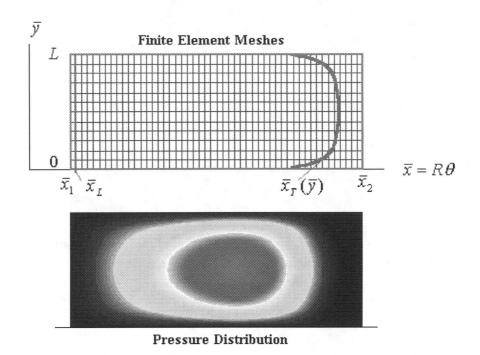

Figure 6.7-4 Boundary conditions

1. Axial Direction

The boundary conditions along the axial direction are that the pressure is ambient (zero) at the edges of a pad. The pressure along the both sides of the bearing is zero:

$$\bar{y} = 0: \quad P = 0 \Rightarrow P_0 = P_x = P_y = P_{\dot{x}} = P_{\dot{y}} = 0 \tag{6.7-17a}$$

$$\bar{y} = L: \quad P = 0 \Rightarrow P_0 = P_x = P_y = P_{\dot{x}} = P_{\dot{y}} = 0 \tag{6.7-17b}$$

If the bearing is symmetric with respect to the centerplane, the pressure distribution is symmetric about the centerplane (no flow to each side) and half bearing model can be used for computational efficiency. At centerplane, the boundary condition is:

$$\bar{y} = L/2: \quad \frac{\partial P}{\partial \bar{y}} = 0 \Rightarrow \frac{\partial P_0}{\partial \bar{y}} = \frac{\partial P_x}{\partial \bar{y}} = \frac{\partial P_y}{\partial \bar{y}} = \frac{\partial P_{\dot{x}}}{\partial \bar{y}} = \frac{\partial P_{\dot{y}}}{\partial \bar{y}} = 0 \tag{6.7-17c}$$

2. Circumferential Direction

For a plan cylindrical bearing with a complete 2π arc. The cyclic boundary condition is applied at ($\bar{x}_1 = 0$) and ($\bar{x}_2 = 2\pi$). For grooves bearings with the lobe arc extending from $\bar{x}_1 = R\theta_1$ to $\bar{x}_2 = R\theta_2$, the pressure along the grooves equals to the oil supply pressure. For hydrodynamic bearings, the supply pressure can be neglected. The boundary conditions are:

$$\bar{x} = \bar{x}_1: \quad P = 0 \Rightarrow P_0 = P_x = P_y = P_{\dot{x}} = P_{\dot{y}} = 0 \qquad (6.7\text{-}18a)$$

$$\bar{x} = \bar{x}_2: \quad P = 0 \Rightarrow P_0 = P_x = P_y = P_{\dot{x}} = P_{\dot{y}} = 0 \qquad (6.7\text{-}18b)$$

Since typical lubricants cannot stand large and continuous negative pressure without rupturing, film cavitations occur and the film pressure boundaries are smaller or equal to the pad physical boundaries. If the pad geometric boundaries are given by $\bar{x}_1$ and $\bar{x}_2$, and the film boundaries are established by $\bar{x}_L$ (leading edge) and $\bar{x}_T$ (trailing edge) as illustrated in Figure 6.7-4, we have:

$$\bar{x}_1 \le \bar{x}_L < \bar{x}_T \le \bar{x}_2 \qquad (6.7\text{-}19)$$

Note that $\bar{x}_1$ and $\bar{x}_2$ are the leading and trailing edges of the pad and $\bar{x}_L$ and $\bar{x}_T$ are the leading and trailing edges of the pressure film. Three sets of circumferential boundary conditions are commonly mentioned in the literatures. The Sommerfeld condition allows the negative pressures in the film and produces unrealistic results such as 90 degrees of attitude angle and unstable rotor motion. The Gumbel (half Sommerfeld) condition is similar to Sommerfeld condition except that all negative pressures are neglected. Although this approach violates the flow continuity at the trailing edge of the film and is physically inappropriate, however, due to its simplicity to apply, some engineers have used it and it leads to analytical solutions for a plain cylindrical bearing with short or long bearing assumptions. Reynolds (Swift-Stieber) condition is more realistic condition, which satisfies the flow continuity requirements and represents the actual condition in a film. For a plain cylindrical bearing, the Reynolds circumferential boundary condition is simply that the pressure film starts slightly ahead of the maximum film thickness location and ends at some point beyond the minimum film thickness in a very gentle manner with the pressure and its gradients are zero. For a lobed bearing, the boundary condition in the circumferential direction can be very complicated. The Reynolds condition is difficult to apply analytically, but it is easy to implement numerically and it should be used in the bearing analysis.

Figure 6.7-5 illustrates the pressure profiles of three different boundary conditions for a plain journal bearing with L=3 in., D=3 in., C_b=0.005 in., Ω=4000 rpm, W=750 Lbf, μ=2.0E-06 Reyns, without turbulence effect. The results are obtained by **DyRoBeS-BePerf**. Note that the Sommerfeld and Gumbel boundary conditions provided in the program are mainly for illustrative purposes; Reynolds boundary conditions should always be selected in the practical bearing design.

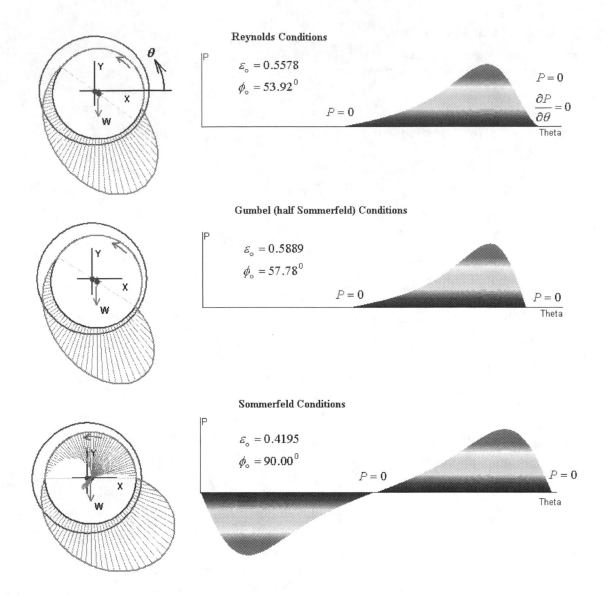

Figure 6.7-5 Three circumferential boundary conditions

For a lobed bearing, due to the bearing oil grooves, preloads, offset, and possible pressure dams, the locations of the film boundaries can have many possibilities. If the lobe is full loaded and no film rupture exists, the film pressure is generated over the entire lobe area and the essential boundary conditions at both pad edges ($P(\bar{x}_1) = 0$ and $P(\bar{x}_2) = 0$) are used. If the lobe is fully unloaded, then there is no need to solve the differential equation and the Reynolds equation does not apply in the cavitation zone. The lubricant breaks up into small streamers, part fluid and part vapor or air, to fill the clearance gap. If the lobe is partially loaded, then film rupture should be considered. The leading edge of the film ($\bar{x}_L$) is usually taken at the maximum film thickness if it is located within the lobe, otherwise, the lubricant supply groove (the leading edge of the pad $\bar{x}_1$) will be taken as the leading edge of the film. The trailing edge of the film ($\bar{x}_T$) is either at a lubricant supply groove (the trailing edge of the pad $\bar{x}_2$) when it is located in a convergent region of the film, or at a transition boundary curve located in the divergent film region where the film rupture occurs. If the film rupture occurs before the lubricant supply groove, the trailing edge boundary, where the pressure and pressure gradients are zero, has to be determined by iterative procedure and it is a function of axial coordinate ($\bar{x}_T = \bar{x}_T(\bar{y})$). The boundary conditions for this case are:

$$\bar{x} = \bar{x}_L : \quad P = 0 \Rightarrow P_0 = P_x = P_y = P_{\dot{x}} = P_{\dot{y}} = 0 \tag{6.7-20a}$$

$$\bar{x} = \bar{x}_T : \quad P_0 = 0, \quad \frac{\partial P_0}{\partial \bar{x}} = \frac{\partial P_0}{\partial \bar{y}} = 0, \quad \text{and} \quad P_x = P_y = P_{\dot{x}} = P_{\dot{y}} = 0 \tag{6.7-20b}$$

Since Hermite interpolation functions are used and the pressure derivatives are also the degrees-of-freedom, the determination of trailing edge boundary curve if required typically takes 1 to 3 iterations.

3. Internal Boundary

One advantage of the Finite Element Method is the ability of handling the bearing with clearance discontinuities such as the pressure dam and pocket bearings as described before. For the boundary nodes around the dam or pocket, two global node numbers are assigned for the same physical node as shown in Figure 6.7-6 in a one-dimensional case.

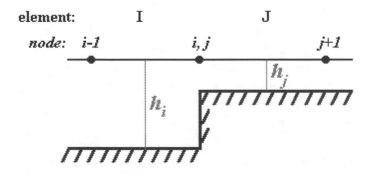

Figure 6.7-6 Boundary condition for film discontinuity

Node i and node j are located at the same physical position (boundary node). However, node i has a film thickness of h_i and is a node for element I, and node j has a film thickness of h_j and is a node for element J. Since node i and j are the same boundary node, the pressure continuity has to be maintained, however, not the slope:

$$P_i = P_j \tag{6.7-21a}$$

$$\frac{\partial P_i}{\partial \bar{x}} \neq \frac{\partial P_j}{\partial \bar{x}} \tag{6.7-21b}$$

Since the pressure gradients are not continuous at the boundary node, the pressure gradients therefore are not specified. That is, we introduce two more degrees-of-freedom (pressure gradients) at the boundary node. It is similar to a hinge connecting two structure bodies where the displacement is continuous and slopes are not.

6.7.6 Solutions for the Primary Unknowns

First, we need to solve the steady state pressure distribution, P_0, with the assembled global equations and prescribed boundary conditions described before. In the process of solving the steady state pressure equations, any negative pressure values are set equal to zero. If the film rupture occurs in the pad, the transition boundary curve is determined by an iterative procedure. With high order interpolation functions and pressure gradients are also the degrees-of-freedom, this boundary curve is established typically in 1 to 3 iterations. The steady state forces are then obtained by integrating the steady state pressure distribution. At static equilibrium, the forces acting on the journal must be balanced. Therefore, iteration in solving the steady state Reynolds equation is required to determine the static equilibrium position. If heat balance is required, then another outer loop iterative procedure is needed since the lubricant properties are dependent upon the temperature.

6.7.7 Computation of Derived Quantities

Once the static equilibrium is found and the steady state pressure distribution has been obtained, the bearing static performance, such as minimum film thickness, maximum film pressure, frictional power loss, flow rate, etc. can be easily determined from previous discussions. Also, the film boundaries established from solving the static pressure distribution are used for the solutions of the four perturbed pressure equations with the same numerical procedure employed in solving the static pressure equation. However, no boundary iteration is required and resulting negative pressures are accepted for these perturbed pressures. The bearing dynamic stiffness and damping coefficients are obtained by integrating these perturbed pressure distributions.

Example 6.1: Pressure Dam Bearing Analysis

A pressure dam bearing has been analyzed by using *DyRoBeS-BePerf*. This bearing was modified from the original two-axial groove bearing by adding a pressure dam in the upper lobe and an axial relief groove in the lower lobe due to the rotor instability problem. The relief track is very deep compared to the bearing clearance such that the pressure vanishes at the groove. Note that, in general, the relief track in the loaded pad should be avoided due to the high operating eccentricity ratio unless the bearing is lightly loaded. The bearing is sketched in Figure 6.7-7. The pertinent bearing parameters and analysis inputs are shown in Figures 6.7-8 and 6.7-9.

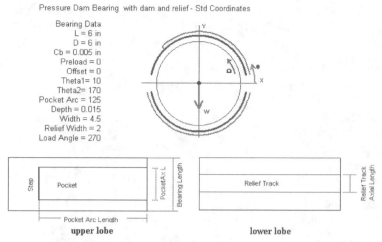

Figure 6.7-7 Pressure dam bearing example

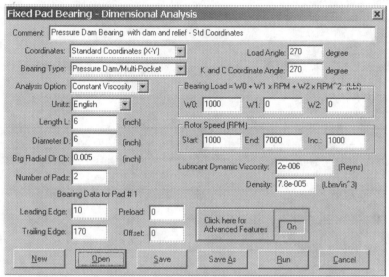

Figure 6.7-8 Bearing parameters

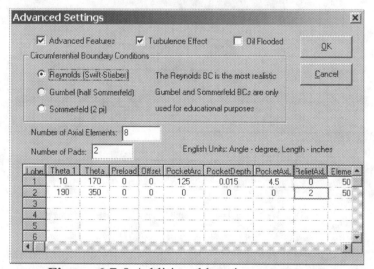

Figure 6.7-9 Additional bearing parameters

This is a very lightly loaded bearing.   The static and dynamic characteristics of the bearing at 6000 rpm are shown in Figure 6.7-10.

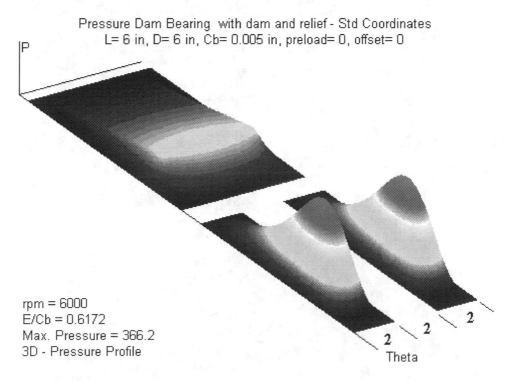

Pressure Dam Bearing with dam and relief - Std Coordinates
L= 6 in, D= 6 in, Cb= 0.005 in, preload= 0, offset= 0
Speed = 6000 rpm
Load = 1000 Lbf
W/LD = 27.7778 psi
Vis. = 2E-06 Reyns
Sb = 2.592
E/Cb = 0.6172
Att. = 56.03 deg
hmin = 1.914 mils
Pmax = 366.165 psi
Hp = 18.0709 hp
Stiffness (Lbf/in)
 2.176E+06   2.918E+06
 3.289E+04   1.670E+06
Damping (Lbf-s/in)
 6.505E+03   2.713E+03
 2.715E+03   2.979E+03
Critical Journal Mass
 16.39

Pressure Dam Bearing with dam and relief - Std Coordinates
L= 6 in, D= 6 in, Cb= 0.005 in, preload= 0, offset= 0

rpm = 6000
E/Cb = 0.6172
Max. Pressure = 366.2
3D - Pressure Profile

Figure 6.7-10 Results at 6000 rpm for centered relief

Very often, the relief track is not in the center, but on both sides for easy manufacturing. The positive value of *Relief Axial* in the data input indicates the centered relief track as shown in Figures 6.7-9 and 6.7-10. The negative value is for the relief track on both sides. Note that the relief value is the sum of both sides. Figures 6.7.11 shows the static and dynamic characteristics of the bearing at 6000 rpm if –2 is entered in the *Relief Axial*. It shows that the bearing with side relief has lower eccentricity (higher film thickness) and lower critical journal mass compared to the bearing with centered relief.

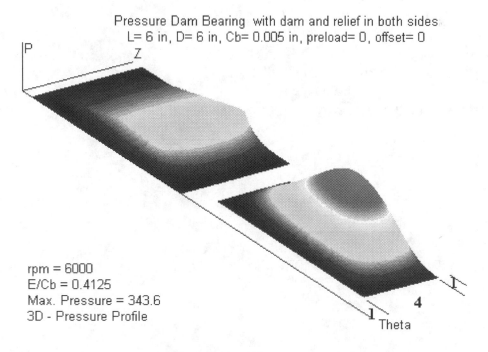

Pressure Dam Bearing with dam and relief in both sides
L= 6 in, D= 6 in, Cb= 0.005 in, preload= 0, offset= 0
Speed = 6000 rpm
Load = 1000 Lbf
W/LD = 27.7778 psi
Vis. = 2E-06 Reyns
Sb = 2.592
E/Cb = 0.4125
Att. = 69.70 deg
hmin = 2.937 mils
Pmax = 343.57 psi
Hp = 15.8343 hp
Stiffness (Lbf/in)
 1.763E+06   3.084E+06
-4.795E+05   1.655E+06
Damping (Lbf-s/in)
 8.146E+03   2.634E+03
 2.638E+03   3.743E+03
Critical Journal Mass
 14.27

Pressure Dam Bearing with dam and relief in both sides
L= 6 in, D= 6 in, Cb= 0.005 in, preload= 0, offset= 0

rpm = 6000
E/Cb = 0.4125
Max. Pressure = 343.6
3D - Pressure Profile

Figure 6.7-11 Results at 6000 rpm for side relief

6.8 Linear and Nonlinear Analyses of a Journal Bearing

A single journal bearing system taken from Kirk and Gunter (1970) is used here to demonstrate the linear and nonlinear analyses and the concepts of linearized dynamic coefficients. The nonlinear analysis is examined first, where the nonlinear fluid film forces are employed in the equations of motion. The motion of the journal is obtained by numerical integration of the governing nonlinear equations. The static equilibrium position can be obtained by neglecting the external forces and only remaining the static load (gravity force). When the rotor speed exceeds the instability threshold, the equilibrium position becomes an equilibrium orbit in which the journal performs periodic motion and forms a closed orbit. The whirling frequency for the self-excited motion is slightly less than one-half of the rotor speed. When the rotor speed is below the instability threshold, under dynamic conditions the journal oscillates around the equilibrium position. Although the complete motion of the journal cannot be predicted by using linear analysis, the locations of the critical speeds and instability threshold can be estimated by utilizing the linearized bearing coefficients in the linear analysis.

Example 6.2: Single Journal Bearing System

For a single journal bearing system as shown in Figure 6.8-1, the rotor is assumed to be rigid and symmetric, which is usually not practical. However, it provides a wealth of fundamental concepts in rotordynamics and bearing analysis. The physical parameters of the system are summarized below:

Journal Mass (M) = 50 Lbm
Journal Unbalance (me) = 0, 0.1, 0.3, 0.5, 1.0 oz-in
Bearing Diameter (D) = 2 in
Bearing Length (L) = 0.5 in
Bearing Radial Clearance (C_b) = 0.002 in
Oil Viscosity (μ) = 1.0E-06 Reyns
Bearing Load due to Gravity ($W=Mg$) = 50 Lbf

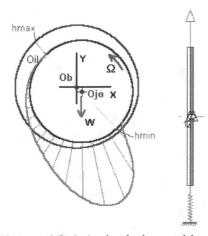

Figure 6.8-1 A single journal bearing

As described in Chapter 3, a dummy shaft element and geometric constraints are used to model this 2 DOF system.

6.8.1 Non-Linear Simulation

The bearing data for the non-linear analysis are entered in the bearing tab as shown in Figure 6.8-2:

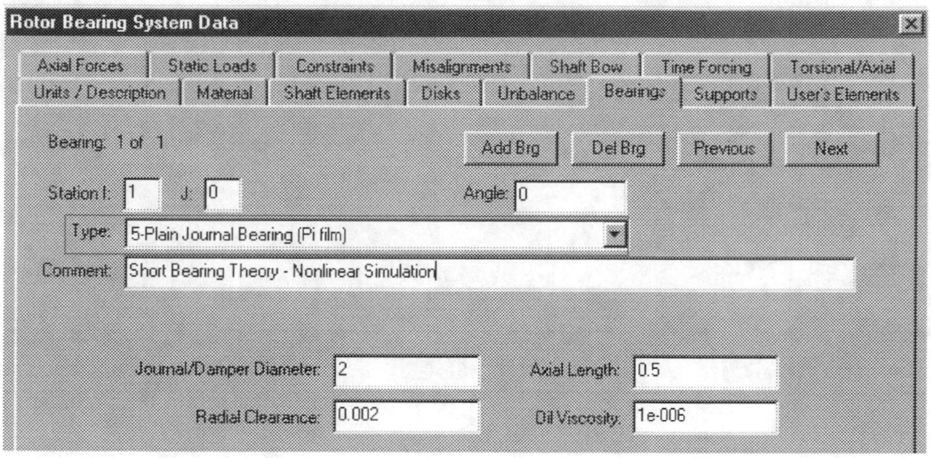

Figure 6.8-2 Bearing data for nonlinear analysis

There are three forces acting on the journal in this example and they are: the nonlinear fluid film forces, mass unbalance, and gravity load. The equations of motion for the journal can be obtained from Newton's Law:

$$M\ddot{x} = F_x(\dot{x}, \dot{y}, x, y) + me\Omega^2 \cos(\Omega t)$$

$$M\ddot{y} = F_y(\dot{x}, \dot{y}, x, y) + me\Omega^2 \sin(\Omega t) - W$$

where W is the static load due to journal weight in the negative Y direction. F_x and F_y are the nonlinear fluid film forces caused by the hydrodynamic fluid film pressure. The film pressure distribution is obtained by solving the Reynolds equation, Eq. (6.5-1), for a given displacements (x,y) and velocities $(\dot{x}, \dot{y})$ at every time instant. Since the $L/D=0.25$ in this example, the short bearing assumption will be proven to be a valid approximation from the following results.

Analytical procedures for the solution of nonlinear differential equations in rotordynamics are not feasible. Due to the high-speed digital computer and numerical integration techniques, a large amount of knowledge in understanding the nonlinear rotor behavior comes from the numerical simulation. For this single journal bearing system, the journal equilibrium position can be determined by setting the unbalance force (*me*) to be zero or simply "*uncheck*" the unbalance box in the analysis option. The system becomes an autonomous system with zero unbalance force. Figure 6.8-3 shows the analysis option at 4000 rpm with gravity force only. The gravity forces are calculated by specifying the gravity constant in the analysis option. In this example, the gravity constant is in the negative Y direction. The details of the numerical integration techniques are discussed in Chapter 7. Only simulated results are presented here.

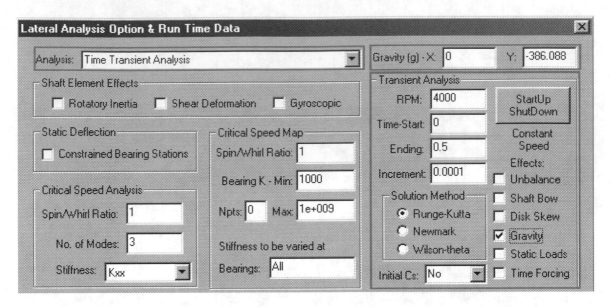

Figure 6.8-3 Analysis input

Figure 6.8-4 shows the solutions of this autonomous system with zero unbalance force (free oscillation) at rotor speeds of 2000, 4000, 6000, 8000 and 10,000 rpm. For every speed listed above, the numerical solution converges to a single point in which the journal is at rest. This point is referred to as *journal static equilibrium position* and is defined by the journal eccentricity and attitude angle. The attitude angle is measured from the load vector. It should be noted that the gravity load is vertically down in the negative Y direction, however, the journal equilibrium position is not in line with the load vector. Instead, the equilibrium position has a displacement in the X direction, which produces a non-zero attitude angle. This coupling between the force vector and displacements is so unique and makes the study of fluid film bearing so fascinating. These cross coupling terms are also responsible for the occurrence of the instability, self-excitation, or oil whirl of the rotor systems.

It is also evident from Figure 6.8-4 that as speed increases, the journal eccentricity ratio decreases and the attitude angle increases. The amount of damping present in the system can be determined from the measurement of the ratio of two successive amplitudes; that is the rate of decay of free oscillations. The larger the damping, the greater the rate of decay is and the shorter time is required to reach the equilibrium position for the same initial conditions. Figure 6.8-4 shows that the system damping decreases as the rotor speed increases. When the rotor speed reaches and exceeds a threshold speed, the equilibrium point no longer exists, and instead there is an equilibrium motion in which the journal performs periodic motion and forms a closed orbit. The motion does not grow indefinitely as linear theory predicts, instead, the journal is whirling in a closed and bounded orbit known as the *limit cycle*. That is, the system becomes unstable in the linear sense and linear theory is no longer valid and nonlinear equations govern the motion.

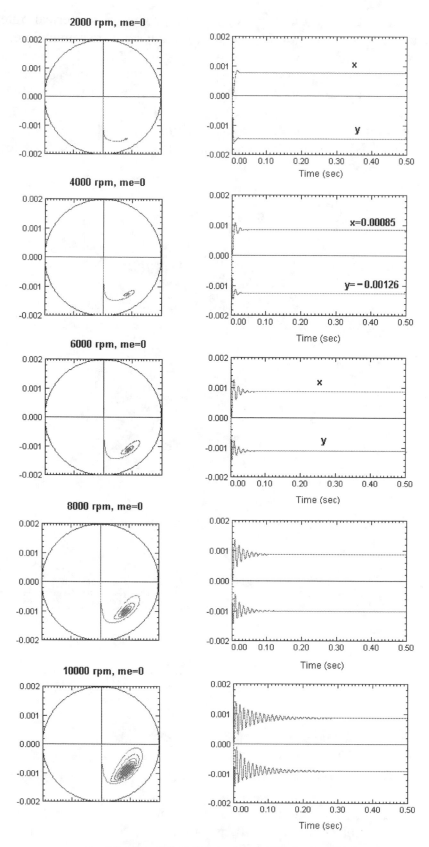

Figure 6.8-4 Journal equilibrium positions

Figure 6.8-5 shows the equilibrium motions (orbits) for rotor speeds of 12,000 and 14,000 rpm. By using FFT (Fast Fourier Transform) analysis of the equilibrium orbits, Figure 6.8-6 shows that the journal whirls with a frequency of 5713 cpm at a rotor speed of 12,000 rpm (whirl/spin ratio=0.476). Figure 6.8-7 shows that the journal whirls with a frequency of 6885 cpm at a rotor speed of 14,000 rpm (whirl/spin ratio=0.492). This whirling frequency is commonly referred to as *"half whirl frequency"* since it is close to the one-half of the rotor spin speed. It is also commonly called *"sub-synchronous"* vibration. Since there is no external excitations; therefore, it is also called *"self-excitation"*.

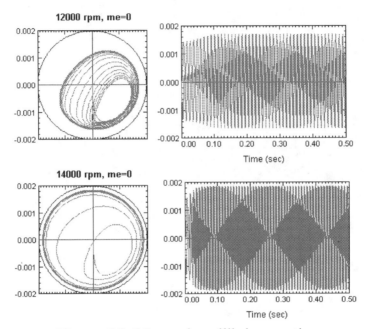

Figure 6.8-5 Journal equilibrium motion

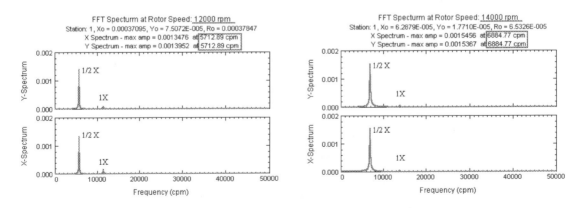

Figure 6.8-6 FFT at 12,000 rpm   **Figure 6.8-7** FFT at 14,000 rpm

Figure 6.8-8 summarizes the journal equilibrium positions or equilibrium orbits with zero unbalance from a rotor speed of 1,000 rpm to 15,000 rpm with an increment of 1,000 rpm. It shows that the journal equilibrium position is a single point when rotor speed is below 10,000 rpm and the equilibrium whirling orbit increases dramatically from 11,000 rpm to 12,000 rpm. At rotor speed of 15000 rpm, the journal motion basically is bounded by the bearing clearance geometry. However, in the "real world", a babbitted bearing will not survive under this high vibration induced heat and stress.

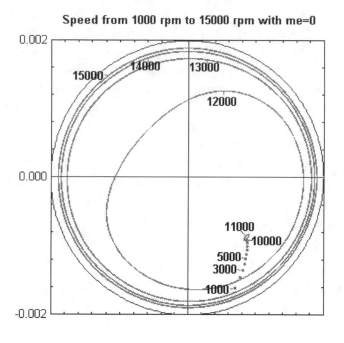

Figure 6.8-8 Journal equilibrium motion

Under dynamic conditions with unbalance force, the journal oscillates around the equilibrium position for rotor speed below the instability threshold. Figure 6.8-9 shows the journal motion at 4,000 rpm with various unbalances included, in additional to the gravity force.

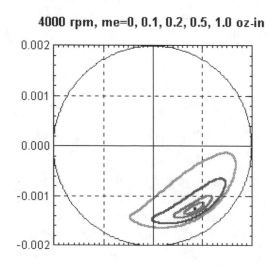

Figure 6.8-9 Journal steady state orbits

The rotor steady state orbit is nearly elliptical for small unbalance and the orbit size increases with the unbalance. However, the orbit size does not increase linearly with the unbalance as the linear theory predicts. As the unbalance increases, the orbit becomes a banana shape. From the FFT analysis, it shows that the journal under unbalance excitation is mainly whirling at 4,000 rpm that is synchronous with the rotor speed. This synchronous vibration is commonly referred to as "1X" vibration. As the orbit increases with the unbalance, 2X vibration component shows up and distorts the elliptical orbit. In the linear theory, the input (force) and output (response) are related linearly; that is, when the unbalance is doubled, the response is doubled. In the non-linear system, this linear relationship between the unbalance and response is no longer proportional, as illustrated in Figure 6.8-9. Only when the vibration amplitudes are small, the force-response relation can be approximated by the linear theory.

Figure 6.8-10 shows that journal orbits from 1,000 rpm to 10,000 rpm with an increment of 1,000 rpm for an unbalance of 0.1 oz-in. It shows that journal whirls around the equilibrium position and the orbit peaks around 5,000 rpm and decreases to almost a constant size for higher speeds.

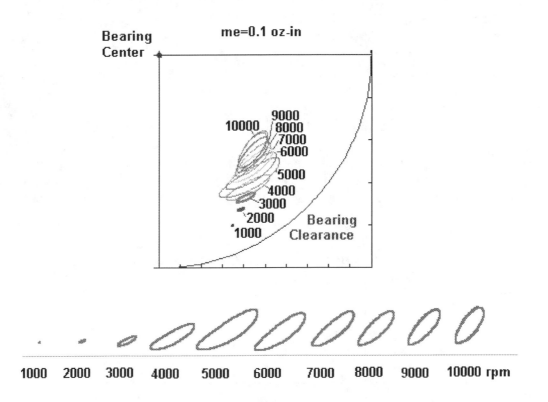

Figure 6.8-10 journal orbits versus speed

A cascade or waterfall plot is shown in Figure 6.8-11 for the X displacement from a rotor speed of 1,000 rpm to 15,000 rpm with an unbalance of 0.1 oz-in. It shows that only synchronous vibration (1X) exists when rotor speed is below 10,000 rpm and the orbit peaks around 5,000 rpm. The rotor speed, where the synchronous vibration (1X) reaches a peak, is commonly referred to as the "*Critical Speed*". The sub-synchronous vibration component starts to show up at 11,000 rpm and increases dramatically from 12,000 to 13,000 rpm. This is commonly referred to as the "*Instability Threshold*". As a common rule of thumb, the onset of instability occurs at twice the critical speed for a rigid rotor as demonstrated in this example. For flexible rotors, as illustrated in the later example, the onset of instability occurs below twice the critical speed.

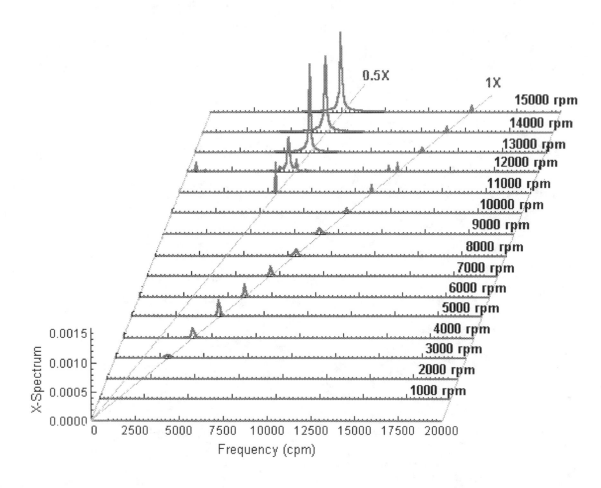

Figure 6.8-11 FFT spectrum (waterfall or cascade plot)

Although it is not advisable to operate the rotor system near and beyond the instability threshold, it is interesting to study the unbalance effects on the rotor in this regime. Figure 6.8-12 shows the journal motion at 10,000 rpm with unbalance force of 0, 0.1, 0.2, 0.4, 0.6, and 0.8 oz-in. As shown before, the journal rests at an equilibrium position at zero unbalance. With small unbalance, the journal oscillates around its equilibrium position with small synchronous (1X) vibration amplitude. As unbalance increases from 0.2 to 0.4, the sub-synchronous (1/2X) vibration appears and dominates the journal motion. However, when the unbalance increases further, the synchronous vibration dominates the motion and the sub-synchronous vibration is suppressed. Although, increase the synchronous vibration can suppress the sub-synchronous vibration, it is not advised to do so. The machine cannot be continuously operated under high vibration regardless the whirling frequency is synchronous or sub-synchronous.

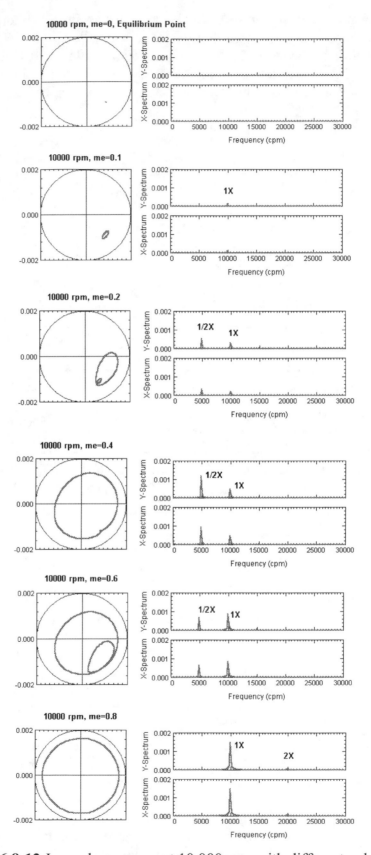

Figure 6.8-12 Journal responses at 10,000 rpm with different unbalances

Figure 6.8-13 shows the journal motion at 12,000 rpm with unbalance of 0, 0.2, and 0.8 oz-in.   It shows the journal whirls at near one-half the rotor speed (whirl/spin ratio=0.476) when unbalance is zero; that is the equilibrium position becomes an equilibrium motion.   Again, when unbalance increases, the synchronous vibration increases, and the sub-synchronous vibration is suppressed when the synchronous vibration is very large.   In diagnostics of rotating machinery, engineers commonly refer the predominant sub-synchronous vibration as Instability.

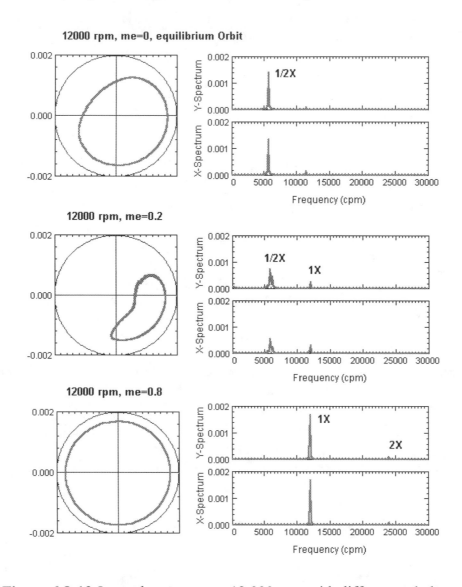

Figure 6.8-13 Journal responses at 12,000 rpm with different unbalances

The non-linear simulation is extremely useful to study the rotor dynamics when large amplitudes are expected and when rotor is operated near and beyond the instability threshold.   During the design stage, non-linear simulation can be tedious and not practical.   The dynamic characteristics of interests in the design stage of a rotor system are: the locations of the critical speeds and system instability threshold.   Fortunately, both questions can be answered by the linear analysis.

6.8.2 Linear Analysis

As illustrated in the non-linear simulation, under dynamic conditions, the journal oscillates around the static equilibrium position with small amplitudes (x,y) if the excitation is small and the rotor speeds are below the instability threshold. At static equilibrium, the nonlinear fluid film forces can be linearized as:

$$\begin{Bmatrix} F_x \\ F_y \end{Bmatrix} = \begin{Bmatrix} F_{xo} \\ F_{yo} \end{Bmatrix} - \begin{bmatrix} K_{xx} & K_{xy} \\ K_{yx} & K_{yy} \end{bmatrix} \begin{Bmatrix} x \\ y \end{Bmatrix} - \begin{bmatrix} C_{xx} & C_{xy} \\ C_{yx} & C_{yy} \end{bmatrix} \begin{Bmatrix} \dot{x} \\ \dot{y} \end{Bmatrix} + (high\ order\ terms)$$

and

$$\begin{Bmatrix} F_{xo} \\ F_{yo} \end{Bmatrix} = \begin{Bmatrix} 0 \\ W \end{Bmatrix}$$

By substituting the linearized fluid film forces into the equations of motion, the journal motion around the static equilibrium position becomes:

$$M\ddot{x} + C_{xx}\dot{x} + C_{xy}\dot{y} + K_{xx}x + K_{xy}y = me\Omega^2 \cos(\Omega t)$$

$$M\ddot{y} + C_{yx}\dot{x} + C_{yy}\dot{y} + K_{yx}x + K_{yy}y = me\Omega^2 \sin(\Omega t)$$

Note that (x,y) are small displacements around the static equilibrium position in the linear analysis. The two planes of motions are coupled by the damping and stiffness. The eight linearized damping and stiffness coefficients can be obtained from **DyRoBeS-BePerf**. The complete 2-D Reynolds equation is solved by application of boundary conditions and integration of pressure distribution as discussed before. It is an iterative process until the convergence criterion is satisfied. Once the static equilibrium is found, the eight bearing dynamic coefficients (stiffness and damping) are obtained by solving the perturbed pressure equations. The plain cylindrical bearing is sketched in Figure 6.8-14 and the data input for the bearing analysis in **DyRoBeS-BePerf** is shown in Figure 6.8-15.

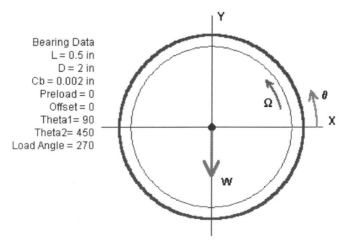

Figure 6.8-14 Plain cylindrical bearing

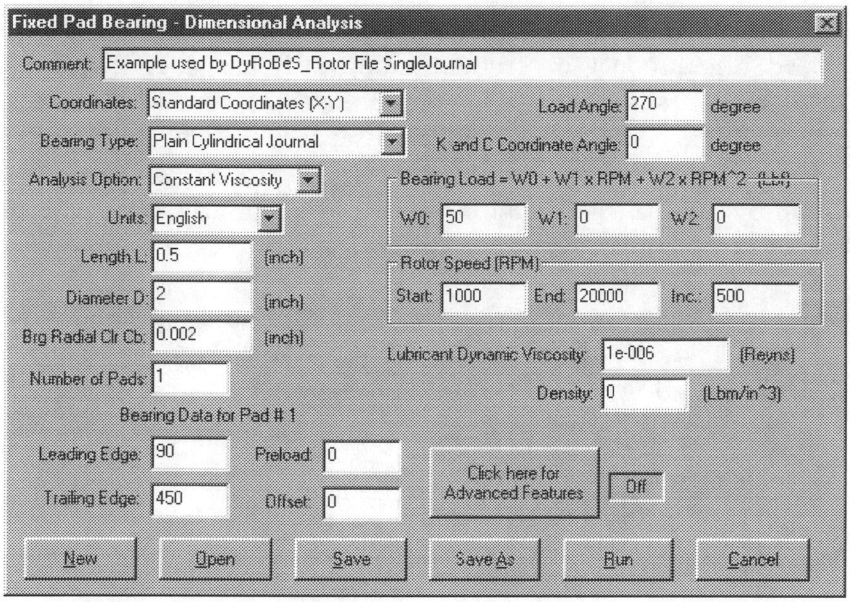

Figure 6.8-15 Bearing data

Although ***DyRoBeS-BePerf*** calculates many bearing performance data, only journal static equilibrium locus and bearing coefficients are presented below. The equilibrium positions, as shown in Figure 6.8-16, obtained from the solution of static Reynolds equation without knowing the rotor equations, are in agreement with the non-linear dynamic simulation results presented previously in Figure 6.8-8. In bearing analysis, the static equilibrium position is determined by applying the static load. Some parts of equilibrium positions will never be reached due to the self-excitation as demonstrated in the non-linear simulation.

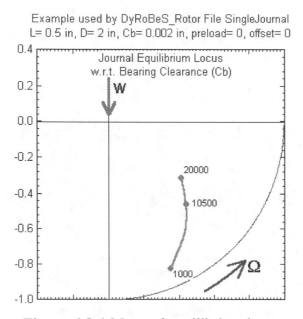

Figure 6.8-16 Journal equilibrium locus

The linearized bearing damping and stiffness coefficients are shown in Figure 6.8-17. For higher speed, the product of cross-coupling stiffness coefficients, ($KxyxKyx$) is negative. This contributes the positive energy to the system forward precessional modes and tends to destabilize the rotor system as discussed in Chapter 3. The linearized damping and stiffness coefficients calculated by **DyRoBeS-BePerf** are saved into a data file and imported by **DyRoBeS-Rotor** for linear rotordynamics analysis as demonstrated in Figure 6.8-18.

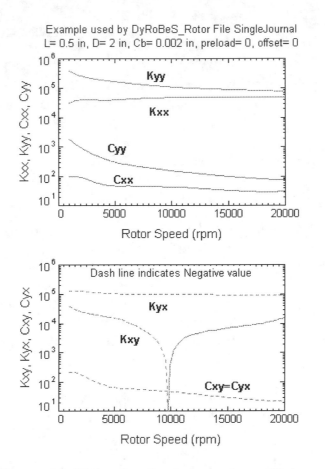

Figure 6.8-17 Bearing linear dynamic coefficients

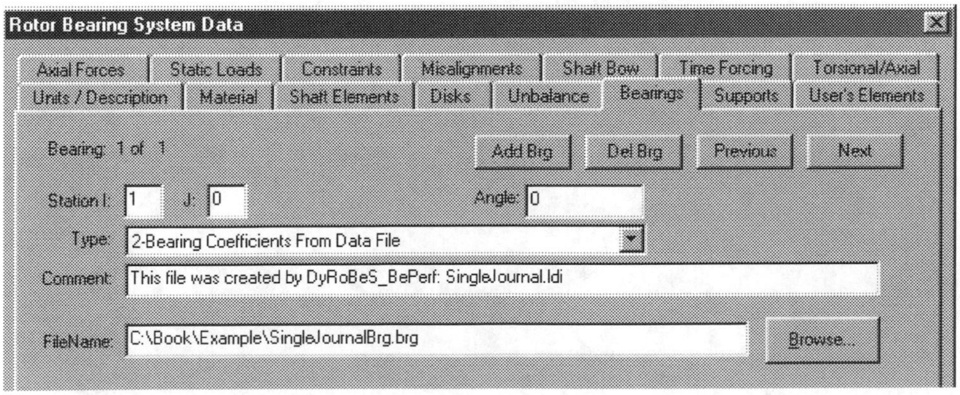

Figure 6.8-18 Import bearing coefficients from data file

Figure 6.8-19 shows the Whirl Speed Map and Stability Map from the whirl speed and stability analysis. The Whirl Speed Map shows that the damped critical speed due to synchronous unbalance excitation is around 4,000 rpm. The backward mode is a real non-vibratory mode with zero whirl frequency from zero to around 5,000 rpm and becomes a vibratory mode after 5,000 rpm. The forward mode is excited by the unbalance excitation at around 4,000 rpm. The Stability Map shows that the instability threshold is around 12,000 rpm where the forward mode becomes unstable in the linear sense. It also shows that the damping factor is about 0.2 at the damped critical speed around 4,000 rpm. The rotor speed, where the peak response occurs due to unbalance, was discussed in Chapter 3 and can be estimated by:

$$\Omega_{peak} = \frac{\omega_n}{\sqrt{1-2\xi^2}} = \frac{\omega_d}{\sqrt{1-2\xi^2}\sqrt{1-\xi^2}}$$

The peak response due to unbalance is predicted to be around 4,300 rpm.

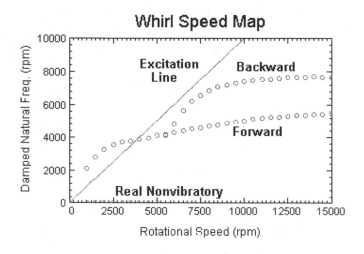

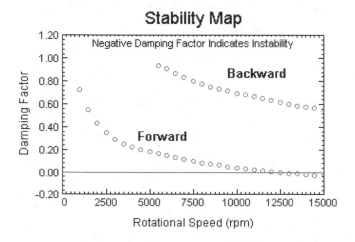

Figure 6.8-19 Whirl speed map and stability map

The steady state unbalance response (*me*= 0.1 oz-in) is shown in Figure 6.8-20. Although the response is shown from zero to 15,000 rpm, the "actual" response above instability threshold (around 12,000 rpm) cannot be predicated by linear analysis and nonlinear simulation should be used in this regime. The journal orbits at various speeds obtained from the linear analysis are compared with those obtained from the nonlinear simulation in Figure 6.8-21. They are in good agreement for small vibrations. It should be noted that the linear analysis is based on the linearization around the static equilibrium position and the response is around the static equilibrium position, not the bearing center.

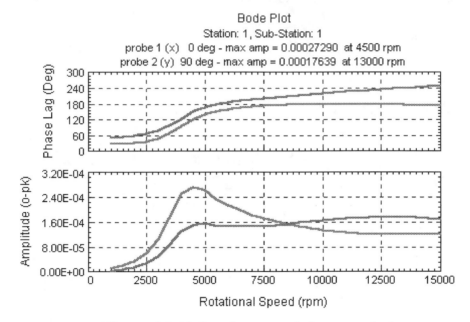

Figure 6.8-20 Steady state unbalance analysis

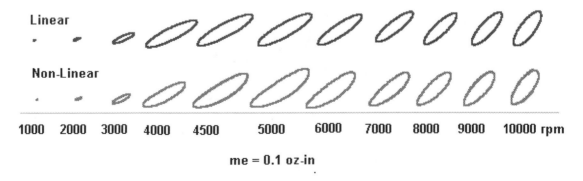

Figure 6.8-21 Comparison of liner and nonlinear orbits

Figure 6.8-22 shows the journal orbits for various unbalance at 4,000 rpm. In the linear analysis, the input (force) and output (response) are related linearly. As the unbalance force increases, the response orbit increases linearly. However, the actual journal motion is constrained by the bearing clearance. Therefore, the orbit for *me*=0.5 is not physically possible. It indicates that the linear analysis is valid only for small vibration around the equilibrium position. For large vibration, nonlinear simulation should be used.

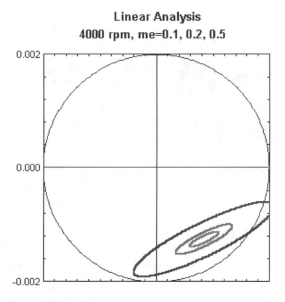

Linear Analysis
4000 rpm, me=0.1, 0.2, 0.5

Figure 6.8-22 Steady state unbalance response orbits

From the nonlinear transient simulation and linear analysis, we learn that:

1. The linear analysis is very useful in the determination of critical speeds and instability threshold. For small vibration around the equilibrium position, the linear analysis gives reasonable results. The linear analysis is fast and very useful in the preliminary rotor design stage.

2. The non-linear analysis gives accurate results for both small and large vibrations. It can be tedious in the determination of critical speeds and instability threshold.

6.9 Tilting Pad Journal Bearings

In modern turbomachinery, there has been a marked increase in operating speed to improve the aerodynamic performance and decrease in shaft-bearing diameter to minimize frictional power loss. As described before, the fluid film bearings are available in a variety of configurations and are widely used in rotating machinery due to their favorable fatigue life and damping characteristics compared to rolling element bearings. In addition to carrying static loads, fluid film journal bearings are often the major source of damping which can attenuate resonant response. However, the non-zero cross-coupling stiffness coefficients existing in the fixed profile journal bearings can introduce a major destabilizing effect. This can cause the rotor system to be in a very destructive self-excited state. The tilting pad journal bearings, as shown in Figure 6.9-1, have been widely used in rotating machinery when the rotor is extremely lightly loaded or in vertical arrangement, or when the rotor operating speed is higher than twice the first critical speed, due to their virtually inherent stability characteristics. Applications are high-speed compressors, pumps, and turbines.

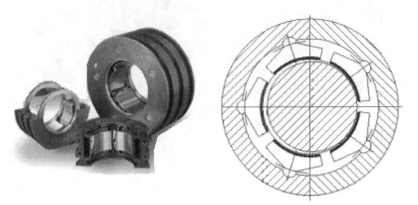

Figure 6.9-1 Tilting pad bearing

However, the tilting pad journal bearings with moving parts are mechanically complex and generally have lower damping and softer stiffness than fixed geometry journal bearings. Furthermore, the manufacturing tolerance stack-up existing in the tilting pad journal bearing, as illustrated by Chen et. al. (1994), can result in a wide range of bearing clearance and preload. This variance can be pronounced for relatively small sized bearings, which have been commonly used in high-speed applications.

Recently, the flexible (flexural) pad journal bearing, as shown in Figure 6.9-2, has been gaining attention in bearing design for high speed rotating machinery. It is a one-piece design similar to that of the conventional tilting pad journal bearing without the complexity of the moving parts. Therefore, the pivot wear, manufacturing tolerance stack-up, and the unloaded pad flutter problems associated with the conventional tilting pad bearing can be eliminated. However, due to the flexibility of the support web, the pad is not free to pitch and the destabilizing tangential oil film force always exists even when the pad inertia is neglected. The rotor-bearing stability must be carefully examined when the flexible pad journal bearing is used. Hence, it is very important to effectively predict the bearing dynamic coefficients of the flexible pad bearings for use in the study of rotor-bearing dynamic characteristics. A comparison of the stack-up tolerance for a typical tilting pad journal bearing and flexible pad journal bearing is listed in the Table 6.9-1 for reference.

Unlike the fixed profile journal bearings that the lobes are fixed, the pads in the tilting/flexible pad bearings are flexible and have additional two translational degrees-of-freedom (radial and tangential directions) and one rotational degree-of-freedom (pitch motion) for each individual pad. These pad motions are governed by the elasticity equations once the pressure distribution over the pad is known. Therefore, at the static equilibrium position, the Reynolds equation must be solved simultaneously with the pad elasticity equations in the determination of the pressure distribution. For a typical 5 pads tilting-pad bearing, there will be a total of 17 nonlinear equations (2 from Reynolds equation, and 3 from elasticity equations for each pad) to be solved in order to determine the static equilibrium position. This can be very time consuming and not practical due to the manufacturing tolerance errors in this type of bearings.

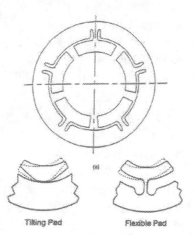

Figure 6.9-2 Flexible pad bearing

Calculation of Minimum and Maximum Preload.

Conventional Tilt Pad Bearing	Flexible Pivot Bearing
Journal Diameter$_{min}$ = 0.8198	Journal Diameter$_{min}$ = 0.8198
Journal Diameter$_{max}$ = 0.8200	Journal Diameter$_{max}$ = 0.8200
Pad Set Bore$_{min}$ = 0.8230	Pad Set Bore$_{min}$ = 0.8224
Pad Set Bore$_{max}$ = 0.8238	Pad Set Bore$_{max}$ = 0.8228
Pad Machined Bore$_{min}$ = 0.8234	Pad Machined Bore$_{min}$ = 0.8232
Pad Machined Bore$_{max}$ = 0.8243	Pad Machined Bore$_{max}$ = 0.8236

$$Preload = m = 1 - \frac{C_b}{C_p} = 1 - \frac{D_b - D_j}{D_b - D_j}$$

$$Preload_{min} = 1 - \frac{D_{bmax} - D_{jmin}}{D_{pmin} - D_{jmin}}$$

$$Preload_{min} = \frac{D_{bmax} - D_{bmax}}{D_{pmax} - D_{jmin}}$$

$$Preload_{max} = 1 - \frac{D_{bmin} - D_{jmax}}{D_{pmax} - D_{jmax}}$$

$$Preload_{max} = \frac{D_{jmin} - D_{bmin}}{D_{pmin} - D_{jmax}}$$

$$m_{min} = 1 - \frac{0.8238 - 0.8198}{0.8234 - 0.8198}$$

$$m_{min} = \frac{0.8236 - 0.8228}{0.8236 - 0.8198}$$

$$m_{min} = \boxed{-0.1000}$$

$$m_{min} = \boxed{0.2105}$$

$$m_{max} = 1 - \frac{0.8230 - 0.8200}{0.8243 - 0.8200}$$

$$m_{max} = \frac{0.8232 - 0.8224}{0.8232 - 0.8198}$$

$$m_{max} = \boxed{0.3023}$$

$$m_{max} = \boxed{0.2352}$$

$$m_{nom} = 1 - \frac{0.8234 - 0.8199}{0.82385 - 0.8199}$$

$$m_{nom} = \frac{0.8234 - 0.8226}{0.8234 - 0.8199}$$

$$m_{nom} = \boxed{0.1139}$$

$$m_{nom} = \boxed{0.2285}$$

Table 6.9-1 Comparison of preload

The pad assembly method for the calculation of tilting pad journal bearing dynamic coefficients was first proposed by Lund (1964). Since then, numerous publications have been presented in the calculation of tilting pad journal bearing coefficients based on the pad assembly method with different numerical techniques (Nicholas, et al., 1979) and inclusion of various effects (Rouch, 1983; Lund and Pedersen, 1987; Kirk and Reedy, 1988). The pad assembly method provides an effective and fast means of calculating tilting pad bearing coefficients by assembling single pad data rather than a time consuming iterative solution for all the pads. The single pad data can be generated separately and stored for future assembly process and parametric study. The fast assembly process makes this method attractive and essential to the practical design engineers in the design parametric study. Isothermal and laminar flow with negligible fluid inertia is assumed in this analysis. A comprehensive literature review on the turbulence, fluid inertia, and thermal effects in fluid film bearings has been documented by Szeri (1980). The computational method for turbulence and thermal effects has also been documented by Someya (1988) and for inertia effect by Reinhardt and Lund (1975). These effects on the bearing dynamic coefficients, in general, are less than the manufacturing tolerance errors existing in the tilting pad journal bearings and uncertainties from other sources (e.g., lubricant properties, flow, actual oil entry temperature, etc.). In the design process, the bearing coefficients at various operating conditions, such as minimum and maximum anticipated oil temperatures, possible ranges of clearance and preload due to manufacturing tolerance, and variation in the bearing static loads, are required for the rotordynamics study. Therefore, an effective, fast, and approximate solution is useful and indispensable.

Based on the pad assembly method, Armentrout and Paquette (1992) have included the web rotational stiffness in the calculation of flexible pad bearing coefficients. However, the pad radial and tangential movements under dynamic conditions were ignored. It is known that this additional two translational flexibility can have a great influence on the bearing dynamic coefficients, especially the damping characteristics. A general method for calculating the dynamic stiffness and damping coefficients of flexible pad journal bearings is described in the following section. The effects of pad translations and rotation and the mass/inertia properties of the pad are included. The conventional tilting pad bearing coefficients can be easily obtained by using the same procedure with elimination of the pad tangential motion. The computational algorithm can be easily implemented into existing tilting pad bearing programs based upon the pad assembly method to evaluate the flexure pad bearing coefficients.

Mathematical Model

The static equilibrium of the journal center is described by a bearing eccentricity (e_b) and attitude angle (ϕ_b), both reference from the bearing center. Under dynamic conditions, the journal oscillates at small displacements (Δx, Δy) around its static equilibrium position, shown in Figure 6.9-3. For small vibrations, the linearized fluid film forces acting on the journal are obtained by summing over all the pads:

$$\begin{Bmatrix} F_x \\ F_y \end{Bmatrix} = \sum_p \begin{Bmatrix} F_x \\ F_y \end{Bmatrix}_p = \sum_p \begin{Bmatrix} F_{xo} + dF_x \\ F_{yo} + dF_y \end{Bmatrix}_p$$

$$= \begin{Bmatrix} F_{xo} \\ F_{yo} \end{Bmatrix} - \begin{bmatrix} K_{xx} & K_{xy} \\ K_{yx} & K_{yy} \end{bmatrix} \begin{Bmatrix} \Delta x \\ \Delta y \end{Bmatrix} - \begin{bmatrix} C_{xx} & C_{xy} \\ C_{yx} & C_{yy} \end{bmatrix} \begin{Bmatrix} \Delta \dot{x} \\ \Delta \dot{y} \end{Bmatrix}$$

$$(6.9\text{-}1)$$

The static equilibrium force, F_{xo}, should be equal to the static load W in the opposite direction if the X axis is aligned with the load vector and F_{yo} is zero.

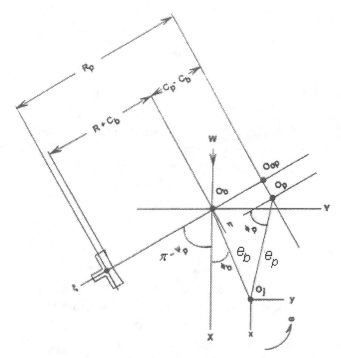

Figure 6.9-3 Coordinate systems

Single Pad Analysis

In order to calculate the dynamic coefficients of a complete bearing, it is necessary to understand the dynamic behavior of the individual pad. A fixed local coordinate system (ξ,η) is defined for each pad with its origin in the bearing center. The ξ-axis passes through the pivot point and the η-axis is perpendicular to the ξ-axis. The ξ-axis is also the static loading line for each pad. For a given journal position, the corresponding pad Sommerfeld number and fixed pad coefficients are calculated by treating the individual pad as a partial arc bearing. The fixed pad coefficients are tabulated as a function of non-dimensional pivot film thickness to be readily used in the bearing assembly process. For small vibration and under dynamic loading, the journal center oscillates with amplitudes of (ξ,η), the pad center of curvature oscillates with amplitudes of (ξ_{op},η_{op}), the pad also moves in the ξ-η plane with two translational (ξ_p,η_p) and one rotational (γ_p) degrees-of-freedom due to the flexibility of the support web. The dynamic forces acting on the journal center, expressed in the fixed local coordinates, are:

$$\left\{ \begin{array}{c} dF_\xi \\ dF_\eta \end{array} \right\} = -\begin{bmatrix} K_{\xi\xi} & K_{\xi\eta} \\ K_{\eta\xi} & K_{\eta\eta} \end{bmatrix} \left\{ \begin{array}{c} \xi - \xi_{op} \\ \eta - \eta_{op} \end{array} \right\} - \begin{bmatrix} C_{\xi\xi} & C_{\xi\eta} \\ C_{\eta\xi} & C_{\eta\eta} \end{bmatrix} \left\{ \begin{array}{c} \dot\xi - \dot\xi_{op} \\ \dot\eta - \dot\eta_{op} \end{array} \right\} \tag{6.9-2}$$

The motion of the pad can be described by the following equations of motion:

$$\begin{bmatrix} m_{\xi\xi} & m_{\xi\eta} & m_{\xi\gamma} \\ m_{\eta\xi} & m_{\eta\eta} & m_{\eta\gamma} \\ m_{\gamma\xi} & m_{\gamma\eta} & m_{\gamma\gamma} \end{bmatrix}_p \left\{ \begin{array}{c} \ddot\xi_p \\ \ddot\eta_p \\ \ddot\gamma_p \end{array} \right\} + \begin{bmatrix} K_{\xi\xi} & K_{\xi\eta} & K_{\xi\gamma} \\ K_{\eta\xi} & K_{\eta\eta} & K_{\eta\gamma} \\ K_{\gamma\xi} & K_{\gamma\eta} & K_{\gamma\gamma} \end{bmatrix}_p \left\{ \begin{array}{c} \xi_p \\ \eta_p \\ \gamma_p \end{array} \right\} = \left\{ \begin{array}{c} -dF_\xi \\ -dF_\eta \\ -R_p dF_\eta \end{array} \right\} \tag{6.9-3}$$

where the pad mass/inertia and stiffness matrices can be obtained from any finite element method and commercial software by using condensation techniques (e.g. Guyan Reduction Method) to reduce the degrees-of-freedom and retaining only three (3) pad degrees-of-freedom; two translations and one rotation. For small vibration, the displacements are related by the following expressions:

$$\xi_{op} = \xi_p \tag{6.9-4}$$

$$\eta_{op} = \eta_p + R_p\gamma_p \tag{6.9-5}$$

Consider harmonic motions with a frequency of ω. The assumed frequency ω is used for the calculation of reduced bearing dynamic coefficients. Normally, the shaft rotational speed is selected to be the reduction frequency. Thus, the reduced bearing coefficients are called the synchronously reduced coefficients. It is convenient and desirable to introduce the impedance notation in complex form as:

$$Z = K + j\omega C \tag{6.9-6}$$

By substituting Eqs. (6.9-4), (6.9-5), and (6.9-6) into Eq. (6.9-2), the dynamic forces can be expressed as:

$$\left\{ \begin{array}{c} dF_\xi \\ dF_\eta \end{array} \right\} = -\begin{bmatrix} Z_{\xi\xi} & Z_{\xi\eta} \\ Z_{\eta\xi} & Z_{\eta\eta} \end{bmatrix} \left\{ \begin{array}{c} \xi - \xi_p \\ \eta - \eta_p - R_p\gamma_p \end{array} \right\}$$

$$= -\begin{bmatrix} Z_{\xi\xi} & Z_{\xi\eta} \\ Z_{\eta\xi} & Z_{\eta\eta} \end{bmatrix} \left\{ \begin{array}{c} \xi \\ \eta \end{array} \right\} + \begin{bmatrix} Z_{\xi\xi} & Z_{\xi\eta} \\ Z_{\eta\xi} & Z_{\eta\eta} \end{bmatrix} \left\{ \begin{array}{c} \xi_p \\ \eta_p + R_p\gamma_p \end{array} \right\} \tag{6.9-7}$$

By introducing the dynamic stiffness notation $\hat K = K - \omega^2 m$, the pad equations of motion, Eq. (6.9-3), can be written as:

$$\begin{bmatrix} \hat{K}_{\xi\xi} & \hat{K}_{\xi\eta} & \hat{K}_{\xi\gamma} \\ \hat{K}_{\eta\xi} & \hat{K}_{\eta\eta} & \hat{K}_{\eta\gamma} \\ \hat{K}_{\gamma\xi} & \hat{K}_{\gamma\eta} & \hat{K}_{\gamma\gamma} \end{bmatrix}_p \begin{Bmatrix} \xi_p \\ \eta_p \\ \gamma_p \end{Bmatrix} = \hat{K}_p \begin{Bmatrix} \xi_p \\ \eta_p \\ \gamma_p \end{Bmatrix} = \begin{Bmatrix} -dF_\xi \\ -dF_\eta \\ -R_p dF_\eta \end{Bmatrix} \tag{6.9-8}$$

or

$$\begin{Bmatrix} \xi_p \\ \eta_p \\ \gamma_p \end{Bmatrix} = \begin{bmatrix} \hat{A}_{\xi\xi} & \hat{A}_{\xi\eta} & \hat{A}_{\xi\gamma} \\ \hat{A}_{\eta\xi} & \hat{A}_{\eta\eta} & \hat{A}_{\eta\gamma} \\ \hat{A}_{\gamma\xi} & \hat{A}_{\gamma\eta} & \hat{A}_{\gamma\gamma} \end{bmatrix}_p \begin{Bmatrix} -dF_\xi \\ -dF_\eta \\ -R_p dF_\eta \end{Bmatrix} = \hat{A}_p \begin{Bmatrix} -dF_\xi \\ -dF_\eta \\ -R_p dF_\eta \end{Bmatrix} \tag{6.9-9}$$

where

$\hat{A}_p = \hat{K}_p^{-1}$ is defined as the pad dynamic flexibility matrix.

Introducing two transformation matrices:

$$\begin{Bmatrix} \xi_p \\ \eta_p + R_p\gamma_p \end{Bmatrix} = \begin{bmatrix} 1 & 0 & 0 \\ 0 & 1 & R_p \end{bmatrix} \begin{Bmatrix} \xi_p \\ \eta_p \\ \gamma_p \end{Bmatrix} \tag{6.9-10}$$

and

$$\begin{Bmatrix} -dF_\xi \\ -dF_\eta \\ -R_p dF_\eta \end{Bmatrix} = \begin{bmatrix} -1 & 0 \\ 0 & -1 \\ 0 & -R_p \end{bmatrix} \begin{Bmatrix} dF_\xi \\ dF_\eta \end{Bmatrix} \tag{6.9-11}$$

By substituting Eqs. (6.9-10) and (6.9-11) into Eq. (6.9-9), gives the following relationship:

$$\begin{Bmatrix} \xi_p \\ \eta_p + R_p\gamma_p \end{Bmatrix} = -\begin{bmatrix} \hat{A}_{\xi\xi} & \hat{A}_{\xi\eta} + R_p\hat{A}_{\xi\gamma} \\ \hat{A}_{\eta\xi} + R_p\hat{A}_{\gamma\xi} & \hat{A}_{\eta\eta} + R_p\left(\hat{A}_{\eta\gamma} + \hat{A}_{\gamma\eta}\right) + R_p^2\hat{A}_{\gamma\gamma} \end{bmatrix}_p \begin{Bmatrix} dF_\xi \\ dF_\eta \end{Bmatrix}$$
$$= -A_p^* \begin{Bmatrix} dF_\xi \\ dF_\eta \end{Bmatrix} \tag{6.9-12}$$

where A_p^* is a (2x2) matrix referred to as the reduced pad dynamic flexibility matrix which relates the pad motion and the fluid film dynamic forces.

Substituting Eq. (6.9-12) into Eq. (6.9-7) and rearranging it, the dynamic forces become:

$$\begin{Bmatrix} dF_\xi \\ dF_\eta \end{Bmatrix} = -\left(I + ZA_p^*\right)^{-1} Z \begin{Bmatrix} \xi \\ \eta \end{Bmatrix} = -Z^* \begin{Bmatrix} \xi \\ \eta \end{Bmatrix} \tag{6.9-13}$$

where I is the identity matrix, Z is the fluid film impedance matrix containing the fixed pad coefficients, and A_p^* is the reduced pad dynamic flexibility matrix containing only the structural properties of the flexible pad. The reduced effective impedance matrix is of the form:

$$Z^* = \begin{bmatrix} Z_{\xi\xi}^* & Z_{\xi\eta}^* \\ Z_{\eta\xi}^* & Z_{\eta\eta}^* \end{bmatrix} = \begin{bmatrix} K_{\xi\xi}^* & K_{\xi\eta}^* \\ K_{\eta\xi}^* & K_{\eta\eta}^* \end{bmatrix} + j\omega \begin{bmatrix} C_{\xi\xi}^* & C_{\xi\eta}^* \\ C_{\eta\xi}^* & C_{\eta\eta}^* \end{bmatrix} \tag{6.9-14}$$

The reduced effective stiffness and damping coefficients for a single flexible pad are determined from the real and imaginary parts of the reduced effective impedance coefficients. By adjusting the reduced pad dynamic flexibility matrix A_p^*, several special cases are discussed below:

1) Tilting pad journal bearing with rigid pivot

In this case, the pad is free to tilt and no translational motion is allowed (Lund, 1964; Nicholas et al., 1979). This is the most common case for high-speed turbomachinery with small journal diameters. The dynamic behavior of the pad can be expressed in mathematical form:

$$\xi_p = 0; \quad \eta_p = 0; \quad \text{and} \quad \left(K_{\gamma\gamma}\right)_p = 0 \tag{6.9-15}$$

The reduced pad dynamic flexibility matrix becomes:

$$A_p^* = \begin{bmatrix} 0 & 0 \\ 0 & 1 \big/ \left(-\omega^2 \tilde{I}_p\right) \end{bmatrix} \tag{6.9-16}$$

where $\tilde{I}_p = \dfrac{I_p}{R_p^2}$ and I_p is the pad inertia. The reduced effective impedance of the pad can be easily calculated from Eq. (6.9-13):

$$Z^* = \left(I + ZA_p^*\right)^{-1} Z = \begin{bmatrix} Z_{\xi\xi} - \dfrac{Z_{\xi\eta} Z_{\eta\xi}}{Z_{\eta\eta} - \omega^2 \tilde{I}_p} & Z_{\xi\eta} - \dfrac{Z_{\xi\eta} Z_{\eta\eta}}{Z_{\eta\eta} - \omega^2 \tilde{I}_p} \\ \dfrac{-\omega^2 \tilde{I}_p Z_{\eta\xi}}{Z_{\eta\eta} - \omega^2 \tilde{I}_p} & \dfrac{-\omega^2 \tilde{I}_p Z_{\eta\eta}}{Z_{\eta\eta} - \omega^2 \tilde{I}_p} \end{bmatrix} \tag{6.9-17}$$

For small oil lubricated bearings, the pad inertia is negligible and the reduced effective impedance matrix becomes:

$$Z_{\xi\xi}^* = Z_{\xi\xi} - \frac{Z_{\xi\eta} Z_{\eta\xi}}{Z_{\eta\eta}}; \qquad Z_{\xi\eta}^* = Z_{\eta\xi}^* = Z_{\eta\eta}^* = 0 \qquad (6.9\text{-}18)$$

The dynamic force from the pad is acting along the ξ-axis and the tangential force vanishes which implies inherent stability of the bearing.

2) Tilting pad journal bearing with flexible pivot

The effect of pivot flexibility on the dynamic coefficients was investigated by several researchers (Rouch, 1983; Lund and Pedersen, 1987; Kirk and Reedy, 1988) and the results have shown that the total damping can be reduced by the effect of the pivot stiffness. In this case, the pad tangential movement is fixed and the radial movement is constrained by pivot stiffness. The pad is free to tilt with zero angular stiffness.

$$\eta_p = 0; \quad \text{and} \quad \left(K_{\gamma\gamma}\right)_p = K_p \qquad (6.9\text{-}19)$$

The reduced pad dynamic flexibility matrix becomes:

$$A_p^* = \begin{bmatrix} \dfrac{1}{\left(K_p - \omega^2 m_p\right)} & 0 \\ 0 & \dfrac{1}{\left(-\omega^2 \tilde{I}_p\right)} \end{bmatrix} \qquad (6.9\text{-}20)$$

The reduced effective impedance of the pad can be easily calculated from Eq. (6.9-13):

$$\mathbf{Z}^* = \frac{1}{\Delta} \begin{bmatrix} \left(K_p - \omega^2 m_p\right)\left(Z_{\eta\eta} - \omega^2 \tilde{I}_p\right) & \left(K_p - \omega^2 m_p\right)\left(-Z_{\xi\eta}\right) \\ \left(-\omega^2 \tilde{I}_p\right)\left(-Z_{\eta\xi}\right) & \left(-\omega^2 \tilde{I}_p\right)\left(Z_{\xi\xi} + K_p - \omega^2 m_p\right) \end{bmatrix} \begin{bmatrix} Z_{\xi\xi} & Z_{\xi\eta} \\ Z_{\eta\xi} & Z_{\eta\eta} \end{bmatrix}$$

$$(6.9\text{-}21)$$

where

$$\Delta = \left(Z_{\xi\xi} + K_p - \omega^2 m_p\right)\left(Z_{\eta\eta} - \omega^2 \tilde{I}_p\right) - \left(Z_{\xi\eta} Z_{\eta\xi}\right) \qquad (6.9\text{-}22)$$

If the pad inertia is negligible ($m_p=I_p=0$), the effective impedances are:

$$Z_{\xi\xi}^* = \frac{K_p\left(Z_{\xi\xi} Z_{\eta\eta} - Z_{\xi\eta} Z_{\eta\xi}\right)}{\left(Z_{\xi\xi} + K_p\right)\left(Z_{\eta\eta}\right) - \left(Z_{\xi\eta} Z_{\eta\xi}\right)}; \quad Z_{\xi\eta}^* = Z_{\eta\xi}^* = Z_{\eta\eta}^* = 0 \qquad (6.9\text{-}23)$$

Again, the tangential dynamic force is zero and the pad has only radial stiffness and damping. The reduced effective impedance can be considered as the oil film impedance in series with the pivot stiffness (Lund and Pedersen, 1987).

The pivot stiffness for a variety of pivot configurations presented by Kirk and Reedy (1988) is listed below for reference:

For a sphere on a flat plat (point contact)

$$K_p = 0.968 \left(E^2 D_p W_p \right)^{1/3} \tag{6.9-24}$$

For a sphere in a sphere (point contact)

$$K_p = 0.968 \left(\frac{E^2 D_h D_p W_p}{D_h - D_p} \right)^{1/3} \tag{6.9-25}$$

For a line contact

$$K_p = \frac{\pi E L}{2\left(1 - v^2\right)\left[-\frac{1}{3} + \ln\left(\frac{4EL\left(D_h - D_p\right)}{2.15^2 \, W_p} \right) \right]} \tag{6.9-26}$$

Where the Poisson's ratio is $v_h = v_p = v = 0.3$ and the Young's modulus is $E_h = E_p = E$. The D_p and D_h are the pivot diameter and pivot housing diameters, respectively. W_p is the static load on the pad and must be determined after the journal equilibrium position has been found. The pivot stiffness is the slope of the load-deflection curve and is a function of W_p. For the unloaded pad, the pivot stiffness is zero.

All the above pivot stiffnesses can be summarized in a general configuration. Consider a general case of two bodies in contact as shown in Figure 6.9-4 (Hamrock, 1991):

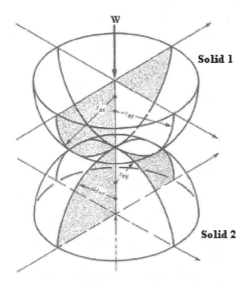

Figure 6.9-4 Two bodies in contact

The deflection in the axis of loading under the load W is (Young, 1989):

$$\delta = \lambda \sqrt[3]{\frac{W^2 C_E^2}{D_k}}$$ (6.9-27)

where

$$C_E = \frac{1-v_1^2}{E_1} + \frac{1-v_2^2}{E_2}$$ (6.9-28)

$$D_k = \frac{1.5}{\frac{1}{r_{1x}} + \frac{1}{r_{1y}} + \frac{1}{r_{2x}} + \frac{1}{r_{2y}}}$$ (6.9-29)

λ is a function of curvatures and is given in reference (Young, 1989). The curvature radii are positive if the center of curvature lies within the given body and negative otherwise. The pivot tangent stiffness is:

$$K_p = \frac{dW}{d\delta} = \frac{3}{2\lambda} \sqrt[3]{\frac{D_k W}{C_E^2}}$$ (6.9-30)

Any special cases can be derived from this generalized equation. For example, a sphere in a sphere socket (point contact) with

$$r_{1x} = r_{1y} = \frac{D_p}{2} \ , \ r_{2x} = r_{2y} = \frac{D_h}{2}, \ E = E_1 = E_2, \ v = v_1 = v_2 = 0.3$$

The generalized Eq. (6.9-30) reduces into the Eq. (6.9-25)

3) Simplified flexible pad bearing

For a simplified flexible pad bearing, the pad can be treated as a lumped inertia in the free end of a cantilever beam. The pad dynamic stiffness matrix is then:

$$\hat{K}_p = \begin{bmatrix} \hat{K}_{\xi\xi} & 0 & 0 \\ 0 & \hat{K}_{\eta\eta} & \hat{K}_{\eta\gamma} \\ 0 & \hat{K}_{\gamma\eta} & \hat{K}_{\gamma\gamma} \end{bmatrix}_p$$ (6.9-31)

where

$$(\hat{K}_{\xi\xi})_p = (AE/L) - \omega^2 m_p$$ (6.9-32)

$$(\hat{K}_{\eta\eta})_p = \left(\frac{12EI}{L^3} - \frac{6W_p}{5L}\right) - \omega^2 m_p \tag{6.9-33}$$

$$(\hat{K}_{\gamma\eta})_p = (\hat{K}_{\eta\gamma})_p = \left(\frac{-6EI}{L^2} - \frac{W_p}{10}\right) \tag{6.9-34}$$

$$(\hat{K}_{\gamma\gamma})_p = \left(\frac{4EI}{L} - \frac{2W_p L}{15}\right) - \omega^2 I_p \tag{6.9-35}$$

The geometric stiffness caused by the axial load on the pad is included in the above equation. The reduced pad dynamic flexibility matrix calculated from Eq. (6.9-13) gives:

$$A_p^* = \begin{bmatrix} \dfrac{1}{\hat{K}_{\xi\xi}} & 0 \\ 0 & \dfrac{1}{\Delta_p}\left(\hat{K}_{rr} - 2R_p\hat{K}_{\eta\gamma} + R_p^2\hat{K}_{\eta\eta}\right) \end{bmatrix} \tag{6.9-36}$$

where

$$\Delta_p = \hat{K}_{\eta\eta}\,\hat{K}_{\gamma\gamma} - \hat{K}_{\eta\gamma}^2 \tag{6.9-37}$$

Pad Assembly

The pad assembly is a straightforward process. The single pad data described in the previous section are calculated using the fixed local coordinates (ξ,η) for each individual pad. However, the complete bearing dynamic characteristics are usually described using the fixed global coordinates (x,y). To assemble the pads, the single pad data are transformed into the global coordinate system by the following coordinate transformation:

$$\begin{Bmatrix} x \\ y \end{Bmatrix} = \begin{bmatrix} \cos\psi_p & -\sin\psi_p \\ \sin\psi_p & \cos\psi_p \end{bmatrix}_p \begin{Bmatrix} \xi \\ \eta \end{Bmatrix}_p \tag{6.9-38}$$

where ψ_p is the angle measured from x to ξ in the direction of rotation. Therefore, the complete bearing static equilibrium forces are:

$$\begin{Bmatrix} F_{xo} \\ F_{yo} \end{Bmatrix} = \begin{Bmatrix} -W \\ 0 \end{Bmatrix} = \sum_p \begin{bmatrix} \cos\psi_p & -\sin\psi_p \\ \sin\psi_p & \cos\psi_p \end{bmatrix}_p \begin{Bmatrix} -W_p \\ 0 \end{Bmatrix}$$

$$= \sum_p \begin{Bmatrix} -W_p\cos\psi_p \\ -W_p\sin\psi_p \end{Bmatrix} \tag{6.9-39}$$

The determination of the static equilibrium position is an iterative process as discussed previously. Once the static equilibrium position (e_b, ϕ_b) is established, the dynamic coefficients can be obtained. For a conventional tilting pad bearing, the pads are identical and the pivots are arranged symmetrical with respect to the load line, then the attitude angle is zero and no cross-coupling terms (stiffness and damping) exist.

The complete bearing impedance coefficients in the global coordinates are:

$$\begin{bmatrix} Z_{xx} & Z_{xy} \\ Z_{yx} & Z_{yy} \end{bmatrix} = \sum_p \begin{bmatrix} \cos\psi_p & -\sin\psi_p \\ \sin\psi_p & \cos\psi_p \end{bmatrix}_p \begin{bmatrix} Z^*_{\xi\xi} & Z^*_{\xi\eta} \\ Z^*_{\eta\xi} & Z^*_{\eta\eta} \end{bmatrix}_p \begin{bmatrix} \cos\psi_p & \sin\psi_p \\ -\sin\psi_p & \cos\psi_p \end{bmatrix}_p \qquad (6.9\text{-}40)$$

The complete bearing stiffness and damping coefficients used in the rotor dynamics calculation are determined from the real and imaginary parts of the impedance coefficients:

$$Z_{i,j} = K_{i,j} + j\omega C_{i,j} \qquad (i = x, y; \, j = x, y) \qquad (6.9\text{-}41)$$

In order to assemble the complete bearing, an important geometric relationship between the bearing eccentricity and pad eccentricity is required. At static conditions, the pad equilibrium position in the ξ coordinate is obtained from Figure 6.9-3 and is normalized with respect to the pad radial clearance C_p:

$$\frac{e_p}{C_p}\cos\varphi_p = \frac{e_b}{C_p}\cos(\pi - \psi_p + \phi_b) + \left(\frac{C_p - C_b}{C_p}\right) \qquad (6.9\text{-}42)$$

where the left hand side of Eq. (6.9-42) can also be expressed as a function of the non-dimensional pivot film thickness:

$$\frac{e_p}{C_p}\cos\varphi_p = 1 - \left(\frac{h_p}{C_p}\right) \qquad (6.9\text{-}43)$$

and the second term in the right hand side of Eq. (6.9-42) is the definition of the pad preload.

Design Consideration

The benefit of using tilting pad journal bearings is due to their virtually inherent stability characteristics. The drawbacks are: cost, increased complexity, and the need for more axial and radial space than the fixed profile type journal bearings. The typical number of pads is 4 pads (72 degrees arc) or 5 pads (57 degrees arc) and up to 12 pads for large bearings. Four pads configuration is commonly used for small shaft diameter and high load applications. The bearing is typically oriented such that the load is either between or on the pivots. Due to the manufacturing and stack-up tolerance, a positive preload should be specified to avoid the possibility of having the negative preload. Typical preload for the tilting pad bearing is from 0.15 to 0.75. The typical value for the pivot offset is from 0.5 to 0.65. A pivot offset other than 0.5 should not be used for machines that reverse rotation can occur. Only pivot offset of 0.5 allows for bi-directional operation. Pivot offset below 0.5 is not recommended. The minimum bearing clearance ratio recommended by Waukesha Bearings is shown in Figure 6.9-5 and recommended by The Glacier Metal Company is shown in Figure 6.9-6.

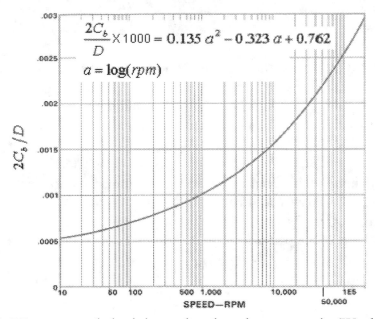

Figure 6.9-5 Recommended minimum bearing clearance ratio (Wsukesha Bearings)

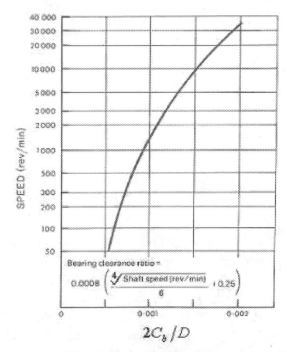

Figure 6.9-6 Recommended minimum bearing clearance ratio (The Glacial Metal)

Example 6.3: A Tilting Pad Journal Bearing

A 5-pads and centrally pivoted tilting pad bearing is presented in this example. This bearing was analyzed by Shapiro and Colsher (1977) and then by Jones and Martin (1979). The pertinent parameters and bearing configuration are listed in the Figure 6.9-7 and Table 6.9-2. Three load directions are analyzed: load on pivot, load between pivots, and load on an arbitrary angle.

Case 1: load on pivot

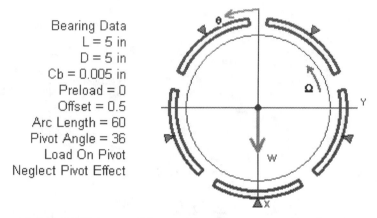

Figure 6.9-7 Tilting pad journal bearing example – load on pivot

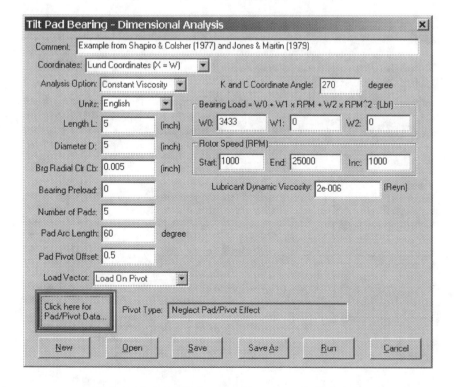

Table 6.9-2 Tilting pad bearing data input

Figure 6.9-8 shows the journal equilibrium locus for rotor speed from 1,000 to 25,000 rpm. It shows that the attitude angles are zero, which implies inherent stability of the bearing. The bearing performance at 5,000 rpm is shown in Figure 6.9-9. The results are in agreement with the previous two publications. Figure 6.9-10 shows the direct stiffness and damping coefficients verses speed. The cross-coupling terms are zero in this case.

Figure 6.9-8 Journal equilibrium locus

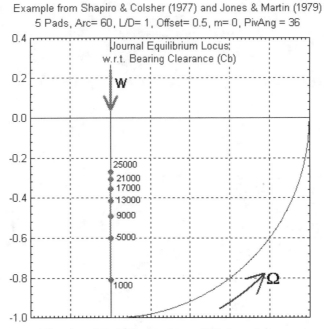

Example from Shapiro & Colsher (1977) and Jones & Martin (1979)
5 Pads, Arc= 60, L/D= 1, Offset= 0.5, m= 0, PivAng = 36
Speed = 5000 rpm
Load = 3433 Lbf
W/LD = 137.32 psi
Vis. = 2E-06 Reyns
Sb = 0.30343
E/Cb = 0.6000
Att. = 0.00 deg
hmin = 1.744 mils
Pmax = 541.948 psi
Hp = 8.60247 hp
Stiffness (Lbf/in)
 2.264E+05   0.000E+00
 0.000E+00   3.008E+06
Damping (Lbf-s/in)
 1.867E+03   0.000E+00
 0.000E+00   6.173E+03
Critical Journal Mass
Stable

Figure 6.9-9 Bearing performance at 5,000 rpm – load on pivot

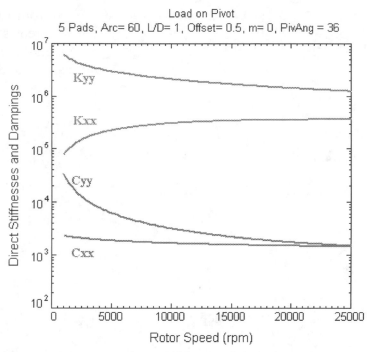

Figure 6.9-10 Bearing stiffness and damping – load on pivot

Case 2: load between pivots

Figure 6.9-11 is the same bearing with load between pivots. Figure 6.9-12 shows the bearing performance at 5,000 rpm. Note that the pad tilt angle and preload have been exaggerated for illustration purposes.

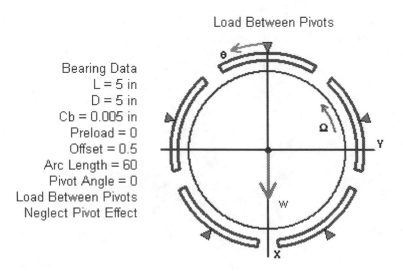

Figure 6.9-11 Tilting pad journal bearing – load between pivots

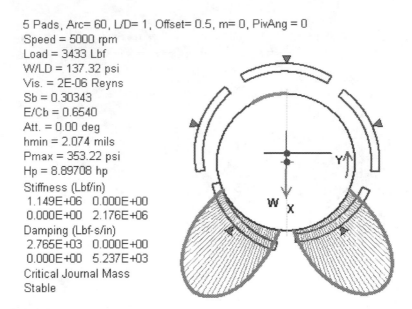

5 Pads, Arc= 60, L/D= 1, Offset= 0.5, m= 0, PivAng = 0
Speed = 5000 rpm
Load = 3433 Lbf
W/LD = 137.32 psi
Vis. = 2E-06 Reyns
Sb = 0.30343
E/Cb = 0.6540
Att. = 0.00 deg
hmin = 2.074 mils
Pmax = 353.22 psi
Hp = 8.89708 hp
Stiffness (Lbf/in)
 1.149E+06   0.000E+00
 0.000E+00   2.176E+06
Damping (Lbf-s/in)
 2.765E+03   0.000E+00
 0.000E+00   5.237E+03
Critical Journal Mass
Stable

Figure 6.9-12 Bearing performance at 5,000 rpm – load between pivots

Case 3: load on an arbitrary angle

Figure 6.9-13 shows the bearing performance at 5,000 rpm for the load vector between the pivot and center of the oil groove. That is 18 degree from the pivot in the direction of rotation. Note that since the load is not symmetric with respect to the pivots, the bearing attitude angle is not zero and cross-coupling terms are not zero.

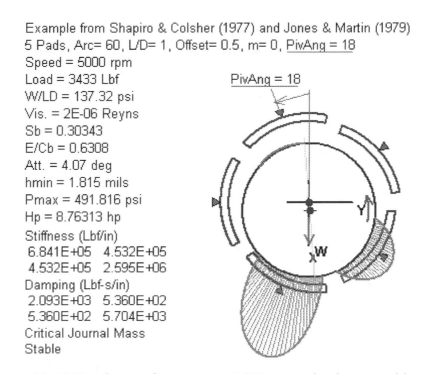

Example from Shapiro & Colsher (1977) and Jones & Martin (1979)
5 Pads, Arc= 60, L/D= 1, Offset= 0.5, m= 0, PivAng = 18
Speed = 5000 rpm
Load = 3433 Lbf
W/LD = 137.32 psi
Vis. = 2E-06 Reyns
Sb = 0.30343
E/Cb = 0.6308
Att. = 4.07 deg
hmin = 1.815 mils
Pmax = 491.816 psi
Hp = 8.76313 hp
Stiffness (Lbf/in)
 6.841E+05   4.532E+05
 4.532E+05   2.595E+06
Damping (Lbf-s/in)
 2.093E+03   5.360E+02
 5.360E+02   5.704E+03
Critical Journal Mass
Stable

PivAng = 18

Figure 6.9-13 Bearing performance at 5,000 rpm – load on an arbitrary angle

The results for three different cases are summarized below:

At 5,000 rpm	Load on Pivot	Load between Pivots	Load on an angle
$\varepsilon = e / C_b$	0.600	0.654	0.6308
ϕ (degree)	0	0	4.07
h_{min} (mils)	1.744	2.074	1.815
P_{max} (psi)	541.9	353.2	491.8
hp (hp)	8.6	8.9	8.8
k_{xx} (Lbf/in)	2.264E05	1.149E06	6.841E05
k_{yy} (Lbf/in)	3.008E06	2.176E06	2.595E06
C_{xx} (Lbf-s/in)	1.867E03	2.765E03	2.093E03
C_{yy} (Lbf-s/in)	6.173E03	5.237E03	5.704E03
$k_{xy} = k_{yx}$	0	0	4.532E05
$C_{xy} = C_{yx}$	0	0	5.360E02

6.10 The Infinitely Short Cylindrical Journal Bearing

Many phenomena in the fluid film bearings are explained by using a plain cylindrical journal bearing where the closed form solutions are available. The Reynolds equation in the fixed coordinate system is given in the previous sections and listed below for reference:

$$\frac{\partial}{\partial \bar{x}}\left(G_{\bar{x}} \frac{h^3}{\mu} \frac{\partial P}{\partial \bar{x}}\right) + \frac{\partial}{\partial \bar{y}}\left(G_{\bar{y}} \frac{h^3}{\mu} \frac{\partial P}{\partial \bar{y}}\right) = \frac{U}{2} \frac{\partial h}{\partial \bar{x}} + \frac{\partial h}{\partial t} \tag{6.10-1}$$

where $\bar{x} = R\theta$ is the circumferential coordinate and $\bar{y}$ is the axial coordinate. Note that angle θ is measured from the load vector. To obtain the closed form solutions, it is convenient to introduce the rotating coordinate θ^* as shown in Figure 6.10-1:

$$\theta = \theta^* + \phi \tag{6.10-2}$$

where ϕ is the attitude angle.

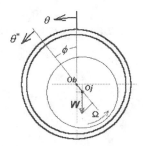

Figure 6.10-1 Coordinate systems

The film thickness for a plain cylindrical bearing is:

$$h = C + e\cos(\theta - \phi) = C + e\cos\theta^* = C(1 + \varepsilon\cos\theta^*) \qquad (6.10\text{-}3)$$

The maximum film thickness ($h_{\max} = C + e$) occurs at $\theta^* = 0$, and the minimum film thickness ($h_{\min} = C - e$) occurs at $\theta^* = \pi$. $\varepsilon = \dfrac{e}{C}$ is the eccentricity ratio which ranges from zero to one. When the eccentricity ratio is zero, the journal center coincides with the bearing center and the film thickness is a constant value that equals to the bearing radial clearance C. When the eccentricity ratio has a value of one, the shaft contacts the bearing surface and the minimum film thickness is zero.

From the film thickness expression Eq. (6.10-3), we have

$$\frac{\partial h}{\partial t} = \dot{e}\cos\theta^* + e\dot{\phi}\sin\theta^* \qquad (6.10\text{-}4)$$

The first term represents the change in film thickness due to radial motion along the journal-bearing line of centers, and the second term represents the change in film thickness due to the precession of the journal center about the bearing center with angular velocity of $\dot{\phi}$, where $\dot{\phi}$ represents the rotation of the line of centers relative to the fixed coordinate system. By using the following identities:

$$\frac{\partial h}{\partial t^*} = \dot{e}\cos\theta^* \qquad\qquad \frac{\partial h}{\partial \theta^*} = -e\sin\theta^* \qquad (6.10\text{-}5)$$

The Reynolds equation in the rotating reference coordinate becomes:

$$\frac{\partial}{\partial \overline{x}^*}\left(G_{\overline{x}}\frac{h^3}{\mu}\frac{\partial P}{\partial \overline{x}^*}\right) + \frac{\partial}{\partial \overline{y}}\left(G_{\overline{y}}\frac{h^3}{\mu}\frac{\partial P}{\partial \overline{y}}\right) = \left(\frac{\Omega}{2} - \dot{\phi}\right)\frac{\partial h}{\partial \theta^*} + \frac{\partial h}{\partial t^*} \qquad (6.10\text{-}6)$$

where $\overline{x}^* = R\theta^*$ is the rotating circumferential coordinate and $\overline{y}$ is the axial coordinate. This is the Reynolds equation in the rotating coordinates as usually quoted in many literatures. Since most of the plain cylindrical bearings in practical applications have a L/D ratio less than 0.5, short bearing assumption is frequently used. This means that the pressure gradient in the axial direction is much greater than in the circumferential direction and the first term in the Reynolds equation can be neglected. Then, the pressure distribution can be obtained by integrating the second-order differential equation with the prescribed essential boundary conditions in the axial direction:

$$P(\theta^*,0) = 0$$
$$P(\theta^*,L) = 0 \qquad (6.10\text{-}7)$$

which yields

$$P(\theta^*, \bar{y}) = \frac{6\mu}{h^3}\left[\left(\dot{\phi} - \frac{\Omega}{2}\right)e\sin\theta^* + \dot{e}\cos\theta^*\right]\left(\bar{y}^2 - L\bar{y}\right) \qquad (6.10\text{-}8)$$

for laminar flow with $G_{\bar{y}} = 1/12$.

The radial and tangential components of the hydrodynamic force acting on the journal are obtained by the integrating the pressure distribution.

$$F_r = \int_{\theta_1^*}^{\theta_2^*}\left(\int_0^L P(\theta^*, \bar{y})\cos\theta^*\, d\bar{y}\right)R\, d\theta^* = -\mu R\left(\frac{L}{C}\right)^3\left[\left(\dot{\phi} - \frac{\Omega}{2}\right)e\, I_3^{11} + \dot{e}\, I_3^{02}\right] \qquad (6.10\text{-}9)$$

$$F_t = \int_{\theta_1^*}^{\theta_2^*}\left(\int_0^L P(\theta^*, \bar{y})\sin\theta^*\, d\bar{y}\right)R\, d\theta^* = -\mu R\left(\frac{L}{C}\right)^3\left[\left(\dot{\phi} - \frac{\Omega}{2}\right)e\, I_3^{20} + \dot{e}\, I_3^{11}\right] \qquad (6.10\text{-}10)$$

where

$$I_n^{lm} = \int_{\theta_1^*}^{\theta_2^*}\frac{\sin^l\theta^*\cos^m\theta^*}{(1 + \varepsilon\cos\theta^*)^n}\, d\theta^* \qquad (6.10\text{-}11)$$

The closed form expression for the above integration is available once the limits of integration are known (Booker, 1965). For a π film assumption (Gumbel, or half Sommerfeld condition) at the circumferential coordinate, the pressure film begins at $\theta^* = \theta_1^*$ and ends at $\theta^* = \theta_2^* = \theta_1^* + \pi$ where the pressure is zero at these two boundaries. In view of the pressure equation, the positive pressures occur in the region $\theta_1^* \le \theta^* \le \theta_1^* + \pi$ where θ^* is given by:

$$\left[\left(\dot{\phi} - \frac{\Omega}{2}\right)e\sin\theta^* + \dot{e}\cos\theta^*\right] \le 0 \qquad (6.10\text{-}12)$$

and at θ_1^*, the pressure is zero:

$$\tan\theta_1^* = \frac{-\dot{e}}{e\left(\dot{\phi} - \frac{\Omega}{2}\right)} \qquad (6.10\text{-}13)$$

At the leading edge of the film, $\theta^* = \theta_1^*$, the slope of the pressure must be positive:

$$\left(\frac{\partial P}{\partial\theta^*}\right)_{\theta^* = \theta_1^*} > 0 \Rightarrow \left[\left(\dot{\phi} - \frac{\Omega}{2}\right)e\cos\theta_1^* - \dot{e}\sin\theta_1^*\right] < 0 \qquad (6.10\text{-}14)$$

This defines the quadrant for θ_1^*.

The nonlinear fluid film forces acting on the rotor in the (X, Y) coordinates which are used in the integration of the rotor motion are obtained through the coordinate transformation:

$$F_x = F_r \cos\phi - F_t \sin\phi$$
$$F_y = F_r \sin\phi + F_t \cos\phi \qquad (6.10\text{-}15)$$

and

$$F_x = F_x(e, \phi, \dot{e}, \dot{\phi}) = F_x(x, y, \dot{x}, \dot{y})$$
$$F_y = F_y(e, \phi, \dot{e}, \dot{\phi}) = F_y(x, y, \dot{x}, \dot{y}) \qquad (6.10\text{-}16)$$

where $(x, y, \dot{x}, \dot{y})$ are journal displacements and velocities. The above nonlinear fluid film forces are valid for any vibrations within the bearing clearance circle and are commonly used in the plain cylindrical bearing nonlinear simulation. Since the rotor motions are commonly specified by $(x, y, \dot{x}, \dot{y})$ and the bearing forces are expressed by $(e, \phi, \dot{e}, \dot{\phi})$, the relationships are:

$$e = \left(x^2 + y^2\right)^{1/2} \qquad\qquad \phi = \arctan\left(\frac{y}{x}\right)$$

$$\dot{e} = \frac{x\dot{x} + y\dot{y}}{e} \qquad\qquad \dot{\phi} = \frac{x\dot{y} - y\dot{x}}{e^2} \qquad (6.10\text{-}17)$$

The linearization of the fluid film forces under short bearing assumption is seldom used since the linearized 2-dimensional Reynolds equation can be solved from previous discussion. They are presented below for reference purposes only. Under static load conditions, the static equilibrium position of the journal center is defined by $(e_0$ and $\phi_0)$ and the limits of integration are from $\theta_1^* = 0$ to $\theta_2^* = \pi$. The film forces are:

$$F_r = -\mu R \left(\frac{L}{C}\right)^3 \left[\left(\dot{\phi} - \frac{\Omega}{2}\right) e \left(\frac{-2\varepsilon}{\left(1 - \varepsilon^2\right)^2}\right) + \dot{e}\left(\frac{\pi\left(1 + 2\varepsilon^2\right)}{2\left(1 - \varepsilon^2\right)^{5/2}}\right)\right] \qquad (6.10\text{-}18)$$

$$F_t = -\mu R \left(\frac{L}{C}\right)^3 \left[\left(\dot{\phi} - \frac{\Omega}{2}\right) e \left(\frac{\pi}{2\left(1 - \varepsilon^2\right)^{3/2}}\right) + \dot{e}\left(\frac{-2\varepsilon}{\left(1 - \varepsilon^2\right)^2}\right)\right] \qquad (6.10\text{-}19)$$

Note that unlike the previous nonlinear equations, which are valid for any motion, the above forces are only valid for small motions around the static equilibrium position. At static equilibrium, $\dot{e} = \dot{\phi} = 0$, assuming the static load is in the X direction, the forces acting on the journal are balanced:

$$F_{r0} = -W \cos\phi_0 \tag{6.10-20}$$
$$F_{t0} = W \sin\phi_0 \tag{6.10-21}$$

By substituting the static forces with ($\dot{e} = \dot{\phi} = 0$) into the above equations, the equilibrium position can be determined by the following equations:

$$\frac{\mu R}{2}\left(\frac{L}{C}\right)^3\left(\frac{\Omega e_0}{W}\right)\left[\frac{\sqrt{16\varepsilon_0^2 + \pi^2\left(1-\varepsilon_0^2\right)}}{2\left(1-\varepsilon_0^2\right)^2}\right] = 1 \tag{6.10-22}$$

and

$$\tan\phi_0 = \frac{\pi\sqrt{1-\varepsilon_0^2}}{4\varepsilon_0} \tag{6.10-23}$$

The linearized bearing dynamic coefficients can then be obtained by the first order Taylor's expansion of the film reactive forces, Eqs. (6.10-18) and (6.10-19), with a small amplitude motion (Δe and $e_0\Delta\phi$) around the static equilibrium position. The linearized bearing coefficients in the radial and tangential directions are:

$$\begin{bmatrix} k_{rr} & k_{rt} \\ k_{tr} & k_{tt} \end{bmatrix} = \frac{\mu R}{2}\left(\frac{L}{C}\right)^3\Omega\begin{bmatrix} \dfrac{4\varepsilon_0(1+\varepsilon_0^2)}{\left(1-\varepsilon_0^2\right)^3} & \dfrac{\pi}{2\left(1-\varepsilon_0^2\right)^{3/2}} \\ \dfrac{-\pi(1+2\varepsilon_0^2)}{2\left(1-\varepsilon_0^2\right)^{5/2}} & \dfrac{2\varepsilon_0}{\left(1-\varepsilon_0^2\right)^2} \end{bmatrix} \tag{6.10-24}$$

$$\begin{bmatrix} c_{rr} & c_{rt} \\ c_{tr} & c_{tt} \end{bmatrix} = \frac{\mu R}{2}\left(\frac{L}{C}\right)^3\Omega\begin{bmatrix} \dfrac{\pi(1+2\varepsilon_0^2)}{\left(1-\varepsilon_0^2\right)^{5/2}} & \dfrac{-4\varepsilon_0}{\left(1-\varepsilon_0^2\right)^2} \\ \dfrac{-4\varepsilon_0}{\left(1-\varepsilon_0^2\right)^2} & \dfrac{\pi}{\left(1-\varepsilon_0^2\right)^{3/2}} \end{bmatrix} \tag{6.10-25}$$

It shows that the stiffness matrix in general is not symmetric and the damping matrix is symmetric. The linearized bearing coefficients in the x and y coordinates are obtained by a coordinate transformation presented before.

$$\begin{bmatrix} k_{xx} & k_{xy} \\ k_{yx} & k_{yy} \end{bmatrix} = \begin{bmatrix} \cos\phi_0 & -\sin\phi_0 \\ \sin\phi_0 & \cos\phi_0 \end{bmatrix}\begin{bmatrix} k_{rr} & k_{rt} \\ k_{tr} & k_{tt} \end{bmatrix}\begin{bmatrix} \cos\phi_0 & \sin\phi_0 \\ -\sin\phi_0 & \cos\phi_0 \end{bmatrix} \tag{6.10-26}$$

The damping matrix has the same transformation.

6.11 Squeeze Film Dampers

Squeeze film dampers mounted on rolling element bearings are commonly used in high-speed gas turbine engines and power turbines to attenuate the unbalance response and bearing transmitted forces. One feature introduced by a squeeze film damper is the introduction of support flexibility and damping in the bearing/support structure. This translates to lower transmitted forces and longer bearing life, particularly for machinery that are designed to operate at super critical speeds. Figure 6.11-1 represents a typical squeeze film damper application in which a squeeze film damper is mounted on rolling element bearings with centering spring. Figure 6.11-2 shows the damper is mounted outside of a fluid film bearing. The damper is constrained from rotating ($\Omega=0$). That is, the damper is free to whirl motion, but not rotation. Thus, the journal and bearing speeds are zero.

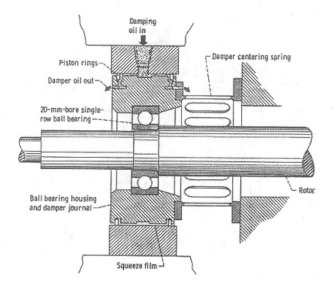

Figure 6.11-1 A typical squeeze film damper mounted on ball bearings with centering spring (Gunter et. al.)

Figure 6.11-2 A squeeze film damper mounted on tilting pad bearing (Photograph courtesy of Rotating Machinery Technology, Inc.)

For short bearing assumption, the pressure equation and fluid film forces presented in the previous session are also valid for the squeeze film dampers with $\Omega=0$. Again, the closed form is available for the pressure integration once the limits of integration are known. For a π film assumption (Gumbel, or half Sommerfeld condition) at the circumferential coordinate, the pressure film begins at $\theta^* = \theta_1^*$ and ends at $\theta^* = \theta_2^* = \theta_1^* + \pi$ where the pressure is zero at these two boundaries. From the pressure equation with $\Omega=0$, the positive pressures occur in the region $\theta_1^* \leq \theta^* \leq \theta_1^* + \pi$ where θ^* is given by:

$$\left[e\dot{\phi}\sin\theta^* + \dot{e}\cos\theta^* \right] \leq 0 \tag{6.11-1}$$

and at θ_1^*, the pressure is zero:

$$\tan\theta_1^* = \frac{-\dot{e}}{e\dot{\phi}} \tag{6.11-2}$$

At the leading edge of the film, $\theta^* = \theta_1^*$, the slope of the pressure must be positive:

$$\left(\frac{\partial P}{\partial \theta^*} \right)_{\theta^*=\theta_1^*} > 0 \implies \left[e\dot{\phi}\cos\theta_1^* - \dot{e}\sin\theta_1^* \right] < 0 \tag{6.11-3}$$

This defines the quadrant for θ_1^*.

The nonlinear squeeze film damper forces acting on the rotor in the (X,Y) coordinates which are used in the integration of the rotor motion are obtained through the coordinate transformation:

$$F_x = F_r \cos\phi - F_t \sin\phi \tag{6.11-4a}$$
$$F_y = F_r \sin\phi + F_t \cos\phi \tag{6.11-4b}$$

Again, the above nonlinear damper forces are valid for any vibrations within the damper clearance and are commonly used in the damper nonlinear simulation.

For steady state circular precession motion ($\dot{e}=0$) and π film assumption (Gumbel boundary condition), the nonlinear squeeze film damper forces are obtained by integrating the film pressure from $\theta_1^* = \pi$ to $\theta_2^* = 2\pi$:

$$F_r = -\mu R \left(\frac{L}{C} \right)^3 \left[\dot{\phi} \left(\frac{-2\varepsilon}{\left(1-\varepsilon^2\right)^2} \right) \right] e = -k_d \, e \tag{6.11-5}$$

and

$$F_t = -\mu R \left(\frac{L}{C} \right)^3 \left[\left(\frac{\pi}{2\left(1-\varepsilon^2\right)^{3/2}} \right) \right] e\dot{\phi} = -c_d \, e\dot{\phi} \tag{6.11-6}$$

The force in the radial direction appears as a secant stiffness coefficients times a displacement and it acts on the journal in a direction opposite to the journal displacement. The force in the tangential direction appears as a secant damping coefficient times a linear velocity and it acts on the journal in a direction opposite to the journal motion. However, these coefficients are nonlinear and amplitude-dependent. They cannot be used directly in the linear analysis and nonlinear iteration procedure for the steady state solution is required. Since the steady state circular precession motion is assumed, the system can be considered as an isotropic system and the stiffness and damping coefficients are the same in any direction. This can be illustrated from the following discussion.

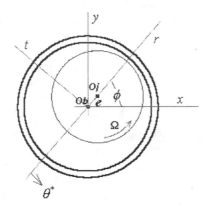

Figure 6.11-3 Coordinate systems

From the Figure 6.11-3, we have:

$$x = e\cos\phi$$
$$y = e\sin\phi$$

$$(6.11\text{-}7)$$

For steady state circular motion (e=constant), the velocities are:

$$\dot{x} = -e\dot{\phi}\sin\phi = -\dot{\phi}\, y$$
$$\dot{y} = e\dot{\phi}\cos\phi = \dot{\phi}\, x$$

$$(6.11\text{-}8)$$

Then the forces in the fixed coordinate system become:

$$F_x = F_r\cos\phi - F_t\sin\phi = -k_d x - c_d \dot{x}$$
$$F_y = F_r\sin\phi + F_t\cos\phi = -k_d y - c_d \dot{y}$$

$$(6.11\text{-}9)$$

Therefore, the stiffness and damping are the same in any directions and the damper is isotropic. Since the rotating reference frame is commonly used in analyzing the steady state circular response of isotropic systems, the forces in the rotating reference frame are:

$$\begin{Bmatrix} F_{x'} \\ F_{y'} \end{Bmatrix} = -\begin{bmatrix} k_d & -\Omega c_d \\ \Omega c_d & k_d \end{bmatrix} \begin{Bmatrix} x' \\ y' \end{Bmatrix}$$

$$(6.11\text{-}10)$$

For the steady state response, the following table summarizes the equivalent stiffness and damping for the cases of circular synchronous motion about the origin and pure radial motion with no precession for the conditions of cavitation (π film) and no cavitation (2π film). The details of derivation can be found in Barrett & Gunter (1975).

Bearing	Film	Motion	Stiffness	Damping
Short Bearing	π film	Circular Synchronous Precession	$\dfrac{2\mu R L^3 \varepsilon \omega}{C^3\left(1-\varepsilon^2\right)^2}$	$\dfrac{\mu R L^3 \pi}{2C^3\left(1-\varepsilon^2\right)^{3/2}}$
	2π film		0	$\dfrac{\mu R L^3 \pi}{C^3\left(1-\varepsilon^2\right)^{3/2}}$
	π film	Pure Radial Squeeze Motion	0	$\dfrac{\mu R L^3 \pi\left(2\varepsilon^2+1\right)}{2C^3\left(1-\varepsilon^2\right)^{5/2}}$
	2π film		0	$\dfrac{\mu R L^3 \pi\left(2\varepsilon^2+1\right)}{C^3\left(1-\varepsilon^2\right)^{5/2}}$
Long Bearing	π film	Circular Synchronous Precession	$\dfrac{24\mu R^3 L \varepsilon \omega}{C^3\left(2+\varepsilon^2\right)\left(1-\varepsilon^2\right)}$	$\dfrac{12\mu R^3 L\pi}{C^3\left(2+\varepsilon^2\right)\left(1-\varepsilon^2\right)^{1/2}}$
	2π film		0	$\dfrac{24\mu R^3 L\pi}{C^3\left(2+\varepsilon^2\right)\left(1-\varepsilon^2\right)^{1/2}}$

Where

R = damper radius
L = damper axial length
C = radial clearance
ω = whirl speed
μ = oil viscosity
ε = eccentricity ratio

Figure 6.11-4 represents the two typical squeeze film dampers with a deep circumferential central groove. The groove is assumed of sufficient depth that the oil pressure is equalized with the oil supply pressure. In case (a), the damper has end seals and the fluid film pressure has two half of identical profiles. The pressure vanishes at the deep center groove and the pressure gradient is zero at both ends due to the end seals. The effective length of the damper is L in this case. In case (b), the damper has no end seals and the fluid film pressure has two identical profiles and vanishes at the ends and at the center deep groove. The effective length is $0.63L$ in this case. If the groove depth is only two to three times the radial damper clearance, then the profile does not vanish at the center groove and the effective length may be increased to $0.8L$.

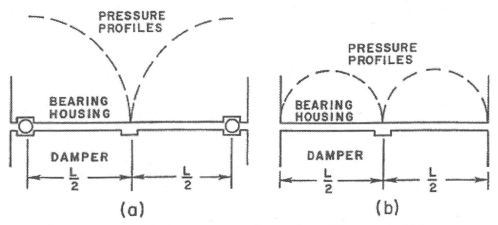

Figure 6.11-4 Damper configurations (Gunter et. Al.)

Note that for the circular synchronous motion, the equivalent stiffness term is a highly nonlinear function of eccentricity and may lead to a nonlinear jump phenomenon under high rotor unbalance. Caution must be taken while designing the damper, since it can significantly either improve or degrade the dynamic characteristics of the rotor system.

Example 6.4: A Single Squeeze Film Damper System

A single squeeze film damper system (2DOF) taken from Gunter et. al. (1977) is used here to demonstrate the nonlinear analysis of a squeeze film damper. The same system has also been studied by Taylor and Kumar (1980) on the unbalance effect and Chen et. al. (1988) on the parametric studies. The damper has a centering spring and the physical parameters of the system are summarized below:

Journal Mass (M) = 33.43 kg
Damper Diameter (D) = 129.6 mm
Damper Length (L) = 22.7 mm
Damper Radial Clearance (C) = 0.1 mm

Oil Viscosity (μ) = 2.66 CentiPoise = 2.66E-03 N-s/m^2
Centering Spring Stiffness = 21540 N/mm

Short bearing and π film assumptions are employed in this example. The steady state centered circular responses for three different unbalance forces are shown in Figure 6.11-5. It shows that the jump phenomenon occurs when the unbalance force is greater than 0.5 kg-mm. A parametric study of the damper variables other than unbalance force, such as damper clearance, radius, and length, on the transmitted load has been presented in Chen et. al. (1988).   Figure 6.11-6 shows the steady state stiffness and damping coefficients calculated by using the tool provided by **DyRoBeS**.   It shows the stiffness and damping increase significantly as the eccentricity ratio increases.

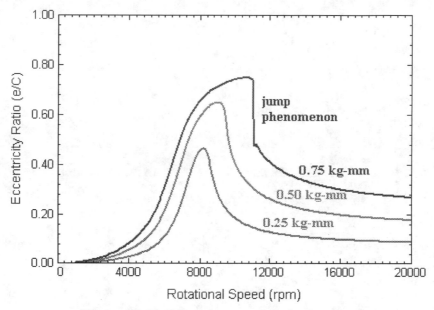

Figure 6.11-5 Steady state unbalance response

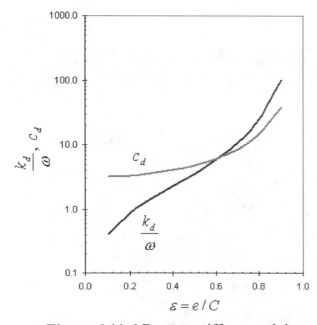

Figure 6.11-6 Damper stiffness and damping

For rotordynamics study, the damper data is entered under **Bearings** tab as shown in Table 6.11-1. The damper stiffness and damping for a given eccentricity and speed can also be calculated by using the design tool provided under **TOOLS** menu as shown in Table 6.11-2.

Table 6.11-1 Damper data used in rotordynamics study

Table 6.11-2 Damper stiffness and damping calculation for a given eccentricity

Figure 6.11-7 shows the steady state orbits at 8,000 rpm for two different side loads, F_y=0 and −1000 N with 0.5 kg-mm unbalance. These results were obtained from the Time Transient Analysis. With zero side load, the result is in agreement with that obtained from the Steady State Response Analysis as shown in Figure 6.11-5. Since the steady state Response Analysis assumes the centered circular orbit, it cannot be used to predict the response with non-zero side loads. The Time Transient Analysis has to be used for this calculation. Another approach for the non-centered circular steady state response is to utilize the Trigonometric Collocation Method as presented by Nelson, Chen, and Nataraj (1989). However, it requires good engineering knowledge in modal analysis in order to use that method.

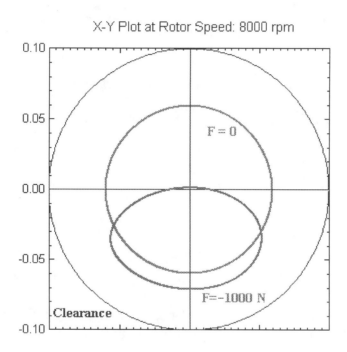

Figure 6.11-7 Steady state orbits with two different side loads

6.12 Floating Ring Bearings

Floating ring bearing has been widely used in the turbocharger applications where the rotor is light and runs at very high speed. Floating Ring Bushing can be treated as two fluid film (plain cylindrical) bearings in series. The inner film bearing has two rotating surface (shaft and bushing). The outer film bearing has only one rotating surface (bushing). Additional two degrees-of-freedom are introduced for each floating ring bearing due to its non-zero ring mass. Figure 6.12-1 shows the floating ring bearing and its model.

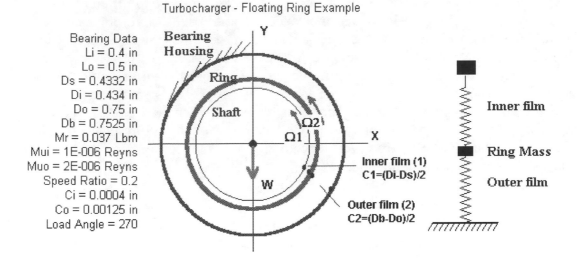

Figure 6.12-1 Floating ring bearing

The radial clearances for the inner and outer films are:

$$C_1 = \frac{D_i - D_s}{2}$$ (6.12-1)

$$C_2 = \frac{D_b - D_o}{2}$$ (6.12-2)

where D_s is the shaft diameter, D_i is the inner diameter of the ring, D_o is the outer diameter of the ring, and D_b is the bearing diameter. The journal rotates with an angular speed of Ω_1 and the ring rotates with an angular speed of Ω_2. Thus, for the inner film, the velocity U in the Reynolds equation is the sum of the shaft velocity and the ring velocity:

$$U_1 = R_s\Omega_1 + R_i\Omega_2 \cong R_s(\Omega_1 + \Omega_2) = R_1(\Omega_1 + \Omega_2) \qquad \text{since } C_i << R_s \qquad (6.12-3)$$

For the outer film, the velocity U in the Reynolds equation is the velocity of the ring:

$$U_2 = R_o\Omega_2 = R_2\Omega_2$$ (6.12-4)

It is convenient to use the subscript "1" to represent the parameters in the inner film and subscript "2" to represent the parameters in the outer film. It is a very straightforward procedure for the nonlinear transient analysis by coupling the rotor equations with two Reynolds equations. For the linear analysis, the linearized coefficients for the inner and outer films need to be determined. At static state condition, the journal center equilibrium position is defined by (e, ϕ) with respect to the bearing center and by (e_1, ϕ_1) relative to the ring center as shown in Figure 6.12-2. The ring center equilibrium position is defined by (e_2, ϕ_2) with respect to the bearing center.

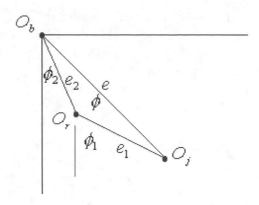

Figure 6.12-2 Relationships for the bearing, journal, and ring centers

Under steady conditions, the friction torques acting on the ring from the inner and outer films must balance. The friction torque acting on the ring from the inner film can be obtained from the shear stress equation:

$$T_1 = \iint R_1 \tau_{1,\bar{z}=0} \, d\bar{x}d\bar{y} \qquad\qquad (6.12\text{-}5)$$

where

$$\tau_{1,\bar{z}=0} = C_f \frac{\mu_1 R_1 (\Omega_1 - \Omega_2)}{h} - \frac{h}{2}\frac{\partial P}{\partial x} \qquad\qquad (6.12\text{-}6)$$

The friction torque acting on the ring from the outer film is:

$$T_2 = \iint R_2 \tau_{2,\bar{z}=h} \, d\bar{x}d\bar{y} \qquad\qquad (6.12\text{-}7)$$

where

$$\tau_{2,\bar{z}=h} = C_f \frac{\mu_2 R_2 \Omega_2}{h} + \frac{h}{2}\frac{\partial P}{\partial x} \qquad\qquad (6.12\text{-}8)$$

At static condition, for a given ring geometry, such as C_1, C_2, R_1, R_2, L_1, L_2, and operating conditions, such as W, Ω_1, μ_1, μ_2, we have a total of 5 unknowns ($\Omega_2, e_1, \phi_1, e_2, \phi_2$) and 5 equations from force and torque balance. Assuming the x-axis is aligned with the load vector W, the 5 equations are:

$$\begin{Bmatrix} F_{x0} \\ F_{y0} \end{Bmatrix}_1 = \begin{Bmatrix} -W \\ 0 \end{Bmatrix} \qquad \text{for inner film} \qquad\qquad (6.12\text{-}9)$$

$$\begin{Bmatrix} F_{x0} \\ F_{y0} \end{Bmatrix}_2 = \begin{Bmatrix} -W \\ 0 \end{Bmatrix} \qquad \text{for outer film} \qquad\qquad (6.12\text{-}10)$$

$$T_1 = T_2 \qquad\qquad \text{torque balance} \qquad\qquad (6.12\text{-}11)$$

Once the equilibrium positions are determined by an iterative procedure, the linearized coefficients can be obtained from the procedures described before. Floating ring bearing is usually used for very high-speed applications and the floating ring is designed to have high ring speed to ensure the proper hydrodynamic lubrication. At very high speed, the eccentricity ratios are very small and the journal, ring, and bearing are nearly concentric. When the eccentricity ratios are zero, the torque balance becomes:

$$T_1 = 2\pi \, \mu_1 R_1^3 \left(\Omega_1 - \Omega_2 \right) L_1 / C_1 = T_2 = 2\pi \, \mu_2 R_2^3 \Omega_2 L_2 / C_2 \qquad (6.12\text{-}12)$$

Thus, the ring speed Ω_2 can be determined from the above equation. Once the ring speed is known, the two Reynolds equations for the inner and outer films can be solved separately and the bearing coefficients can be determined just like the conventional plain cylindrical bearing. It is convenient to define the ring-journal speed ratio as:

$$\nu = \frac{\Omega_2}{\Omega_1} = \frac{1}{1 + \left[\left(\dfrac{\mu_2}{\mu_1} \right) \left(\dfrac{R_2}{R_1} \right)^3 \left(\dfrac{L_2}{L_1} \right) \left(\dfrac{C_1}{C_2} \right) \right]} \qquad (6.12\text{-}13)$$

For the nonlinear analysis, the floating ring bearing data is entered under the **_Bearing_** tab in **_DyRoBeS-Rotor_** as shown in Table 6.12-1. Note that station J is the ring station. The nonlinear fluid film forces for the inner and outer films are dependent on the motions of the journal and ring. Nonlinear simulation can be performed using this input.

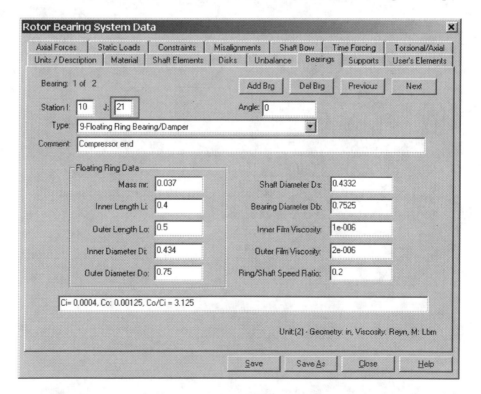

Table 6.12-1 Floating ring bearing data input in **_DyRoBeS-Rotor_**

For the linear analysis, the floating ring bearing linearized coefficients can be calculated by using *DyRoBeS-BePerf* as shown in Table 6.12-2. The linearized bearing coefficients for the inner film and the outer film can be calculated and saved by using *DyRoBeS-BePerf*, and then be imported by *DyRoBeS-Rotor* to be used in the rotordynamics study. The ring mass is modeled as a flexible support. The inner film connects the rotor to the ring (flexible support). The outer film connects the ring to the bearing housing. The bearing housing can be either flexible or rigid in *DyRoBeS*.

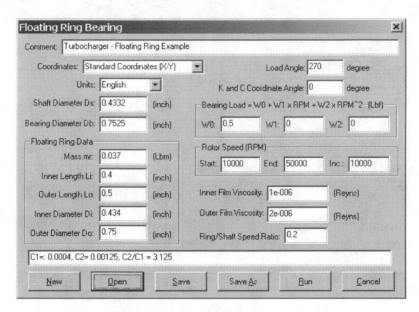

Table 6.12-2 Floating ring bearing analysis in *DyRoBeS-BePerf*

Example 6.5: Turbocharger Supported by Floating Ring Bushings

Figure 6.12-3 shows a turbocharger configuration. The turbocharger runs at 100,000 rpm supported by two floating ring bearings (Gunter and Chen, 2005).   The limit cycle motion is presented in Figure 6.12-4.

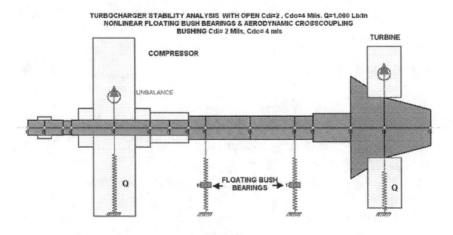

Figure 6.12-3 Turbocharger with floating ring bearings

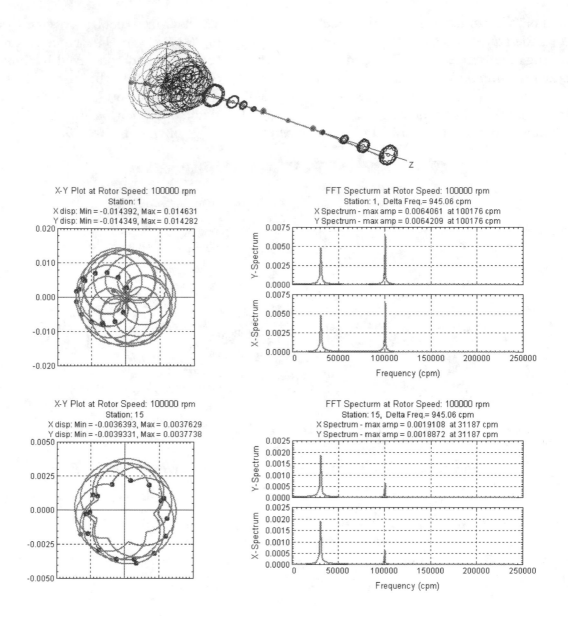

Figure 6.12-4 Limit cycle motions

6.13 Gas Lubricated Bearings

Gas bearings have been used in very high-speed and light load applications. The applications of gas bearings provide maintenance free and oil free clean operating environment. Significant low power consumption due to low viscosity of process gas allows the machines to operate at very high speeds. However, the load carrying capability of gas bearings is very small compared with the conventional oil-lubricated bearings. Gas bearings are applied at temperatures that range from cryogenic to more than a thousand degrees Fahrenheit. The temperature limit is imposed by the bearing material, not the gas. The viscosities of common gases increase with temperature but are comparatively insensitive to moderate temperature and pressure changes. On the contrast, viscosities of liquids vary inversely with temperature and are strongly sensitive to temperature variations as shown Figure 6.13-1 and in the Appendix C. For a gas bearing, the temperature and pressure changes are normally small, constant viscosity model is commonly used in handling the gas bearings.

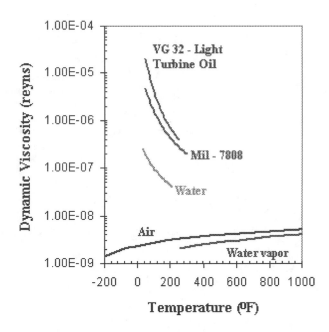

Figure 6.13-1 Viscosity of common fluids

Figure 6.13-2 shows two commonly used foil gas bearings. One has leaf foils and one has bump foils. In order to produce more damping, the bump and leaf foils have many different configurations. Tilting pad gas bearings with spring-loaded diaphragm has also been used in high-speed compressor applications.

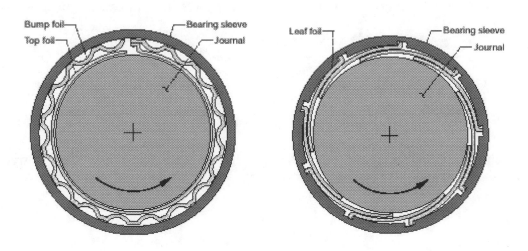

Figure 6.13-2 Gas foil bearings
(Photograph courtesy of NASA)

The governing equation for the fluid element is Reynolds equation with compressible ideal gas under isothermal conditions:

$$\frac{\partial}{\partial x}\left(\frac{Ph^3}{12\mu}\frac{\partial P}{\partial x}\right) + \frac{\partial}{\partial y}\left(\frac{Ph^3}{12\mu}\frac{\partial P}{\partial y}\right) = \frac{U}{2}\frac{\partial(Ph)}{\partial x} + \frac{\partial(Ph)}{\partial t} \tag{6.13-1}$$

This compressible Reynolds equation is more difficult to analyze due to the existence of the pressure (P) in each terms compared with the incompressible flow, which makes the problem "non-linear". Again, weak formulation based on variational principle is applied for generating the finite element model for the boundary value problems. Since this is a nonlinear problem, Newton-Raphson's iterative scheme is utilized to solve the pressure increment, or pressure correction. The static pressure distribution at the "$i+1$" th iteration is:

$$P_{i+1} = P_i + \eta_i \tag{6.13-2}$$

η_i is the pressure correction and is obtained from the following linearized equation:

$$I(P_{i+1}) = I(P_i) + \frac{dI(P_i)}{dP}\eta_i = 0 \tag{6.13-3}$$

where $I(P)$ is a functional obtained from the weak formulation

$$I(P) = \int_A v\left[\nabla\cdot\left(\frac{Ph^3}{12\mu}\nabla P - \frac{1}{2}UPh\right)\right] = 0 \tag{6.13-4}$$

Where ∇ is the gradient operator. The Frechet derivative at P_i operating on η_i is (Szeri, 1980):

$$\frac{dI(P)}{dP}\eta = \frac{d}{d\varepsilon}I(P+\varepsilon\eta)\big|_{\varepsilon=0} \tag{6.13-5}$$

By applying Galerkin method ($v = N$), the linearized equation of the pressure correction for a typical fluid element becomes:

$$\boldsymbol{K}\boldsymbol{\eta}_i = \boldsymbol{F} \tag{6.13-6}$$

where

$$K = \int_A \begin{Bmatrix} N \\[6pt] \dfrac{\partial N}{\partial x} \\[6pt] \dfrac{\partial N}{\partial y} \end{Bmatrix}^{\mathrm{T}} \begin{bmatrix} \dfrac{U_{\bar{x}}}{2}\dfrac{\partial h}{\partial x} & \dfrac{U_{\bar{x}}}{2}h & 0 \\[10pt] \dfrac{h^3}{12\mu}\dfrac{\partial P}{\partial x} & \dfrac{Ph^3}{12\mu} & 0 \\[10pt] \dfrac{h^3}{12\mu}\dfrac{\partial P}{\partial y} & 0 & \dfrac{Ph^3}{12\mu} \end{bmatrix} \begin{Bmatrix} N \\[6pt] \dfrac{\partial N}{\partial x} \\[6pt] \dfrac{\partial N}{\partial y} \end{Bmatrix} dA \tag{6.13-7}$$

and

$$F = -\int_A \begin{Bmatrix} N \\[6pt] \dfrac{\partial N}{\partial x} \\[6pt] \dfrac{\partial N}{\partial y} \end{Bmatrix}^{\mathrm{T}} \begin{Bmatrix} \dfrac{U_{\bar{x}}}{2}\dfrac{\partial P}{\partial x}h + \dfrac{U_{\bar{x}}}{2}P\dfrac{\partial h}{\partial x} \\[10pt] \dfrac{Ph^3}{12\mu}\dfrac{\partial P}{\partial x} \\[10pt] \dfrac{Ph^3}{12\mu}\dfrac{\partial P}{\partial y} \end{Bmatrix} dA \tag{6.13-8}$$

Assembling all the fluid elements and applying the appropriate boundary conditions, the static pressure distribution P now can be calculated by utilizing the Newton-Raphson's iterative scheme with the given film thickness. For bearings with flexible supports, such as foil bearings, the film thickness is dependent upon the structural deflection. The structural deflection depends upon the pressure distribution. Therefore, the Reynolds equation must be coupled with the structural equation in the determination of static pressure.

A non-dimensional bearing parameter, Λ, also known as the compressibility number, is commonly used in the gas bearing analysis:

$$\Lambda = \frac{6\mu\Omega}{P_a}\left(\frac{R}{C}\right)^2 = 12\pi\,\frac{\mu N_s}{P_a}\left(\frac{R}{C}\right)^2 \tag{6.13-9}$$

where P_a is the ambient pressure. This compressibility number is similar to the Sommerfeld number used in the incompressible fluid bearing analysis. As $\Lambda \rightarrow 0$, the solution reduces to the solution of the incompressible flow.

Example 6.6 A Gas Journal Bearing

A plain cylindrical gas journal bearing taken from Raimondi (1961) is presented below. The pertinent parameters are listed and shown in the input data page.

L = 1 in.	D = 1 in.	C = 0.0006 in.
N = 52800 rpm	μ = 2.7E-09 reyns	
P_a = 14.7 psi	P = 12.9 psi =>	W = 12.9 Lbf

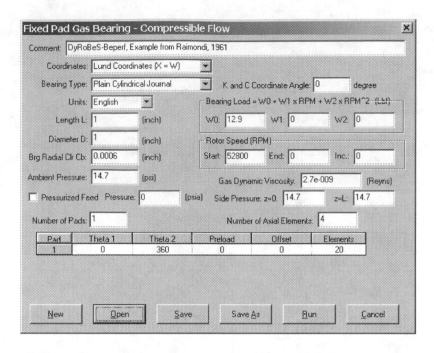

Table 6.13-1 Gas bearing input

The calculated results are listed below:

```
================================================================================

                Rotational Speed =   52800.      rpm
                   Bearing Load =   12.900      Lbf

        Compressibility Number =    4.2315

     Eccentricity Ratio (E/Cb) =     0.598
        Attitude Angle (degree) =    24.78

  Minimum film thickness (h) =   0.00024      in.
         Maximum Pressure (P) =   37.103      psi (Lbf/in^2)
                   at Angle =     18.00      degree
       Friction Power Loss =  0.21235E-01 horsepower

 Bearing Coefficients, x axis is    0.00 degrees from the X axis
 Dimensional Coefficients (units: K= Lbf/in,  C= Lbf-sec/in)
 Kxx,Kxy,Kyx,Kyy =   70131.      7498.2      14049.      42523.
 Cxx,Cxy,Cyx,Cyy =  2.3705     -0.88981     0.76221     2.1260
================================================================================
```

The static performance is summarized below for comparison purposes:

	DyRoBeS-BePerf	Raimondi
ε	0.598	0.60
ϕ	24.78	24.5
h_{min} (in)	0.00024	0.00024
Hp (hp)	0.0212	0.0212
P_{max} (psia)	37.1	37.6

6.14 Rolling Element Bearings

One of the most commonly used bearing types is the rolling element bearing. Typically, the stiffness is derived from the load-deflection curve in the contact zones between rollers (balls) and inner, outer races. It is highly dependent upon the working preload of the bearing, which is strongly influenced by the installation methods, clearance, and tolerance. The situation is further complicated by the operation conditions, such as type of lubricants, supply methods, temperature, and rotational speed. Since the stiffness of a rolling element bearing is proven to be difficult to calculate, therefore, a tolerance is always recommended when using a calculated value and whenever possible test data should always be used. The stiffness for ball bearings is typically found to be on the order of 2.0E07 N/m (1.0E05 Lb/in) to 2.0E08 N/m (1.0E06 Lb/in). For roller bearings, the stiffness is approximately 5-10 times higher. Two major disadvantages of the rolling element bearings compared with the fluid film bearings are the lack of external damping and limited life due to metal-to-metal contact. In rotordynamics study, the rolling element bearing is commonly represented by an isotropic radial stiffness ($k=k_{xx}=k_{yy}$) and very small damping. The damping factor estimated by Kramer (1993) lies in the range of 0.0004 to 0.004.

Gargiulo (1980) has provided a simple way to estimate the rolling element bearing stiffness. This approximation provides a convenient way to estimate the bearing stiffness since it requires minimum data input. The radial deflection and stiffness under a given load are summarized below.

1) Deep-Groove or Angular-Contact Radial Ball Bearings

$$\delta = 46.2E - 06 \sqrt[3]{\frac{F^2}{DZ^2 \cos^5 \alpha}} \tag{6.14-1}$$

$$K = 0.0325E06 \sqrt[3]{DFZ^2 \cos^5 \alpha} \tag{6.14-2}$$

2) Self-Aligning Ball Bearings

$$\delta = 74.0E - 06 \sqrt[3]{\frac{F^2}{DZ^2 \cos^5 \alpha}} \tag{6.14-3}$$

$$K = 0.0203E06 \sqrt[3]{DFZ^2 \cos^5 \alpha} \tag{6.14-4}$$

3) Spherical Roller

$$\delta = 14.5E - 06 \sqrt[4]{\frac{F^3}{L^2 Z^3 \cos^7 \alpha}} \tag{6.14-5}$$

$$K = 0.0921E06 \sqrt[4]{FL^2 Z^3 \cos^7 \alpha} \tag{6.14-6}$$

4) Straight Roller or Tapered Roller

$$\delta = 3.71E - 06 \; \frac{F^{0.9}}{L^{0.8} Z^{0.9} \cos^{1.9} \alpha}$$ (6.14-7)

$$K = 0.300E06 \; F^{0.1} Z^{0.9} L^{0.8} \cos^{1.9} \alpha$$ (6.14-8)

where

δ	Radial Deflection
K	Radial Stiffness (Lbf/in)
F	External Radial Force (Lbf)
D	Ball Diameter (in)
Z	Number of Rolling Elements
L	Roller Effective Length (in)
α	Contact Angle (rad)

Note that the stiffness in the above equations is the tangent stiffness evaluated at the equilibrium position. These linearized bearing calculations are provided in **DyRoBeS-Rotor** under **TOOLS** menu.

Generalized Non-Linear Isotropic Bearing Model

In some cases, the load-deflection curve can be found by expensive tests and experiments. The generalized non-linear isotropic bearing model can be used to simulate the bearing when the load-deflection curve is known. This option also allows for the bearing clearance to be included in the analysis. At a time instance, the shaft center has a displacement of (x, y), as shown in Figure 6.14-1, and the tangential velocity of the shaft, having a radius of R, at the contact point is given by:

$$v_t = R\,\Omega + \left(-\dot{x}\sin\theta + \dot{y}\cos\theta\right) = R\,\Omega + \left(\frac{-\dot{x}y + \dot{y}x}{r}\right)$$ (6.14-9)

and

$$r = \sqrt{x^2 + y^2}$$ (6.14-10)

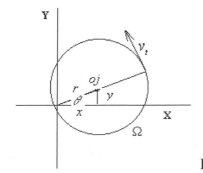

Figure 6.14-1 Contact tangential velocity

Let $(-F_r)$ be the radial restoring force acting on the shaft due to displacement which will be explained in details in the following, and $(-F_t)$ is the tangential force due to Coulomb Friction:

$$-F_t = \mu\,(-F_r)\,sign(v_t) \tag{6.14-11}$$

The total forces acting on the shaft due to displacement (r), friction (μ), and linear viscous damping (C) are:

$$F_x = (-F_r)\cos\theta - (-F_t)\sin\theta - C\dot{x}$$
$$F_y = (-F_r)\sin\theta + (-F_t)\cos\theta - C\dot{y} \tag{6.14-12}$$

Two types of equations are provided for the radial force in **DyRoBeS** and they are:

Case 1: Continuous Force-Deflection Curve

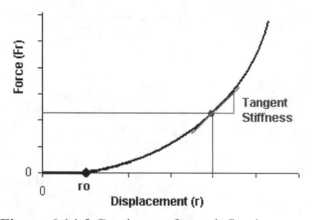

Figure 6.14-2 Continuous force-deflection curve

When $r < r_0$ (deadband or gap)

$F_r = 0,\ F_t = 0,\ F_x = 0,\ F_y = 0$ No forces are acting on the shaft

When $r_0 \leq r$, the force is approximated by a polynomial function

$$F_r = k_0(r-r_0)^{a_0} + k_1(r-r_0)^{a_1} + k_2(r-r_0)^{a_2} + k_3(r-r_0)^{a_3} + k_4(r-r_0)^{a_4}$$
$$F_t = \mu\,F_r\,sign(v_t) \tag{6.14-13}$$

Note that r_0 is zero if no gap or deadband exists.

Case 2: Piecewise Linear Curves

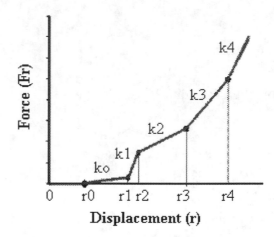

Figure 6.14-3 Piecewise linear force-displacement curve

When $r < r_0$ (deadband or gap)

$F_r = 0, \ F_t = 0, \ F_x = 0, \ F_y = 0$ No forces are acting on the shaft

When $r_0 \leq r < r_1$

$$F_r = k_0 (r - r_0)$$
$$F_t = \mu_0 \, F_r \, sign(v_t)$$

(6.14-14)

When $r_1 \leq r < r_2$

$$F_r = k_0 (r_1 - r_0) + k_1 (r - r_1)$$
$$F_t = \mu_1 \, F_r \, sign(v_t)$$

(6.14-15)

When $r_2 \leq r < r_3$

$$F_r = k_0 (r_1 - r_0) + k_1 (r_2 - r_1) + k_2 (r - r_2)$$
$$F_t = \mu_2 \, F_r \, sign(v_t)$$

(6.14-16)

When $r_3 \leq r < r_4$

$$F_r = k_0 (r_1 - r_0) + k_1 (r_2 - r_1) + k_2 (r_3 - r_2) + k_3 (r - r_3)$$
$$F_t = \mu_3 \, F_r \, sign(v_t)$$

(6.14-17)

and so on. Where k_i is the slope from r_i to r_{i+1}.

This option gives the user the ability to specify the bearing gap, contact stiffness, which is the soft mounting stiffness and the limit stiffness after a given deformation. This is the option that had been utilized for the AMB rotor drop transient analysis presented by Kirk, Gunter and Chen (2004). The following results are taken from the reference paper for the NEDO ISO test rotor drop simulation.

Example 6.7: Transient Rotor Drop Evaluation of AMB Machinery

The NEDO test rotor as modeled has a length of 0.851 m, mass of 25.963 kg and can operate at speeds to 30,000 rpm and higher. The diameter of the shaft at the AMB is 0.063 m and at the auxiliary bearing it is 0.0396 m. The rotor is shown in Figure 6.14-4 with the two bearing locations showing at the auxiliary bearing locations. The free-free mode is found to be 43,152 rpm and the associated mode shape is shown in Figure 6.14-5. The midspan imbalance is 5E-5 kg-m for the following drop case results. Only selected results from the reference are presented here.

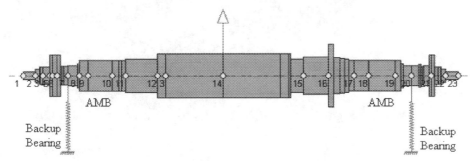

Figure 6.14-4 The NEDO ISO test rotor
(Photograph courtesy of Kirk et al., 2004)

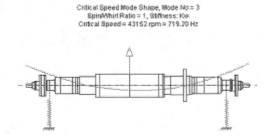

Figure 6.14-5 Test Rotor Free-Free Mode

The results of the rotor drop at 30,000 rpm and decelerated to 10,000 rpm is given in Figure 6.14-6, where the 3D orbits show a well behaved drop, with only a small side movement. This is confirmed by the time traces of the bearing force and response for the first bearing location, station 7, given in Figure 6.14-7. The orbit x-y motions at stations 7 and 20, bearing 1 and 2 respectively, show an initial bounce then rapid decay to a small motion as shown in Figure 6.14-8. A drop at below the free-free mode is considered as a good design for AMB support rotors.

The result of a drop at 50,000 and decelerated to 40,000 rpm, during a one second time period, is given in Figures 6.14-9, 10, 11. The 3D rotor response clearly shows large whirling and the orbit plots for bearing one shows the rotor motion taking the entire clearance gap of the auxiliary bearing. The bearing reaction and response versus time clearly indicate the danger of coming down through the free-free mode, even at the rate of 10,000 rpm per second. The peak force is in excess of 5,000 N, while the static load is only 128 N and numerous peaks of 2,000 N are observed for bearing one. The vertical motion goes from -172 μm to 121 μm. The gap clearance is only 100 μm and the soft spring retainer goes very stiff at 170 μm. A slow unloaded deceleration through this speed range would not be tolerated by the retainer bearings most likely.

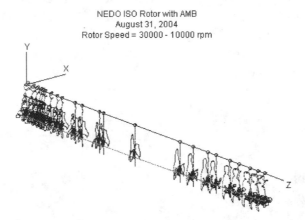

Figure 6.14-6 Test Rotor 3D Station Orbits for Drop from 30,000 to 10,000 rpm

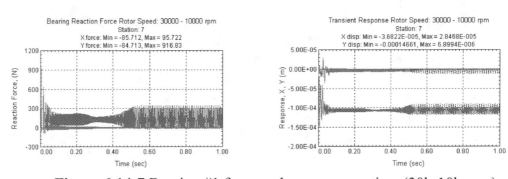

Figure 6.14-7 Bearing #1 force and response vs. time (30k-10k rpm)

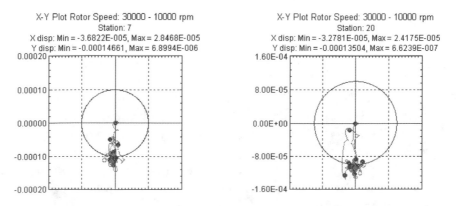

Figure 6.14-8 Bearings #1 and #2 response orbits (30k-10k rpm)

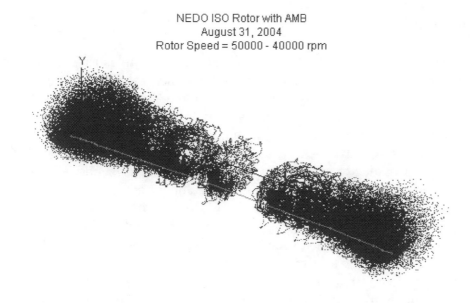

NEDO ISO Rotor with AMB
August 31, 2004
Rotor Speed = 50000 - 40000 rpm

Figure 6.14-9 Test Rotor 3D Station Orbits for Drop from 50 000 to 40 000 rpm

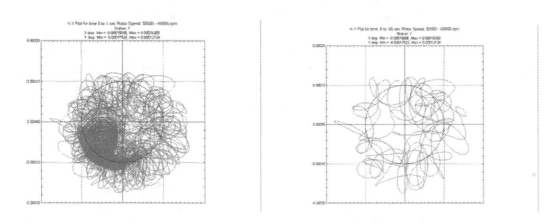

Figure 6.14-10 Bearing #1 response orbit (a) (50k-40k rpm) (b) around 43000 rpm

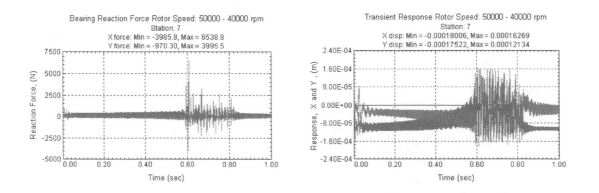

Figure 6.14-11 Bearing #1 force and response vs. time (50k-40k rpm)

6.15 Magnetic Bearings

Active magnetic bearings (AMB) have been used in many industrial applications where the lubricants are not allowed. Recently, due to the rapid growth in the high-speed motor applications, magnetic bearings become a part of the standard package. Two major advantages of the magnetic bearings compared with the fluid film bearings are oil-free and low power loss. However, the load carrying capability is lower, and it requires more space for the backup (touch down, or auxiliary) bearings and installation. The backup bearings must be provided to support the rotor in the event of mechanical disturbance or disruption of the active control system. In turbomachinery applications, the typical AMB stiffness ranges from 30E06 N/m (0.17E06 Lbf/in) to 50E06 N/m (0.286E06 Lbf/in), which is usually less than the stiffness of the same size conventional oil-film bearing. AMB damping is usually designed such that the amplification factor is below 2.5 for the rigid body modes. Both AMB stiffness and damping do not change with rotor rotational speed, but change with frequency of disturbance. In general, it is recommended to operate the machine with AMB below the first free-free bending mode, especially if the bearings or sensors are located close to the shaft nodes. Figure 6.15-1 illustrates the basic principle of AMB. The rotor motion is monitored by a displacement sensor, which is typically located very close to the bearing location. The magnetic bearing produces attractive magnetic forces based on the control schemes. The analog controller has been installed in many applications over the past years. The control logic follows the proportional, integral, and derivative schemes (PID control). However, most of the new magnetic bearing systems installed today, utilize digital controls with many features, such as Automatic Balancing System (ABS), or Automatic Vibration Rejection (AVR).

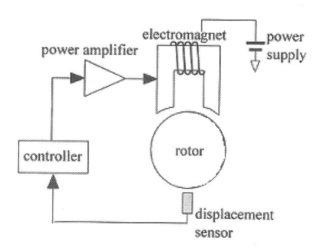

Figure 6.15-1 Basic principle of AMB

The load that can be carried by a magnetic bearing is lower than that of conventional oil-film bearing. This is usually limited by the magnetic saturation of the ferromagnetic material of stator and rotor core, the maximum coil current available from the power amplifier as shown in Figure 6.15-2.

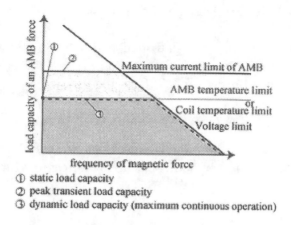

Figure **6.15-2** Limits on AMB (ISO publication)

Two options are used to model the active magnetic bearing in the current **DyRoBeS** (version 9.0). These options are basic and should be used with cautions. The linear Proportional-Integral-Derivative (PID) controller with low pass filter is used in the steady state analysis (Stability and Forced Response Analyses). The nonlinear active magnetic bearing requires more input data and is used in the non-linear transient analysis. For both options, the sensor stations may be different than the bearing stations (sensor non-collocation) and the model may be different for the two bearing axes.

Linear PID Controller with Low Pass Filter

This bearing is modeled as a PID controller in series with a unity gain first order low pass filter (generally used to model the amplifier). Two additional degrees-of-freedom will be added to each of the x and y equations to model the controller states for each bearing. The output of the PID controller at each axis is:

$$C_p x_s + C_i \int x_s dt + C_d \dot{x}_s \qquad (6.15\text{-}1)$$

Where x_s is the displacement at sensor location. The control force at each direction of the bearing location in the S-domain is:

$$F = \left(C_p + C_i \frac{1}{S} + C_d S \right) \left(\frac{2\pi f_c}{S + 2\pi f_c} \right) x_s \qquad (6.15\text{-}2)$$

where

C_p = proportional gain
C_i = integral gain
C_d = derivative gain
f_c = amplifier cut-off frequency

Non-Linear PID Model for Transient Analysis

This bearing is a standard PID controlled active magnetic bearing with sensor non-collocation, gap non-linearity and current saturation effects for the transient analysis only. The control current is determined from the following expression:

$$i_c = C_p x_s + C_i \int x_s dt + C_d \dot{x}_s \tag{6.15-3}$$

The currents supplied to the magnetic bearing are determined from the following:

$$i_1 = i_{b,p} - i_c \tag{6.15-4}$$

$$i_2 = i_{b,n} + i_c \tag{6.15-5}$$

$$\text{if } i < 0, \quad i = 0; \quad \text{if } i > i_{\text{limit}}, \quad i = i_{\text{limit}}$$

The force in the magnetic bearing is:

$$F = F_c \cdot \left[\left(\frac{i_1}{h_1} \right)^2 - \left(\frac{i_2}{h_2} \right)^2 \right] \tag{6.15-6}$$

where

$$h_1 = gap - x_b$$

$$h_2 = gap + x_b$$

$i_{b,p}, i_{b,n}$ = bias currents in positive and negative axes, respectively

x_s, x_b = displacements at sensor and bearing locations

F_c = force constant

Example 6.8: Rotor Levitation on AMB

A rotor system shown in Figure 6.15-3 is supported by two magnetic bearings at stations 2 and 7. Two backup bearings have radial clearances of 0.01 inches at stations 1 and 8. The rotor weighs 28.86 Lbf. The levitation of the rotor system before rotation is illustrated in this example. The bearing parameters are shown in Figure 6.15-4. The bearing axes are arranged with 45 degrees and 135 degrees from the fixed global X and Y axes respectively. The rotor is levitated before operation. The rotor is initially at rest on the backup bearings with a radial clearance of 0.01 inches. The initial conditions are given in file name with an extension .ics. Once the power is on, the rotor is quickly levitated to the bearing centerline as shown in Figure 6.15-5. The bearing forces shown in Figure 6.15-6 can also be calculated by force and moment balance for this simple static determined system. The parameters in the bearing number 1 are varied in this example. Figure 6.15-7 shows the effect of proportional gain on the rotor response. Figure 6.15-8 shows the effect of integral gain on the rotor response.   Figure 6.15-9 shows the effect of derivative gain on the rotor response.

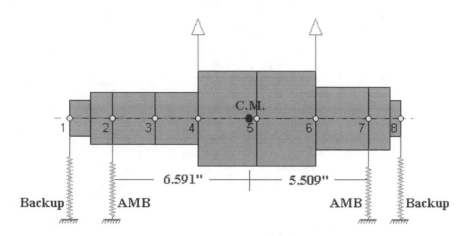

Figure 6.15-3 AMB rotor system

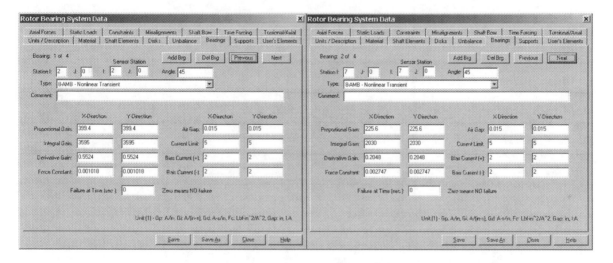

Figure 6.15-4 AMB parameters

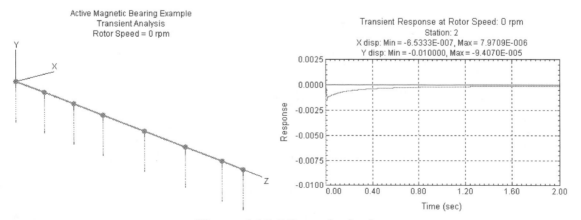

Figure 6.15-5 Rotor levitation

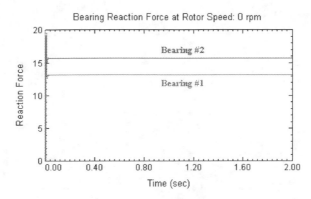

Figure 6.15-6 AMB forces

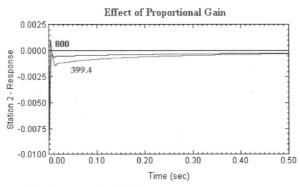

Figure 6.15-7 Effect of proportional gain

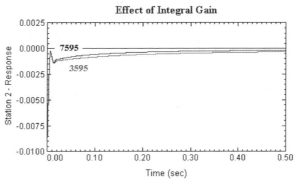

Figure 6.15-8 Effect of integral gain

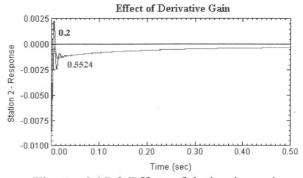

Figure 6.15-9 Effect of derivative gain

6.16 Liquid Annular Seals

The liquid annular seals used in the pumps are known to raise the "dry" critical speeds by a considerable amount. The "dry" critical speeds represent the calculation without process liquid. The "wet" critical speeds, on the other hand, are determined during the pump normal operation with the process liquid inside it. The wet critical speeds are often considerably different (usually higher) from the dry critical speeds due to the fluid interaction forces occurring at annual seals, and impeller-diffuser interfaces. Due to the pressure drop across the seal, the pressure difference develops a strong radial restoring force opposing to the shaft displacement. This effect is known as *Lomakin effect* which can be illustrated in Figure 6.16-1 taken from Florjancic and McCloskey (1991). Considering the rotor has a radial displacement (e) from the seal center and this eccentricity produces a smaller clearance ($C-e$) in the direction of the displacement and a larger clearance ($C+e$) in the opposite direction. Ignoring the effects of shaft rotation and inlet flow preswirl in this illustration. As the flow accelerates into the seal from its initial zero axial fluid velocity to the nominal velocity, a static pressure drop occurs due to the Bernoulli effect. The nominal axial velocity is slower in the region where the clearance is smaller and the pressure is higher in this low flow velocity region from Bernoulli effect. In addition to the pressure drop along the seal, an additional entrance loss occurs at the inlet region due to the development of the flow field into its fully developed form. The entrance loss is larger in the high clearance region than that in the low clearance region. After the initial pressure drop due to the entrance loss, the pressure drop due to the friction loss along the seal is nearly linear. This unsymmetric pressure distribution produces a restoring force acting on the rotor in the opposite direction to the rotor displacement.

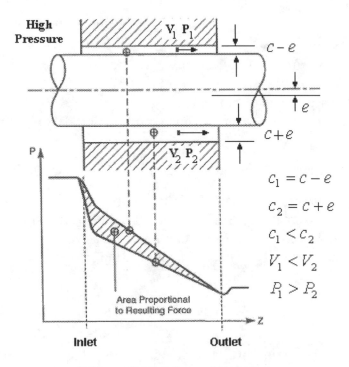

Figure 6.16-1 Lomakin effect

Black and Jenssen (1970) have extended Lomakin's theory in the development of the rotordynamics coefficients by using the bulk flow analysis.   Later Childs (1983) formulated a more complete bulk flow model using Hirs' lubrication equation, which includes the influence of fluid inertia terms and inlet swirl. Childs (1993) has extended his methods even further with more complicated model.

For small motion, the linearized forces acting on the rotor generated in an annular seal are:

$$\begin{Bmatrix} F_x \\ F_y \end{Bmatrix} = - \begin{bmatrix} K_d & k_c \\ -k_c & K_d \end{bmatrix} \begin{Bmatrix} x \\ y \end{Bmatrix} - \begin{bmatrix} C_d & c_c \\ -c_c & C_d \end{bmatrix} \begin{Bmatrix} \dot{x} \\ \dot{y} \end{Bmatrix} - \begin{bmatrix} m_d & 0 \\ 0 & m_d \end{bmatrix} \begin{Bmatrix} \ddot{x} \\ \ddot{y} \end{Bmatrix} \qquad (6.16\text{-}1)$$

The restoring forces are proportional to the displacement, velocity, and acceleration. In this aspect, the seals resemble fluid-film bearings discussed previously with additional inertia term. However, the governing equations for an annular seal are quite different from the "normal" turbulent Reynolds equations used in the bearing analysis.

A fundamental relationship between the total axial pressure drop and the mean axial flow velocity is:

$$\Delta P = \frac{1}{2}\left(1 + \xi + 2\sigma\right)\rho V^2 \qquad (6.16\text{-}2)$$

where

$$\sigma = \lambda \frac{L}{C} \qquad (6.16\text{-}3)$$

ξ is the entrance loss coefficient (typical value = 0.1)

ρ is the fluid density

V is the mean axial flow velocity

L is the seal length

C is the seal radial clearance

λ is the frication loss factor which is a function of Reynolds number

and $\frac{1}{2}\left(1 + \xi\right)\rho V^2$ defines the inlet pressure drop. It shows that the inlet pressure drop is greater in the region where the clearance is larger due to high velocity than the region where the clearance is smaller. Since σ is a nonlinear function of the mean axial velocity, the mean axial velocity must be determined iteratively.

The circumferential velocity of the fluid entering the seal is commonly expressed as a fraction of shaft surface speed ($R\Omega$):

$$u_c = \alpha R\Omega \qquad (6.16\text{-}4)$$

where α is the inlet swirl ratio. In Black's model (1970), the inlet swirl ratio is 0.5. The value can be lower if a swirl break is used. In Childs' model (1983), the inlet swirl ratio can be varied. It can be seen that inlet swirl has a significant influence on the cross-coupled stiffness and that therefore stability can be improved by reducing the inlet swirl.

There are two ways to include the seal dynamic coefficients in ***DyRoBeS***. One explicit approach is using the ***TOOLS*** in the main menu. One can enter the seal data, total pressure drop, speed, and calculation method (Black or Childs), the dynamic coefficients and leakage rate are calculated and shown in the same dialogue box. Since the pressure drop depends upon the rotor speed, it is convenient to include the seal calculation inside the rotordynamics code. Therefore, another implicit approach is to enter the seal data under ***BEARING*** tab in the rotor bearing data editor (bearing type 11), the seal dynamic coefficients then are calculated inside the rotordynamics code for various analyses.

6.16.1 Black Model

The seal dynamic coefficients based on Black's model are well documented in Black & Jenssen (1970), Barrett (1984), and Corbo and Malanoski (2003) and are listed here for reference:

$$K_d = \mu_3 \left(\mu_0 - 0.25 \mu_2 \Omega^2 T^2 \right) \tag{6.16-5}$$

$$k_c = \mu_3 \left(0.5 \mu_1 \, \Omega T \right) \tag{6.16-6}$$

$$C_d = \mu_3 \, \mu_1 \, T \tag{6.16-7}$$

$$c_c = \mu_3 \, \mu_2 \, \Omega T^2 \tag{6.16-8}$$

$$m_d = \mu_3 \, \mu_2 \, T^2 \tag{6.16-9}$$

with

$$\mu_0 = \frac{(1+\xi)\sigma^2}{(1+\xi+2\sigma)^2} \tag{6.16-10a}$$

$$\mu_1 = \frac{(1+\xi)^2 \sigma + (1+\xi)(2.33+2\xi)\sigma^2 + 3.33(1+\xi)\sigma^3 + 1.33\sigma^4}{(1+\xi+2\sigma)^3} \tag{6.16-10b}$$

$$\mu_2 = \frac{0.33(1+\xi)^2(2\xi-1)\sigma + (1+\xi)(1+2\xi)\sigma^2 + 2(1+\xi)\sigma^3 + 1.33\sigma^4}{(1+\xi+2\sigma)^4} \tag{6.16-10c}$$

$$\mu_3 = \frac{\pi R \Delta P}{\lambda} \tag{6.16-10d}$$

$$\lambda = 0.079\, R_a^{-0.25}\left[1+\left(\frac{7R_c}{8R_a}\right)^2\right]^{0.375} \quad \text{Friction Loss Factor} \tag{6.16-10e}$$

$$R_a = \frac{2\rho VC}{\mu} \qquad \text{Axial Reynolds number} \tag{6.16-10f}$$

$$R_c = \frac{\rho R\Omega C}{\mu} \qquad \text{Circumferential Reynolds number} \tag{6.16-10g}$$

$$\mu \qquad \text{is the fluid viscosity}$$

$$T = \frac{L}{V} \tag{6.16-10h}$$

The above parameters (μ_0, μ_1, μ_2) are based on short seal solution. The corrected parameters for finite length seal are (Black & Jenssen, 1970):

$$\mu_{0,c} = \frac{\mu_0}{1+0.28(L/R)^2} \tag{6.16-11a}$$

$$\mu_{1,c} = \frac{\mu_1}{1+0.23(L/R)^2} \tag{6.16-11b}$$

$$\mu_{2,c} = \frac{\mu_2}{1+0.06(L/R)^2} \tag{6.16-11c}$$

Since the friction loss factor is a function of Reynolds number that is dependent upon the velocity, therefore, an iterative process is needed to solve the mean axial velocity.

Example taken from Barrett (1984) is used to demonstrate this calculation. The inputs and calculated results are shown in Figure 6.16-2. The small differences between **DyRoBeS** and previous publication are due to:

1. **DyRoBeS** uses finite length corrected parameters. However, short seal theory was used in the reference. It is known that short seal solutions tend to overestimate the dynamic coefficients of finite length seals.

2. In reference paper, although iteration was used to determine the average axial velocity, however, the initial σ was used to calculate the dynamic coefficients. In **DyRoBeS**, the final converged σ is used in the related calculation.

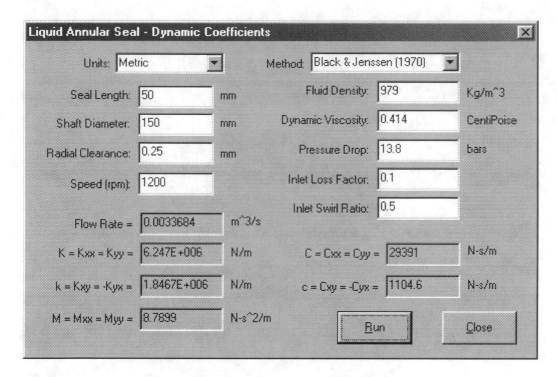

Figure 6.16-2 Liquid seal input and output

6.16.2 Childs Model

Childs (1983) formulated the seal dynamic coefficients based on Hirs' lubrication equation. The fluid inertia terms are included in the momentum equations and the inlet swirl is also included. The short seal theory is used. The coefficients are summarized in the following:

$$K_d = \left(\frac{\pi R \Delta P}{\lambda}\right) \frac{2\sigma^2}{(1+\xi+2\sigma)}\left\{1.25E - \frac{(\Omega T)^2}{4\sigma}\left\{\frac{1}{2}\left(\frac{1}{6}+E\right) + \frac{2v_0}{a}\left[\left(E+\frac{1}{a^2}\right)(1-e^{-a}) - \left(\frac{1}{2}+\frac{1}{a}\right)e^{-a}\right]\right\}\right\}$$

$$k_c = \left(\frac{\pi R \Delta P}{\lambda}\right) \frac{\sigma^2 \Omega T}{(1+\xi+2\sigma)}\left\{\frac{E}{\sigma}+\frac{B}{2}\left(\frac{1}{6}+E\right) + \frac{2v_0}{a}\left\{EB+\left(\frac{1}{\sigma}-\frac{B}{a}\right)\left[\left(1-e^{-a}\right)\left(E+\frac{1}{2}+\frac{1}{a}\right)-1\right]\right\}\right\}$$

$$C_d = \left(\frac{\pi R \Delta P}{\lambda}\right) \frac{2\sigma^2 T}{(1+\xi+2\sigma)}\left[\frac{E}{\sigma}+\frac{B}{2}\left(\frac{1}{6}+E\right)\right]$$

$$c_c = \left(\frac{\pi R \Delta P}{\lambda}\right) \frac{2\sigma \Omega T^2}{(1+\xi+2\sigma)}\left\{\frac{1}{2}\left(\frac{1}{6}+E\right) + \frac{v_0}{a}\left[\left(1-e^{-a}\right)\left(E+\frac{1}{2}+\frac{1}{a^2}\right) - \left(\frac{1}{2}+\frac{e^{-a}}{a}\right)\right]\right\}$$

$$m_d = \left(\frac{\pi R \Delta P}{\lambda} \right) \frac{\sigma \left(\frac{1}{6} + E \right)}{\left(1 + \xi + 2\sigma \right)} T^2$$

(6.16-12)

where

$$\lambda = 0.066 \, R_a^{-0.25} \left(1 + \frac{1}{4b^2} \right)^{0.375} \text{ Friction Loss Factor}$$

$$R_a = \frac{\rho VC}{\mu} \quad \text{Axial Reynolds number (note: it is a half of the Reynolds number}$$

defined by Black)

$$R_c = \frac{\rho R \Omega C}{\mu} \text{ Circumferential Reynolds number}$$

$$b = \frac{R_a}{R_c} = \frac{V}{R\Omega}$$

$$a = \sigma \left(1 + 0.75\beta \right)$$

$$\beta = \frac{1}{1 + 4b^2}$$

$$B = 1 + 4b^2 \beta \left(0.75 \right)$$

$$E = \frac{\left(1 + \xi \right)}{2 \left(1 + \xi + B\sigma \right)}$$

$$T = \frac{L}{V}$$

An inter-stage seal of the High Pressure Hydrogen Turbopump (HPFTP) of the Space Shuttle Main Engine (SSME) used in reference paper is presented below as another example. For an inlet swirl ratio of 0.5 (the inlet circumferential velocity is a half of the surface speed), the inputs and calculated results are listed in Figure 6.16-3. Figure 6.16-4 shows the input and output for an inlet swirl ratio, $u_c = 0$. Note that in the reference paper, $v_0 = 0$ indicates $u_c = 0.5 \, R\Omega$ and $v_0 = -0.5$ indicates $u_c = 0$. Typical value for the Inlet Swirl Ratio is 0.5; that is, the circumferential velocity of the fluid entering the seal is a half of the shaft surface speed. However, a swirl break can be used to lower the swirl ratio in order to improve the rotor stability by decreasing the seal cross-coupling stiffness as shown in the summary Table 6.16-1. It shows that by eliminating the inlet swirl ratio, the direct stiffness is about the same, however, the cross-coupling stiffness has decreased by more than 50%.

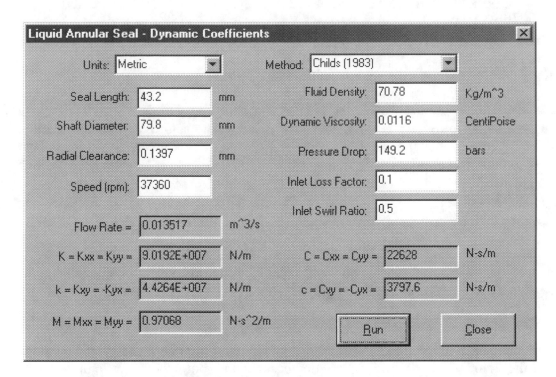

Figure 6.16-3 Seal input and output for 0.5 inlet swirl ratio

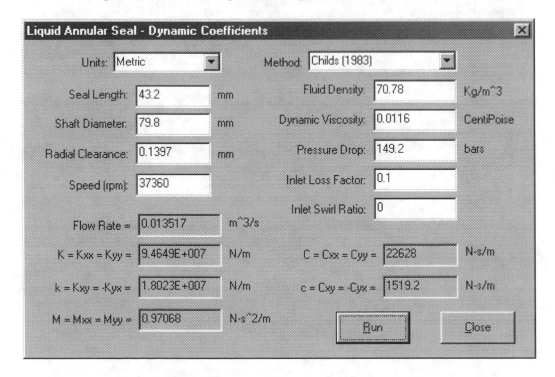

Figure 6.16-4 Seal input and output for zero inlet swirl ratio

	Inlet Swirl Ratio = 0.5	Inlet Swirl Ratio = 0.0
K_d (direct stiffness)	9.0192E07	9.4649E07
k_c (cross-coupling stiffness)	4.4264E07	1.8023E07
C_d (direct damping)	22628	22628
c_c (cross-coupling damping)	3797.6	1519.2

Table 6.16-1 Result summary

Note that two theoretical methods (Black model and Childs Model) for the liquid annual seals are provided in **DyRoBeS** to be used in the rotordynamics calculation. In most publications the theoretical models are compared with the test experiments and good corrections were obtained, although certain differences remain. Further differences are evident in the industrial application; therefore, the precaution must be taken while using these coefficients. A tolerance on these calculated coefficients should be applied when the accurate critical speed and stability margins are critical. A tolerance of +-30% is recommended by Kramer (1993).

6.16.3 Data input under *Bearing* Tab

The seal dynamic coefficients can be calculated internally when performing the rotordynamics analyses. Due to the similarity with bearing data, the seal data are entered under the **Bearing** tab. A brief description of the data input is listed in Figure 6.16-5:

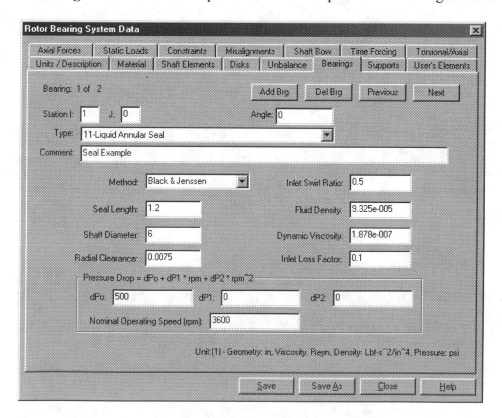

Figure 6.16-5 Seal data entered as a pseudo bearing

The inlet pressure drop as the fluid entering the seal is defined by $\frac{1}{2}(1+\xi)\rho V^2$.

Since the total pressure drop (or the pump discharge pressure) is a function of rotor speed, therefore, a second order polynomial is used to calculate the seal total pressure drop for a given speed in **DyRoBeS**.

6.17 Aerodynamic Cross Coupling Force

The aerodynamic excitation caused by the impeller clearance variation is a destabilizing force with a form:

$$\begin{Bmatrix} F_x \\ F_y \end{Bmatrix} = -\begin{bmatrix} 0 & K_{xy} \\ K_{yx} & 0 \end{bmatrix}\begin{Bmatrix} x \\ y \end{Bmatrix} \tag{6.17-1}$$

Two most commonly used methods are Alford (1965) and Wachel (1983) equations. In general, the value calculated from Alford equation provides a lower limit and the value calculated from Wachel equation provides an upper limit. These equations have been used as design criteria for improved stability of centrifugal compressors by Kirk and Donald (1983).

Alford has proposed the cross-coupled stiffness:

$$K_{xy} = -K_{yx} = \frac{\beta T}{D H} \tag{6.17-2}$$

where

K_{xy}	cross coupled stiffness (Lbf/in)
T	stage torque (lbf-in)
D	blade pitch (mean) diameter (in)
H	blade height (in)
β	efficiency factor (design parameter)

The Alford equation was originally proposed for axial flow turbines and compressors. It has also been also used for centrifugal compressors. The value for the efficiency factor (β) for different types of machines has been suggested by Researchers at Texas A&M University and listed below for reference:

= 0.5	for shrouded axially bladed disks
= 1.5	for un-shrouded axially bladed disks
= 2 - 3	for un-shrouded radial flow impellers
= 5 - 10	for extreme cases, overhung impellers

Wachel has proposed an empirical formula for estimating the aerodynamic cross-coupled stiffness based on the several instability problems occurred in the practical applications:

$$K_{xy} = -K_{yx} = \frac{6300 \; hp \; Mw}{D \; H \; Rpm} \; \cdot \; \frac{\rho_d}{\rho_s}$$ (6.17-3)

where

K_{xy}	cross coupled stiffness (Lbf/in)
hp	stage horsepower
Mw	molecular weight
D	impeller diameter (in)
H	restrictive dimension in flow path (in)
Rpm	stage speed
ρ_d	density of fluid at discharge
ρ_s	density of fluid at suction

These aerodynamic cross-coupling calculations are implemented in **DyRoBeS-Rotor** under the **TOOLS** menu as shown in Figure 6.17-1.

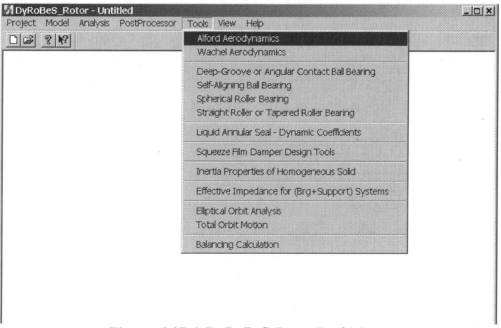

Figure 6.17-1 DyRoBeS-Rotor Tool Menu

System Governing Equations and Analyses

7

7.1 Governing Equations of Motion

The modeling of complicated rotating machinery is not an exact science, but depends on good engineering judgment and practical experience. The modeling process transforms the complex physical system into a representative, hopefully simple, mathematical model. The components of the system under investigation must be identified, and the governing equation for each component is readily available. The assumptions and simplifications made on the formulation of the components must be fully understood in order to properly utilize these equations and interpret the analysis results. The equations of motion that describe the dynamic behavior of the entire system are obtained by assembling the equations of motion of the appropriate components. The governing equation of motion for a general rotor bearing system, with a constant spin speed, Ω, is:

$$M\,\ddot{q}(t) + \left(C_b + \Omega G\right)\dot{q}(t) + \left(K_s + K_b\right)q(t) = Q(\dot{q}, q, \Omega, t) \tag{7.1-1}$$

It can also be written in a more commonly used standard form for vibration problems:

$$M\,\ddot{q}(t) + C\,\dot{q}(t) + K\,q(t) = Q(\dot{q}, q, \Omega, t) \tag{7.1-2}$$

Where q is the system displacement vector to be solved. The mass/inertia matrix M is a positive definite real symmetric matrix contributed by the kinetic energy. The gyroscopic matrix G is a real skew-symmetric matrix, which is contributed by the part of the rotational kinetic energy caused by the gyroscopic moments. The structural stiffness matrix K_s derived from the strain energy is symmetric and conservative in nature. In rare cases, where the axial torque is included in the modeling, the stiffness matrix, due to the axial torque, is non-symmetric (circulatory) and non-conservative. The matrices C_b and K_b are the linearized bearing (or bearing like) dynamic damping and stiffness matrices. In general, they are non-symmetric real matrices, due to the fact that bearings are non-conservative in nature. The forcing vector Q contains all the forcing functions, such as synchronous excitations due to mass unbalance, shaft bow, and skew disks, and constant gravity loads and static loads, and any applied forcing functions due to blade loss, sudden loading, and all the nonlinear interconnection forces.

In some applications, there are needs to study the rotor motion during startup, shutdown, going through critical speeds, or rotor drop for the magnetic bearing systems. In these situations, the angular velocity (spin speed) is no longer a constant and is a function of time. The governing equations of motion for a variable rotational speed system are:

$$
\boldsymbol{M}\ddot{\boldsymbol{q}}(t) + \left[\boldsymbol{C}_b + \dot{\varphi}\,\boldsymbol{G}\right]\dot{\boldsymbol{q}}(t) + \left[\left(\boldsymbol{K}_s + \boldsymbol{K}_b\right) + \ddot{\varphi}\,\boldsymbol{G}\right]\boldsymbol{q}(t)
$$
$$
= \dot{\varphi}^2\,\boldsymbol{Q}_1(\varphi) + \ddot{\varphi}\,\boldsymbol{Q}_2(\varphi) + \boldsymbol{Q}_3(\dot{\boldsymbol{q}}, \boldsymbol{q}, \varphi, \dot{\varphi}, \ddot{\varphi}, t)
$$

(7.1-3)

where $\varphi, \dot{\varphi},$ and $\ddot{\varphi}$ are the angular displacement, angular velocity ($\dot{\varphi} = \Omega$), and angular acceleration (deceleration) of the rotor system. Two more terms introduced in the governing equations due to the speed variation are: circulatory matrix $\ddot{\varphi}\boldsymbol{G}$ and forcing function $\ddot{\varphi}\boldsymbol{Q}_2$. The equations of motion for a variable rotational speed system, Eq. (7.1-3), are nonlinear and only are used in the time transient analysis.

In most applications, the nonlinear interconnection forces are linearized around the static equilibrium position as discussed in Chapter 6 and linear systems are employed in the analysis. In some applications, the nonlinear forces cannot be linearized, such as non-centered squeeze film damper forces. In this case, nonlinear transient analysis is required to study the dynamics of the rotor behavior. The dynamic characteristics of primary interest in the linear rotor-bearing systems are: 1) critical speeds and associated mode shapes and energy distribution, 2) system stability and damped natural frequencies of whirl and associated mode shapes, 3) steady state synchronous response and transmitted loads, and 4) transient response due to general time forcing functions. For the nonlinear systems, the time transient analysis is commonly used to study the dynamic behaviors. Although the determination of rotor static deflection and bearing reactions is not "dynamics" in theory, it is however frequently required in order to perform the bearing analysis and determine the linearized bearing coefficients. Therefore, the following analyses are discussed:

- Static deflection and bearing loads
- Critical speed analysis
- Critical speed map
- Whirl speed and stability analysis
- Steady state synchronous response for general linear systems
- Steady state synchronous response for generalized non-linear isotropic systems
- Time transient analysis

7.2 Static Deflection and Bearing Loads

For a static loading, the system excitation vector $\boldsymbol{Q}$ is constant. The associated static response (deflections) can be obtained from the constrained stiffness matrix and force vector:

$$q = K^{-1}Q \qquad\qquad (7.2\text{-}1)$$

Once the static displacements are obtained from the above equation, the bearing reaction forces, shaft element internal shear forces and bending moments can be determined accordingly.   In **DyRoBeS**, this option calculates the shaft static deflection and bearing/constraint reaction forces for the static determinate and indeterminate problems. The finite element stations, where linear bearings are located, can be either constrained (zero displacements) or flexible with specified linear bearing stiffnesses being used in the calculation.   For speed-dependent bearing coefficients, the bearing stiffness at the lowest speed is used in **DyRoBeS**.   The finite element stations, where the non-linear bearings, dampers and magnetic bearings are located, are constrained.   The external static loads, gravity, constraints, and misalignment are included in this analysis.

Figure 7.2-1 shows the static deflection of a centrifugal compressor shaft due to gravity.   The rotor assembly weighs 26.16 Lb and is supported by two identical bearings with a stiffness of 2.0E06 Lb_f/in.   The maximum deflection of 3.4E-05 inches occurs at the impeller.   The reaction forces at the bearings are 8.01 Lb at bearing number 1 and 18.15 Lb at bearing number 2.   The sum of the bearing reactions is equal to the total rotor assembly weight of 26.16 Lb.

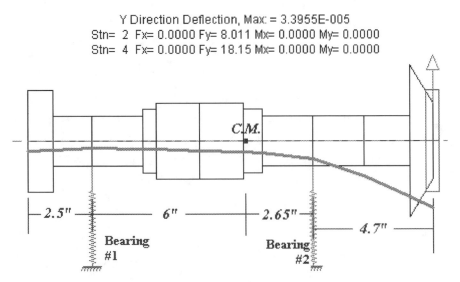

Y Direction Deflection, Max: = 3.3955E-005
Stn= 2 Fx= 0.0000 Fy= 8.011 Mx= 0.0000 My= 0.0000
Stn= 4 Fx= 0.0000 Fy= 18.15 Mx= 0.0000 My= 0.0000

Figure 7.2-1 Rotor static deflection and bearing loads

7.3 Critical Speeds and Modes

The primary consideration in the design of rotor systems is the placement of forward synchronous critical speeds with respect to the operating speed of the machine.   The other dynamic behaviors are all somewhat related to the position of the critical speeds.   When the critical speeds are within the operating speed range, the rotor may experience large synchronous vibration.   When the rotor is operated far above the critical speeds, the rotor may be susceptible to instability and experience high sub-synchronous vibration.   When the rotor is operated far below the rigid body critical speed, it can be sensitive to the operating environments, due to the rigid body motion. Since damping has little effect on the position of critical speeds, the undamped isotropic system is usually used for critical speed calculation. When damping is significant, the equivalent dynamic stiffness should be used in the critical speed analysis.

$$k_d = \sqrt{k^2 + (\omega c)^2}$$
(7.3-1)

Due to the simplifying assumptions applied in the critical speed calculation, extreme care must be taken in the preparation and interpretation of these results.

For undamped rotor systems with isotropic supports, the critical speeds may be determined directly from a reduced eigenvalue problem associated with the system equations expressed in a rotating reference frame. These undamped modes are circular relative to the fixed reference frame, but are constant relative to the rotating reference frame. Therefore, it is convenient to consider only one of the two planes of motion. This procedure will simplify the calculation and reduce computation time. The undamped natural circular whirl speeds and mode shapes can be obtained from the homogeneous form of system equations expressed in the rotating reference frame. Assuming a constant eigensolution, the reduced eigenvalue problem in the (XZ) plane is:

$$\left(K_{XZ} - \omega_i^2 M_{XZ} \right) y_i = 0$$
(7.3-2)

where ω_i^2 $(i = 1,2,....n)$ are called eigenvalues and ω_i are recognized as the natural frequencies of the system or critical speeds in this case. The vectors y_i are known as the associated (right) eigenvectors or mode shapes. By removing the damping matrix and circulatory matrix due to the non-conservative forces, the matrices K_{XZ}, M_{XZ} are all symmetric and are assembled from all the appropriate components, such as shaft elements (e), disks (d), bearings (b), and flexible supports (f):

$$K_{XZ} = \sum_{XZ} K^e + \sum_{XZ} K^d + \sum_{XZ} K^b + \sum_{XZ} K^f$$
(7.3-3)

$$M_{XZ} = \sum_{XZ} \left[M_T^e + (1-2\gamma)M_R^e \right] + \sum_{XZ} \left[\left(M_T^d + M_R^d \right) - \gamma \hat{G}^d \right] + \sum_{XZ} M^f$$
(7.3-4)

and

$$\gamma = \frac{\Omega}{\omega} \qquad \text{is the spin-whirl ratio.}$$

All the matrices are self-explanatory and described in previous chapters. The subscript (XZ) denotes that only (XZ) plane of motion is considered, and the symbol Σ represents the assembly procedure from the elemental to the system level. The spin-whirl ratio in the reduced mass matrix is due to the gyroscopic effect. The spin/whirl ratio is specified for the determination of various types of critical speeds. The typical values and the type of critical speeds are:

$\gamma = 1$ Forward synchronous critical speeds

$\gamma = -1$ Backward synchronous critical speeds

$\gamma = 0$ Planar critical speeds for non-rotating systems

$\gamma = 2$ Half frequency whirl (sub-synchronous critical speeds)

$\gamma = n$ Forward super-synchronous critical speeds

The forward synchronous critical speeds and modes are the most commonly calculated due to mass imbalance in the rotor system. Depending on the choice of the spin-whirl ratio, and the values of the rotor moments of inertia, the matrix M_{xz} may not be positive definite. Under these circumstances, it does not have solutions, which correspond to situation when the gyroscopic effect is such that it prevents intersecting of the excitation line with one or more of the natural frequencies of whirl curves. This phenomenon was explained in Chapters 3 and 4.

Since the matrices M and K are both real and symmetric, the eigenvalues and eigenvectors are real. For a constrained system, the K is positive definite and all the eigenvalues are positive. When the system is unconstrained, which is usually used to determine the free-free modes, the K is positive semidefinite (singular matrix), and some of the eigenvalues are zero, while the remaining ones are positive. Zero eigenvalues occur when the system is unconstrained, in which case the associated modes are called *rigid body modes* and the structure can move as a rigid body without elastic deformation. In order to utilize the standard QR algorithm, an eigenvalue shift scheme is necessary to avoid the computational difficulties for rigid body modes.

Whenever possible, it is desirable to have at least 15% (10% is required in most Standards) separation margin between the operating speed and the critical speeds, to ensure safe and smooth operation. When the critical speeds are within the undesirable range, some parameters need to be adjusted to shift the critical speeds outside this operating range. Bearing stiffnesses, mass properties of the disks, and shaft elements of the rotating assemblies are usually the variables that can be changed to achieve this requirement. Bearing locations have a great influence in positioning the critical speeds, however, they cannot be easily changed in an existing design without major modifications in the layout and general arrangement.

If the bearing stiffnesses are the only variables that can be changed, then one should look to those bearings with high potential energy density. If a particular bearing has very small potential energy density (e.g. less than 5%) of a particular mode, then minor modifications on this bearing have little effect on that mode. In fact, increasing the bearing stiffness will have an adverse effect, due to the reduction in the modal damping. An increase or decrease in the diameter of the shaft elements with large potential energy density can be used to effectively raise or lower the critical speed. Decreasing or increasing the mass of a disk with high translational kinetic energy density can also significantly increase or decrease the critical speed. Caution must be taken to ensure that the center of gravity of the disk will not be moved while changing the mass properties of the disk. Also, decreasing or increasing the polar moment of inertia of a disk, which has high rotational kinetic energy density, can decrease or increase the synchronous critical speed due to the gyroscopic effects. If the kinetic and potential energies of a vibration mode are significant in the flexible bearing supports, changing the support structure can be very effective in shifting that associated frequency. For the support modes, the noncontact displacement probes, measuring the shaft vibration may not be enough for a safe monitoring system and an accelerometer or velocity pickup on the bearing housing may be required.

The potential energy distribution among the rotor assembly, bearings, and flexible supports can also provide information for design purposes. If the rotor assembly (shaft elements) has more than 70% of the total potential energy of a vibration mode, then this vibration mode is characterized as a *flexible rotor mode*. If the rotor assembly (shaft elements) has less than 30% of the total potential energy of a vibration mode, then this vibration mode is characterized as a *rigid rotor mode*. Typically, the first two lowest frequency modes are the rigid rotor modes and the other high frequency modes are flexible rotor modes. For some flexible rotor systems, the rigid body modes do not exist. In some cases, the rigid rotor modes could be overdamped or could not be excited by synchronous excitation, thus they will not be observed in the response data. For rotors operated above the rigid bearing critical speeds (e.g. most high speed compressors), it is desirable to design the bearings so that the potential energy of the critical speed can be evenly distributed among the rotor and bearings. If the critical speed falls into the rigid rotor section, the system will be more susceptible to instability (e.g. oil whirl or whip) and vibration could be large enough in the bearing station to damage the bearing. If the critical speed falls into the flexible rotor section, vibration may be very large at some critical stations due to the lack of damping.

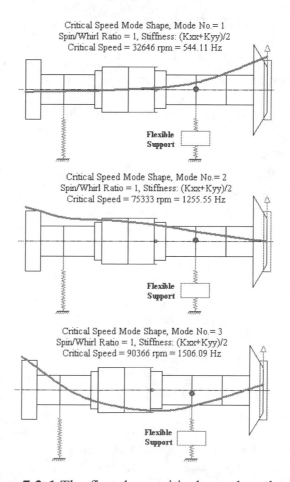

Figure 7.3-1 The first three critical speed mode shapes

Figure 7.3-1 shows the mode shapes for the first three forward synchronous critical speeds of the centrifugal compressor. In this case, a flexible support is assumed at bearing number 2. The compressor design speed is 45,000 rpm. It shows that the compressor is operated above the first critical speed. The associated energy distributions of first two modes are shown in Figure 7.3-2. In the first mode, there is about 61 percent of potential energy in the shaft and about 33 percent in the bearings. The flexible support has about 6 percent potential energy. The majority of the kinetic energy is in the impeller disk. Interestingly, the shaft has about 11 percent potential energy in the second mode and bearings have about 86 percent potential energy. Furthermore almost all the kinetic energy is in the shaft for this mode. These results can also be qualitatively observed from the mode shapes. In the kinetic energy distribution, "T" represents the kinetic energy due to mass translational motion, "R" represents the kinetic energy due to rotatory inertia, and "G" represents the kinetic energy due to gyroscopic moment. It shows the negative kinetic energy for the forward modes comes from the gyroscopic moment.

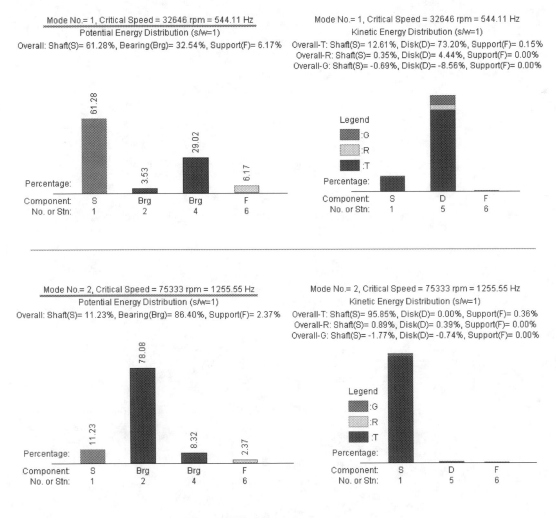

Figure 7.3-2 Energy distribution

Critical Speed Map

The critical speed map is an outgrowth of the critical speed analysis. It calculates the critical speeds for a range of bearing stiffness, assuming the bearings are connected to a rigid foundation. This map is used as a check for the design of the shaft and the bearings, from a stiffness point of view. Figure 7.3-3 shows the classical critical speed map in which the bearing stiffnesses of both bearings are the same and are varied simultaneously. Sometimes only one bearing can be changed and the other bearing stiffnesses are fixed. In this case, the critical speed map can be used as parametric study. Figure 7.3-4 shows the critical speeds verses the stiffness of bearing number 2 while the stiffness of bearing number 1 is fixed at a value of 2.0E06 Lbf/in. **DyRoBeS** allows the user to select the bearings where the bearing stiffnesses are to be varied in the analysis.

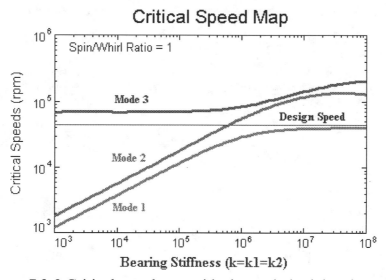

Figure 7.3-3 Critical speed map with change in both bearing stiffnesses

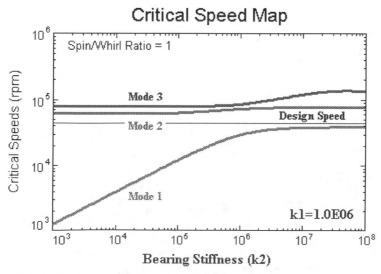

Figure 7.3-4 Critical speed map with change in bearing number 2 stiffness

7.4 Whirl Speeds and Stability Analysis

The whirl speeds and system stability are determined from the homogeneous form of the equations of motion. The linearized equations of motion describing the free vibration of a general linear damped rotor dynamics system are:

$$M\ddot{q}(t) + C\dot{q}(t) + Kq(t) = 0 \qquad (7.4\text{-}1)$$

where all the matrices are real and M is positive definite. The matrices C and K can be any real matrices due to the non-conservative forces of the interconnection components (bearings, seals, etc). Since most of the eigenvalue solvers require the matrix to be in the first order form, it is convenient to write the second order homogeneous equation in the state space form:

$$A\dot{x} + Bx = 0 \qquad (7.4\text{-}2)$$

where the matrices A, B and the vector x are defined as:

$$A = \begin{bmatrix} M & 0 \\ 0 & I \end{bmatrix}, \qquad B = \begin{bmatrix} C & K \\ -I & 0 \end{bmatrix}, \qquad x = \begin{Bmatrix} \dot{q} \\ q \end{Bmatrix} \qquad (7.4\text{-}3)$$

On seeking the eigensolution of the exponential form:

$$x(t) = y\, e^{\lambda t} \qquad (7.4\text{-}4)$$

Introducing the eigensolution into the state space form and dividing through by $e^{\lambda t}$, we obtain the eigenvalues problem:

$$(\lambda A + B)\, y = 0 \qquad (7.4\text{-}5)$$

Since matrix A is a positive definite real symmetric matrix and B is an arbitrary real matrix, the eigenvalue problem can be reduced to the standard format:

$$(-A^{-1}B)\, y = \lambda\, y \qquad (7.4\text{-}6)$$

where

$$(-A^{-1}B) = \begin{bmatrix} -M^{-1}C & -M^{-1}K \\ I & 0 \end{bmatrix} \qquad (7.4\text{-}7)$$

$(-A^{-1}B)$ is an arbitrary real matrix. If the eigenvalues λ are complex, then they must occur in pairs of complex conjugates, as do the eigenvectors y. The eigenvalues and eigenvectors have the forms:

$$\lambda = \sigma + j\omega \qquad\qquad \lambda^* = \sigma - j\omega \tag{7.4-8}$$

$$\boldsymbol{y} = \left\{ \begin{array}{c} \lambda\boldsymbol{u} \\ \boldsymbol{u} \end{array} \right\} \qquad\qquad \boldsymbol{y}^* = \left\{ \begin{array}{c} \lambda\boldsymbol{u}^* \\ \boldsymbol{u}^* \end{array} \right\} \tag{7.4-9}$$

where the real parts of the eigenvalues, σ, are the system *damping exponents*, which are used to determine the system stability. A positive damping exponent indicates system instability in the linear sense. The imaginary parts of the eigenvalues, ω, are the system damped natural frequencies. They are also referred to as *whirl speeds* or *whirl frequencies*. If the damped natural frequency is a positive value, this mode is referred to as a *precessional mode* with an oscillating frequency that equals the damped natural frequency. If the damped natural frequency equals zero, this mode is referred to as a *real mode* or non-oscillating mode.

Very often the logarithmic decrements or damping factors are used to express the degree of the system stability. A negative logarithmic decrement or damping factor indicates system instability. When the value of a logarithmic decrement exceeds 1, that particular mode is considered to be well-damped. The logarithmic decrement for a precessional mode is defined as:

$$\delta = \frac{-2\pi\sigma}{\omega} \tag{7.4-10}$$

The damping factor is:

$$\xi = \frac{\delta}{\sqrt{(2\pi)^2 + \delta^2}} \tag{7.4-11}$$

The undamped natural frequency is:

$$\omega_n = \frac{\omega}{\sqrt{1 - \xi^2}} \tag{7.4-12}$$

In most applications, if the system becomes unstable, the rotor whirls in its first mode (lowest whirl frequency) with a forward precession. It should be noted that the first unstable mode could be either a rigid body mode or a bending mode, depending on the potential energy distribution. The precessional mode solution for each pair of complex conjugate eigenvalues is defined as the sum of the associated complex eigenvectors:

$$\begin{aligned} \overline{\boldsymbol{u}}(t) &= e^{\lambda t}\boldsymbol{u} + e^{\lambda^* t}\boldsymbol{u}^* \\ &= 2e^{\sigma t}\left(\mathrm{Re}(\boldsymbol{u})\cos\omega t - \mathrm{Im}(\boldsymbol{u})\sin\omega t \right) \end{aligned} \tag{7.4-13}$$

The precessional mode solution $\overline{u}$ is real and the motion of each (x,y) coordinate pair describes an exponentially decreasing or increasing orbit for stable and unstable modes, respectively. At the instability threshold or for the undamped systems $(\sigma=0)$, the precessional mode solution is also referred to as the precessional mode shape. It then has the basic harmonic motion form.

$$\overline{u}(t) = u_c \cos\omega t + u_s \sin\omega t \qquad (7.4\text{-}14)$$

The above equation shows that at each finite element station, the rotor translational motions are described as:

$$x(t) = x_c \cos\omega t + x_s \sin\omega t \qquad (7.4\text{-}15a)$$
$$y(t) = y_c \cos\omega t + y_s \sin\omega t \qquad (7.4\text{-}15b)$$

Note that x and y displacements oscillate in the same damped natural frequency ω with different amplitudes and phase angles. This harmonic motion describes an elliptical orbit as discussed in Chapter 2. The direction of precession (whirling) is determined by the rate of precessional angle

$$sign(\dot{\phi}) = sign(x_c y_s - x_s y_c) \qquad (7.4\text{-}16)$$

When $\dot{\phi}$ is positive value, the orbit motion is a forward whirl (precession). When $\dot{\phi}$ is negative value, the orbit motion is a backward whirl (precession). When $\dot{\phi}$ is zero, the orbit motion degenerates to a straight-line path.

In general, the complete rotor whirls either forward or backward. However, the rotor can also have mixed precession, i.e., the rotor can have forward precession and backward precession simultaneously, at different sections of the rotor. Hence, to determine the rotor direction of precession, the direction of whirling should be evaluated at all the finite element stations. The mixed precession occurs commonly in the long flexible rotors supported by the fluid film bearings. In addition to the direction of whirling, the size, shape, and orientation of the elliptical orbits change from station to station as illustrated in Chapter 2.

Figure 7.4-1 shows the Whirl Speed Map and Figure 7.4-2 shows the Stability Map for a centrifugal compressor as shown in Figure 7.2-1. A damping value of 200 Lbf-s/in is assumed in the bearings and both bearings are connected to the rigid ground. From the Whirl Speed Map, it shows that the first forward damped critical speed is around 34,000 rpm and the first backward damped critical speed is around 30,000 rpm. It also shows that the gyroscopic effect has little influence on the second mode. The Stability map shows that the damping factor of the forward mode decreases as the rotor speed increases, and also that the damping factor of the backward mode increases as the rotor speed increases.

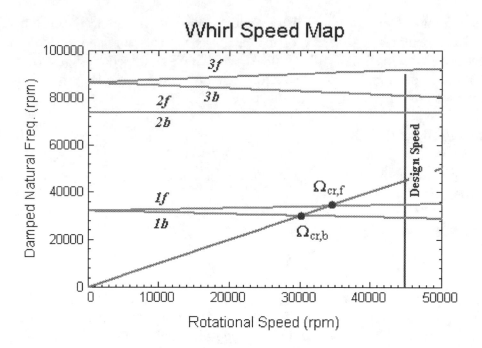

Figure 7.4-1 Whirl speed map with isotropic bearings

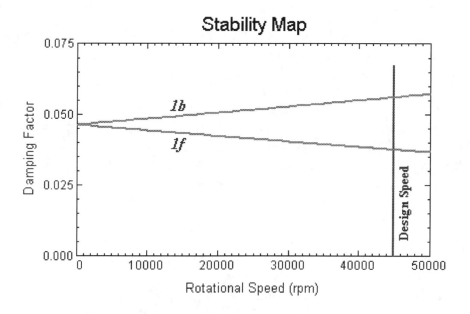

Figure 7.4-2 Stability map with isotropic bearings

In the presence of the gyroscopic effect and the general linearized bearing model, the Whirl Speed Map and Stability Map can be very complicated. The associated precessional mode shapes must be used to identify the modes and properly construct the Whirl Speed Map and Stability Map. In practice, the compressor is supported by two fluid film bearings. To demonstrate the complex nature of the whirl speed map, a compressor supported by two 3-lobe fluid film bearings is employed for this illustration. The Whirl Speed Map is shown in Figure 7.4-3. In this example, the mode number is based on the values of the frequency and not based on the mode shape. The mode shapes for the first four modes at rotor speeds of 20000, 25000, 30000, 35000, 40000, 45000, and 50000 rpm are shown in Figure 7.4-4. At 20000 rpm, the first two forward modes are classic rigid body modes. The third mode is a mixed mode, predominated by the forward precessional motion, is mixed with rigid body and bending motion. The fourth mode is a backward bending mode. At 25000 rpm, the second forward mode has some shaft bending motion, and the fourth mode (backward mode) starts to have forward motion. At 30000 rpm, modes number 3 and 4 switch order. Also the forward and backward modes start to influence each other. Therefore, modes 2 and 3 are mixed modes, meaning, some portion of the shaft has forward precessional motion and part of shaft has backward precessional motion. However for both modes, the motions are predominated by the forward precessional motions. At 40000 rpm, the modes become either pure forward or pure backward modes. The first mode is still a rigid body forward mode, the second mode is a backward bending mode, and the third and fourth modes are forward bending modes. At 45000 rpm, modes 1 and 2 switch order. At 50000 rpm, modes 2 and 3 switch order. Extreme care must be taken while preparing the Whirl Speed Map.

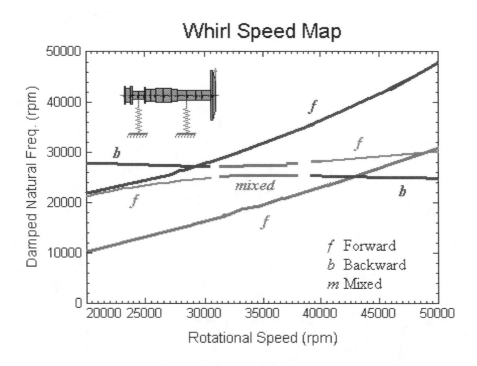

Figure 7.4-3 Whirl speed map

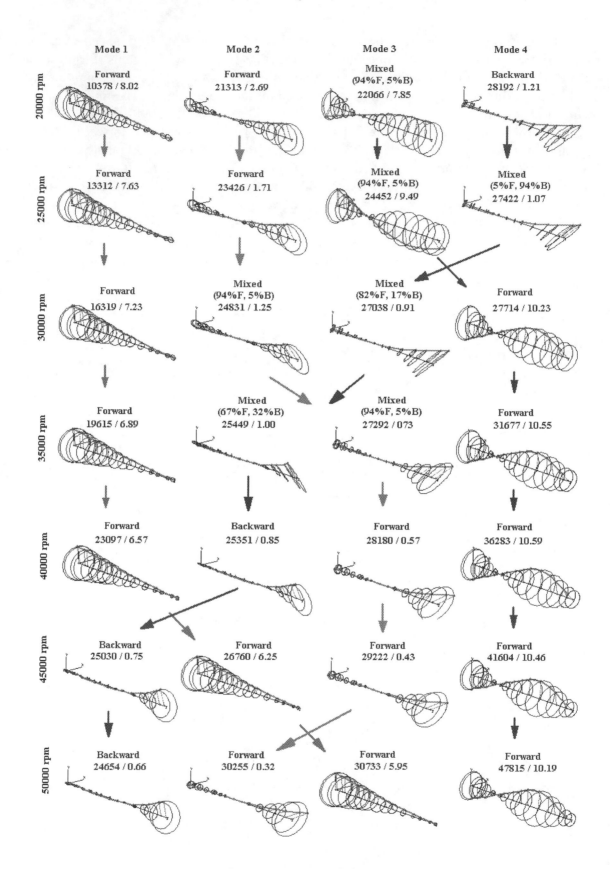

Figure 7.4-4 Mode shapes

7.5 Steady State Synchronous Response

The most common steady-state synchronous excitations include mass unbalance, disk skew, and shaft bow. The rotors always have some amount of residual unbalance no matter how well they are balanced. The unbalance of a rotating assembly is usually determined by using the multi-plane balancing machines. These residual unbalance forces are discrete and located at different planes, with magnitudes of *me* (mass times eccentricity). The unbalance forces are assumed to be independent and each unbalance has the form:

$$
\begin{Bmatrix} F_{u,x} \\ F_{u,y} \\ M_{u,x} \\ M_{u,y} \end{Bmatrix} = \begin{Bmatrix} me\Omega^2\cos(\Omega t + \phi_u) \\ me\Omega^2\sin(\Omega t + \phi_u) \\ 0 \\ 0 \end{Bmatrix}
$$

$$
= \begin{Bmatrix} me\Omega^2\cos\phi_u \\ me\Omega^2\sin\phi_u \\ 0 \\ 0 \end{Bmatrix}\cos\Omega t + \begin{Bmatrix} -me\Omega^2\sin\phi_u \\ me\Omega^2\cos\phi_u \\ 0 \\ 0 \end{Bmatrix}\sin\Omega t
$$

(7.5-1)

If the disk is mounted in a skew position on the shaft, with an angle of τ between the rotor axis and the disk axis and a phase angle of ϕ_τ, the resulting moments are similar to the unbalance force expression:

$$
\begin{Bmatrix} F_{\tau,x} \\ F_{\tau,y} \\ M_{\tau,x} \\ M_{\tau,y} \end{Bmatrix} = \begin{Bmatrix} 0 \\ 0 \\ \tau(I_p - I_d)\Omega^2\cos(\Omega t + \phi_\tau) \\ \tau(I_p - I_d)\Omega^2\sin(\Omega t + \phi_\tau) \end{Bmatrix}
$$

$$
= \begin{Bmatrix} 0 \\ 0 \\ \tau(I_p - I_d)\Omega^2\cos\phi_\tau \\ \tau(I_p - I_d)\Omega^2\sin\phi_\tau \end{Bmatrix}\cos\Omega t + \begin{Bmatrix} 0 \\ 0 \\ -\tau(I_p - I_d)\Omega^2\sin\phi_\tau \\ \tau(I_p - I_d)\Omega^2\cos\phi_\tau \end{Bmatrix}\sin\Omega t
$$

(7.5-2)

The residual shaft bow may be present in the rotor-bearing systems due to many various reasons, including assembly tolerances and uneven thermal distribution. When the residual shaft bow exists in a rotor system, a constant magnitude rotating force that is synchronized with the shaft spin speed, is acting on the rotor system. The excitation force caused by the residual shaft bow is:

$$
F_b = K_s q_b
$$

(7.5-3)

where K_s is the shaft bending stiffness and q_b is the amount of initial shaft displacement in the fixed reference frame due to residual shaft bow. The shaft bow rotates with the rotating reference and is specified in the rotating reference frame. At a typical cross section of the shaft, the displacement relationship between fixed reference frame and rotating reference frame, as presented in Chapter 2, is:

$$\begin{Bmatrix} x \\ y \\ \theta_x \\ \theta_y \end{Bmatrix}_F = \begin{Bmatrix} x' \\ y' \\ \theta'_x \\ \theta'_y \end{Bmatrix}_R \cos\Omega t + \begin{Bmatrix} -y' \\ x' \\ -\theta'_y \\ \theta'_x \end{Bmatrix}_R \sin\Omega t \tag{7.5-4}$$

The synchronous excitation force due to residual shaft bow in the fixed reference frame becomes:

$$F_b(t) = K_s \left[r_c \cos\Omega t + r_s \sin\Omega t \right] \tag{7.5-5}$$

where r is the specified initial residual shaft bow in the rotating reference frame:

$$r_c = \left[x'_1, y'_1, \theta'_{x1}, \theta'_{y1}, \ldots\ldots, x'_n, y'_n, \theta'_{xn}, \theta'_{yn} \right]^T \tag{7.5-6}$$

and

$$r_s = \left[-y'_1, x'_1, -\theta'_{y1}, \theta'_{x1}, \ldots\ldots, -y'_n, x'_n, -\theta'_{yn}, \theta'_{xn} \right]^T \tag{7.5-7}$$

It should be noted that the excitation amplitude of a shaft bow is a constant, while the excitation amplitude of an unbalance or disk skew is a function of the square of spin speed. However, they all produce the synchronous excitation:

$$Q = Q_c \cos\Omega t + Q_s \sin\Omega t \tag{7.5-8}$$

where Ω is the rotor spinning speed in rad/sec.

7.5.1 Steady State Synchronous Response for Linear Systems

For the generalized linear systems, the steady state response has the same form:

$$q = q_c \cos\Omega t + q_s \sin\Omega t \tag{7.5-9}$$

On differentiation of the solution and substitution into the equation of motion, the following set of linear algebra equations apply:

$$\begin{bmatrix} K - \Omega^2 M & \Omega C \\ -\Omega C & K - \Omega^2 M \end{bmatrix} \begin{Bmatrix} q_c \\ q_s \end{Bmatrix} = \begin{Bmatrix} Q_c \\ Q_s \end{Bmatrix} \tag{7.5-10}$$

It is usually convenient, although not necessary, to analyze the steady state synchronous response in a complex form. The complex synchronous excitation and associated steady state response are related by the following algebraic equations:

$$\left(\boldsymbol{K} - \Omega^2 \boldsymbol{M} + j\Omega\boldsymbol{C}\right)\hat{\boldsymbol{q}} = \hat{\boldsymbol{Q}}$$

(7.5-11)

where the complex quantities, $\hat{\boldsymbol{q}}$, $\hat{\boldsymbol{Q}}$, are given by the following expressions:

$$\hat{\boldsymbol{Q}} = \boldsymbol{Q}_c - j\,\boldsymbol{Q}_s$$

(7.5-12)

$$\hat{\boldsymbol{q}} = \boldsymbol{q}_c - j\,\boldsymbol{q}_s$$

(7.5-13)

The steady state response can be calculated by utilizing either real or complex domains. Figure 7.5-1 shows the Bode Plot and Figure 7.5-2 shows the Polar Plots at the impeller for a high-speed compressor. Since generalized bearing coefficients are used, the system is not isotropic and the displacements at the x and y directions are different, resulting in elliptical rotor orbits.

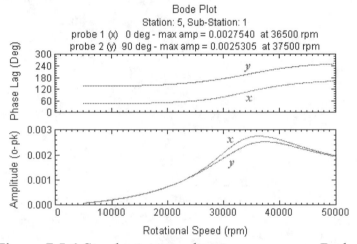

Figure 7.5-1 Steady state synchronous response – Bode plot

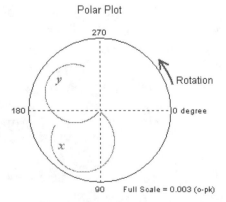

Figure 7.5-2 Steady state synchronous response – Polar plot

7.5.2 Steady State Synchronous Response for Non-Linear Isotropic Systems

For systems with isotropic bearings (linear or nonlinear) and nonlinear squeeze film dampers, the steady state response is circular in the fixed reference frame and constant in the rotating reference frame. The constant steady-state synchronous response, p_o, can be obtained from the equations of motion expressed in the rotating reference frame with the following:

$$\omega = \Omega$$
$$p = p_0, \qquad \dot{p} = \ddot{p} = 0 \tag{7.5-14}$$

The equations of motion in the rotating reference frame reduce to a set of nonlinear equations:

$$\left[K + \Omega \hat{C} - \Omega^2 M \right] p_0 = Q_s + F_n \tag{7.5-15}$$

where p_o is the constant steady state response in the rotating reference frame, Q_s is the synchronous excitation, including mass unbalance, disk skew, and shaft bow in the rotating reference frame. F_n is the nonlinear bearing and damper force vector. Typically the nonlinear force vector is a very sparse vector. The transformed matrix $\hat{C}$ can be obtained by using the coordinate transformation given in Chapter 2 as:

$$\hat{C} = R^T C S \tag{7.5-16}$$

For simplicity of notation, D is defined as the system dynamical stiffness matrix.

$$D = \left[K + \Omega \hat{C} - \Omega^2 M \right] \tag{7.5-17}$$

For a NDOF degrees-of-freedom system, the Eq. (7.5-15) consists of NDOF nonlinear algebraic equations. There are many solution techniques available for the above nonlinear equations. However, as the number of simultaneous equations increases, the difficulties of convergence and computation time have been found to increase significantly. Since there are only a small number of nonlinear bearing and damper coordinates, the following re-ordering procedure presented by McLean and Hahn (1983) reduces the number of required nonlinear equations from NDOF to the number of nonlinear bearing/damper coordinates. The advantage of this re-ordering procedure is its ability to reduce the computation time and minimize the convergence difficulties. The technique is valid for all types of nonlinear isotropic supports, where the motion is circular and the radial and tangential forces are functions of displacements and bearing/damper parameters. For given bearing/damper properties, the nonlinear forces are dependent on the displacements only.

Partitioning equation (7.5-15) into a set of nonlinear bearing/damper coordinates, p_d, and its complement, p_c, yields:

$$\begin{bmatrix} \boldsymbol{D}_{cc} & \boldsymbol{D}_{cd} \\ \boldsymbol{D}_{dc} & \boldsymbol{D}_{dd} \end{bmatrix} \begin{Bmatrix} \boldsymbol{p}_c \\ \boldsymbol{p}_d \end{Bmatrix} = \begin{Bmatrix} \boldsymbol{Q}_c \\ \boldsymbol{Q}_d \end{Bmatrix} + \begin{Bmatrix} 0 \\ \boldsymbol{F}_d(\boldsymbol{p}_d) \end{Bmatrix} \tag{7.5-18}$$

From the upper half of equations

$$\boldsymbol{p}_c = \boldsymbol{D}_{cc}^{-1}\big(\boldsymbol{Q}_c - \boldsymbol{D}_{cd}\,\boldsymbol{p}_d\big) \tag{7.5-19}$$

which when substituted into the lower half yields

$$\boldsymbol{Q}_r + \boldsymbol{D}_r\,\boldsymbol{p}_d = \boldsymbol{F}_d(\boldsymbol{p}_d) \tag{7.5-20}$$

where

$$\boldsymbol{Q}_r = \big(\boldsymbol{D}_{dc}\,\boldsymbol{D}_{cc}^{-1}\boldsymbol{Q}_c\big) - \boldsymbol{Q}_d \tag{7.5-21}$$

and

$$\boldsymbol{D}_r = \boldsymbol{D}_{dd} - \boldsymbol{D}_{dc}\,\boldsymbol{D}_{cc}^{-1}\boldsymbol{D}_{cd} \tag{7.5-22}$$

Now, the nonlinear bearing/damper displacements, $\boldsymbol{p}_{\mathrm{d}}$, can be solved from the small order of nonlinear equations involving nonlinear bearing/damper coordinates only, instead of the original large number of simultaneous equations. Once the nonlinear bearing/damper displacement vector, $\boldsymbol{p}_{\mathrm{d}}$, has been determined from the iterative solution schemes, the complementary displacement, $\boldsymbol{p}_{\mathrm{c}}$, can be obtained directly from Eq. (7.5-19).

7.6 Transient Response

There are two types of transient analysis provided in *DyRoBeS*. One is the rotor system with a constant rotational speed, and the other one is with a variable rotational speed. For a constant rotational speed, the transient analysis is used to determine the steady state response for the non-linear systems or the linear/nonlinear systems subject to sudden excitations. In many applications, there are needs to study the rotor motion during startup, shutdown, going through critical speeds, or rotor drop for the magnetic bearing systems. In these situations, the angular velocity (spin speed) is no longer a constant and is a function of time. The governing equations of motion for a constant rotational speed have been presented in Eq. (7.1-1); the governing equations of motion for a variable rotational speed system are discussed in details here:

$$\begin{aligned} \boldsymbol{M}\ddot{\boldsymbol{q}}(t) + \big[\boldsymbol{C} + \dot{\varphi}\,\boldsymbol{G}\big]\dot{\boldsymbol{q}}(t) + \big[\boldsymbol{K} + \ddot{\varphi}\,\boldsymbol{G}\big]\boldsymbol{q}(t) \\ = \dot{\varphi}^2\,\boldsymbol{Q}_1(\varphi) + \ddot{\varphi}\,\boldsymbol{Q}_2(\varphi) + \boldsymbol{Q}_3(\dot{\boldsymbol{q}}, \boldsymbol{q}, \varphi, \dot{\varphi}, \ddot{\varphi}, t) \end{aligned} \tag{7.6-1}$$

Two more terms introduced in the governing equations due to the speed variation are: circulatory matrix $\ddot{\varphi}G$ and forcing function $\ddot{\varphi}Q_2$. All the linearized damping and stiffness terms are in the damping and stiffness matrices, and all the other interconnection nonlinear forces are included in Q_3. Q_1 and Q_2 are functions of $(\varphi, \dot{\varphi})$ and $(\varphi, \ddot{\varphi})$, respectively. For mass unbalance, Q_1 and Q_2 at station "i" can be obtained from the following expression:

$$\begin{Bmatrix} F_{u,x} \\ F_{u,y} \\ M_{u,x} \\ M_{u,y} \end{Bmatrix} = \dot{\varphi}^2 \begin{Bmatrix} me\cos\varphi \\ me\sin\varphi \\ 0 \\ 0 \end{Bmatrix} + \ddot{\varphi} \begin{Bmatrix} me\sin\phi \\ -me\cos\phi \\ 0 \\ 0 \end{Bmatrix} \qquad (7.6\text{-}2)$$

For a system with a constant rotational speed ($\dot{\varphi} = \Omega, \ddot{\varphi} = 0$), the transient response can be obtained by direct numerical integration algorithms in the physical coordinates or by modal expansion using a linear transformation in the modal (normal) coordinates. The use of modal analysis in the rotor systems was proposed in 70s (Childs, 1976) and early 80s (Li and Gunter 1982, Nelson et al., 1983). The advantage of the modal analysis is the use of modal expansion (truncation) to transform a large order of coupled equations into a small set of uncoupled (if complex modes are used) or coupled (if real modes are used) equations. However, the determination and selection of the retained modes require sound engineering judgment. The transformation between modal and physical coordinates for the nonlinear systems, at every time step, can be very tedious and time consuming since all the nonlinear forces are functions of physical coordinates. For a system with a variable rotational speed ($\dot{\varphi} = \Omega, \ddot{\varphi} \neq 0$), the transient response has to be obtained by direct numerical integration algorithms in the physical coordinates. For simplicity of notion, the equations of motion, equation (7.6-1) can be rewritten as:

$$M\ddot{q}(t) + C\,\dot{q}(t) + K\,q(t) = Q \qquad (7.6\text{-}3)$$

Where the general damping matrix C consists of the damping and gyroscopic matrices, and the general stiffness matrix K consists of the stiffness and circulatory matrices.

Direct Numerical Integration Algorithms

The direct step-by-step numerical integration calculates the transient response for a given constant rotor speed or variable speed and time interval. The system can be linear or non-linear. In structural dynamics, two commonly used direct integration algorithms that solve the original second-order differential equations are: Newmark-β method and Wilson-θ method. By properly selecting the integration parameters in both methods, both methods are unconditionally stable schemes in linear problems. Another group of numerical integration methods, such as Runge-Kutta and Gear's methods, solve the first-order differential equations (state space form) are commonly used by scientists in other fields.

The Newmark-β Method

The most commonly used Newmark method is known as the constant-average-acceleration method. The Newmark constant average acceleration method assumes the average acceleration in the time interval t_i to $t_{i+1}=t_i+\Delta t$, to be the average value of the discrete initial and final accelerations as illustrated in Figure 7.6-1.

$$\ddot{q}(t) = \frac{1}{2}\left(\ddot{q}_i + \ddot{q}_{i+1}\right) \tag{7.6-4}$$

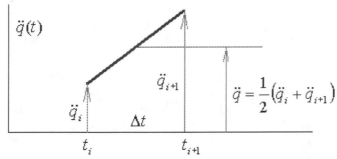

Figure 7.6-1 Constant acceleration method

Successive integration of Eq. (7.6-4) gives the velocity and displacement at time $t_{i+1}=t_i+\Delta t$.

$$\dot{q}_{i+1} = \dot{q}_i + \left(\frac{\Delta t}{2}\right)\left(\ddot{q}_i + \ddot{q}_{i+1}\right) \tag{7.6-5}$$

$$q_{i+1} = q_i + (\Delta t)\dot{q}_i + \left(\frac{\Delta t^2}{4}\right)\left(\ddot{q}_i + \ddot{q}_{i+1}\right) \tag{7.6-6}$$

By rearranging Eqs. (7.6-5) and (7.6-6), the velocity and acceleration at time $t_{i+1}=t_i+\Delta t$ are expressed as functions of displacement, velocity, acceleration at time t_i and displacement at time $t_{i+1}=t_i+\Delta t$.

$$\dot{q}_{i+1} = \left(\frac{2}{\Delta t}\right)\left(q_{i+1} - q_i\right) - \dot{q}_i \tag{7.6-7}$$

$$\ddot{q}_{i+1} = \left(\frac{4}{\Delta t^2}\right)\left(q_{i+1} - q_i\right) - \left(\frac{4}{\Delta t}\right)\dot{q}_i - \ddot{q}_i \tag{7.6-8}$$

The equations of motion, Eq. (7.6-3), are numerically integrated for each time step, beginning with the initial conditions of $q(0)=q_0$, $\dot{q}(0)=\dot{q}_0$. The initial accelerations are obtained from the dynamic equilibrium equation:

$$\ddot{q}_i = M^{-1}\left(Q_i - C\dot{q}_i - Kq_i\right) \tag{7.6-9}$$

The dynamic equilibrium is satisfied at both t_i and t_{i+1}. The equilibrium equations at time t_{i+1} may be written as follows:

$$M\ddot{q}_{i+1} + C\dot{q}_{i+1} + Kq_{i+1} = Q_{i+1} \tag{7.6-10}$$

Substitution of Eqs. (7.6-7) and (7.6-8) into Eq. (7.6-10) gives a set of equations with displacements at time t_{i+1} to be determined.

$$\left(K + \frac{2}{\Delta t}C + \frac{4}{\Delta t^2}M\right)q_{i+1} = \hat{Q}_{i+1} \tag{7.6-11}$$

and

$$\hat{Q}_{i+1} = Q_{i+1} + M\left(\frac{4}{\Delta t^2}q_i + \frac{4}{\Delta t}\dot{q}_i + \ddot{q}_i\right) + C\left(\frac{2}{\Delta t}q_i + \dot{q}_i\right) \tag{7.6-12}$$

Once the displacement at time t_{i+1} is solved from above equation (7.6-11), the velocity and acceleration at time t_{i+1} can be determined from Eqs. (7.6-7) and (7.6-8).

The Wilson-θ Method

The Wilson-θ method is essentially an extension of the linear acceleration method. Acceleration is assumed to be linear from time t_i to time $t_i + \theta\Delta t$, where $\theta \geq 1.0$. For unconditional stability in linear problems, θ must be greater than or equal to 1.37 and the value of 1.4 is usually employed. It has been shown that the optimal value of θ is 1.420815 and this value is used in **DyRoBeS**. The acceleration at time $(t_i + \tau)$ from linear assumption as illustrated in Figure 7.6-2 is given as:

$$\ddot{q}(t_i + \tau) = \ddot{q}(t_i) + \frac{\tau}{\theta\Delta t}\left(\ddot{q}(t_i + \theta\Delta t) - \ddot{q}(t_i)\right) \tag{7.6-13}$$

where τ is the time increment from t_i and $0 \leq \tau \leq \theta\Delta t$. Integrating Eq. (7.6-13), the velocity and displacement at time $(t_i + \tau)$ are:

$$\dot{q}(t_i + \tau) = \dot{q}(t_i) + \tau\,\ddot{q}(t_i) + \frac{\tau^2}{2\theta\Delta t}\left(\ddot{q}(t_i + \theta\Delta t) - \ddot{q}(t_i)\right) \tag{7.6-14}$$

$$q(t_i + \tau) = q(t_i) + \tau\,\dot{q}(t_i) + \frac{\tau^2}{2}\ddot{q}(t_i) + \frac{\tau^3}{6\theta\Delta t}\left(\ddot{q}(t_i + \theta\Delta t) - \ddot{q}(t_i)\right) \tag{7.6-15}$$

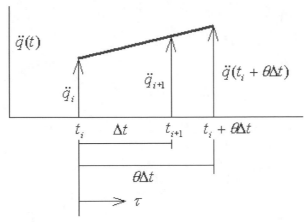

Figure 7.6-2 Linear acceleration method

From Eqs. (7.6-14) and (7.6-15), the velocity and displacement at time $t_i + \theta \Delta t$ are obtained by replacing τ by $\theta \Delta t$:

$$\dot{q}(t_i + \theta \Delta t) = \dot{q}(t_i) + \frac{\theta \Delta t}{2} \left(\ddot{q}(t_i + \theta \Delta t) + \ddot{q}(t_i) \right) \tag{7.6-16}$$

$$q(t_i + \theta \Delta t) = q(t_i) + \theta \Delta t \, \dot{q}(t_i) + \frac{(\theta \Delta t)^2}{6} \left(\ddot{q}(t_i + \theta \Delta t) + 2\ddot{q}(t_i) \right) \tag{7.6-17}$$

From Eq. (7.6-17), the acceleration can be expressed in terms of the displacement at time $t_i + \theta \Delta t$:

$$\ddot{q}(t_i + \theta \Delta t) = \frac{6}{(\theta \Delta t)^2} \left(q(t_i + \theta \Delta t) - q(t_i) \right) - \frac{6}{\theta \Delta t} \dot{q}(t_i) - 2\ddot{q}(t_i) \tag{7.6-18}$$

Substituting Eq. (7.6-18) into Eq. (7.6-16), the velocity is expressed in terms of the displacement at time $t_i + \theta \Delta t$:

$$\dot{q}(t_i + \theta \Delta t) = \frac{3}{(\theta \Delta t)} \left(q(t_i + \theta \Delta t) - q(t_i) \right) - 2\dot{q}(t_i) - \frac{\theta \Delta t}{2} \ddot{q}(t_i) \tag{7.6-19}$$

In order to obtain the solution at time $t_i + \Delta t$, the equilibrium equations at time $t_i + \theta \Delta t$ are considered. The equilibrium equations at time $t_i + \theta \Delta t$ may be written as follows:

$$\boldsymbol{M} \ddot{q}(t_i + \theta \Delta t) + \boldsymbol{C} \dot{q}(t_i + \theta \Delta t) + \boldsymbol{K} q(t_i + \theta \Delta t) = \boldsymbol{Q}(t_i + \theta \Delta t) \tag{7.6-20}$$

Since the accelerations are assumed to be linear, a linearly projected load vector $\boldsymbol{Q}(t_i + \theta \Delta t)$ is:

$$\boldsymbol{Q}(t_i + \theta \Delta t) = \boldsymbol{Q}(t_i) + \theta \left[\boldsymbol{Q}(t_i + \Delta t) - \boldsymbol{Q}(t_i) \right] \tag{7.6-21}$$

The displacements at time $t_i + \theta \Delta t$ are solved using Eq. (7.6-20) after substitution of Eqs. (7.6-18) and (7.6-19), and rearrangement:

$$\left(K + \frac{3}{\theta \Delta t} C + \frac{6}{(\theta \Delta t)^2} M \right) q(t_i + \theta \Delta t) = \hat{Q}(t_i + \theta \Delta t) \tag{7.6-22}$$

and

$$\hat{Q}(t_i + \theta \Delta t) = Q(t_i) + \theta \left[Q(t_i + \Delta t) - Q(t_i) \right]$$
$$+ M \left(\frac{6}{(\theta \Delta t)^2} q_i + \frac{6}{\theta \Delta t} \dot{q}_i + 2 \ddot{q}_i \right) + C \left(\frac{3}{\theta \Delta t} q_i + 2 \dot{q}_i + \frac{\theta \Delta t}{2} \ddot{q}_i \right) \tag{7.6-23}$$

Once the displacement at time $t_i + \theta \Delta t$ is solved from Eq. (7.6-22), the acceleration at $t_i + \theta \Delta t$ is obtained by Eq. (7.6-18). Then, the acceleration, velocity, and displacement at time $t_i + \Delta t$ can be calculated by substituting the acceleration at time $t_i + \theta \Delta t$ into Eqs. (7.6-13), (7.6-14), and (7.6-15), and setting $\tau = \Delta t$.

It should be noted that both Newmark-β and Wilson-θ methods are unconditionally stable in the linear problems. That is, the solution is stable for any value of Δt in linear problems. However, the result may not be accurate if the time step is too large. For the nonlinear problems, an iterative procedure such as Newton-Raphson iteration is required to ensure that the dynamic equilibrium condition is satisfied (Bathe, 1995). It is expected that smaller time steps produce better results. Larger time steps can decrease the accuracy of the solution and also can introduce some unwanted numerical oscillations in the solution. One must exhibit some care when using very small time steps so that the computational times do not become excessive. However, for the highly non-linear case, a small time interval is necessary for the solution convergence. The time step suggested by many researchers in the linear problems is about Tcr/20 or smaller for stability reasons. Tcr is the natural period for the critical frequency. For systems with high order degrees-of-freedom, the Wilson-θ method provides the numerical dissipation to filter out the response of the higher inaccurate modes and might be superior to the average acceleration method (Craig, 1981). These direct integration methods have been discussed in Bathe (1995) and Craig (1981). Due to the similarity of the Newmark-β and the Wilson-θ method, they can be implemented in a single computer subroutine or function.

Gear's and Runge-Kutta Methods

Gear's and Runge-Kutta methods solve the first-order ordinary differential equations (state space form). The standard form for the initial-value problem of first order ordinary differential equations is:

$$\dot{x} = f(x, t) \tag{7.6-24}$$

where

$$\dot{x} = \begin{Bmatrix} \ddot{q} \\ \dot{q} \end{Bmatrix}, \quad x = \begin{Bmatrix} \dot{q} \\ q \end{Bmatrix} \tag{7.6-25}$$

and

$$f(x,t) = F - A^{-1}Bx = \begin{Bmatrix} Q \\ 0 \end{Bmatrix} - \begin{bmatrix} M^{-1}C & M^{-1}K \\ -I & 0 \end{bmatrix} \begin{Bmatrix} \dot{q} \\ q \end{Bmatrix} \tag{7.6-26}$$

There are several solution methods for the first order ordinary differential equations, such as Runge-Kutta method, Adams, and Gear's methods. Runge-Kutta method is the most commonly used in engineering applications. The classical Runge-Kutta method is a fourth-order method. However, there are fifth-order and six-order methods readily available in many numerical computation textbooks. Gear's method is suitable for the stiff equations. The solution methods for the first order differential equations may not be computationally efficient, compared to the solution methods for the second-order differential equations. However, they often produce better results.

Modal Analysis

The system equation of motion for a constant rotational speed in the state space form is:

$$A\dot{x} + Bx = F \tag{7.6-27}$$

where

$$A = \begin{bmatrix} M & 0 \\ 0 & I \end{bmatrix}, \quad B = \begin{bmatrix} C & K \\ -I & 0 \end{bmatrix}, \quad x = \begin{Bmatrix} \dot{q} \\ q \end{Bmatrix}, \quad F = \begin{Bmatrix} Q \\ 0 \end{Bmatrix} \tag{7.6-28}$$

Since the matrix B is generally asymmetric. Thus, in order to obtain a biorthogonality relation between the system eigenvectors, it is necessary to define both the eigenvalue problem (right eigenvector) and the associated adjoint eigenvalue problem (left eigenvector):

$$\left(\lambda_i A + B \right) y_i = 0 \tag{7.6-29}$$

and

$$z_i^T \left(\lambda_i A + B \right) = 0^T \quad \text{or} \quad \left(\lambda_i A^T + B^T \right) z_i = 0 \tag{7.6-30}$$

These two eigenvalue problems yield a set of eigenvalues $\lambda_i, i = 1, 2, \ldots, 2n$ and associated right and left eigenvectors y_i and z_i, respectively. The two sets of eigenvectors satisfy the two following biorthogonality relations:

$$z_j^T A y_i = R_i \delta_{ij} \tag{7.6-31}$$

$$z_j^T B y_i = -\lambda_i R_i \delta_{ij} \tag{7.6-32}$$

where R_i is the system norm and δ_{ij} is the Kronecker delta. The eigenvectors obtained from the state space form are related to the eigenvectors obtained from the original second order eigenvalue problem by the following relations:

$$y = \begin{Bmatrix} \lambda u \\ u \end{Bmatrix} \qquad z = \begin{Bmatrix} \lambda v \\ v \end{Bmatrix} \tag{7.6-33}$$

The primary step in the modal analysis is to introduce the coordinate transformation by using the modal expansion:

$$x(t) = \sum_{i=1}^{2\hat{n} \leq 2n} y_i \eta_i(t) = \hat{Y} \eta \tag{7.6-34}$$

Normally the modes retained in the modal truncation ($2\hat{n}$) are much less than the original large order number of modes ($2n$). Since the modes are complex, it is necessary to retain the complex conjugate pairs in order to have a real response in the physical coordinates. The substitution of the linear transformation into the state space equation and utilization of the biorthogonality relations yields the following set of equations in terms of the complex modal coordinates:

$$R_i \dot{\eta}_i - \lambda_i R_i \eta_i = z_i^T F \qquad i = 1,2,...2\hat{n} \tag{7.6-35}$$

The modal initial conditions can be obtained in the same manner by using the biorthogonality relations:

$$x(0) = \begin{Bmatrix} \dot{q}_0 \\ q_0 \end{Bmatrix} = \sum y_i \eta_i(0) \tag{7.6-36}$$

By multiplying the above equations by $z_j^T A$, the initial conditions in the modal coordinates become:

$$\eta_i(0) = \frac{1}{R_i} z_j^T A x(0) \tag{7.6-37}$$

Note that in the absence of nonlinear forces, i.e., the system is linear and the excitation force $F = F(t)$ is a function of time only, the differential equations in the modal coordinates are decoupled, and may be solved independently by whatever procedures are appropriate for the excitation. However, for nonlinear systems, where $F = F(x,t) = F(\dot{q},q,t)$, the differential equations in the modal coordinates are still coupled, though the order of equations is smaller than the original order by modal truncation. The manipulations of the complex modal coordinates (two sets of complex eigenvectors, and transformation between complex modal coordinates and real physical coordinates) can be very tedious and may not be as computationally efficient as using the direct numerical integration algorithms in the real physical coordinates.

Another approximation is using the assumed modes to reduce the number of equations (Nelson and Chen, 1993). This approach cannot decouple the equations of motion even for the linear problems, but it can reduce the order of equations and thereby reduce the computational effort. The assumed modes approach is implemented by introducing the Rayleigh-Ritz transformation:

$$q = \sum_{i=1}^{\hat{n}} u_i \eta_i = U \eta \qquad (7.6\text{-}38)$$

Usually, $\hat{n}$ is much less than n, therefore, the order of equations is decreased. There are several options in choosing the linearly independent assumed-mode vector u_i. The real eigenvectors obtained from the associated conservative non-gyroscopic system are commonly used in this reduction, since it is very easy to implement. Only the modes whose natural frequencies are within the frequency range of interest are retained. Truncating higher-order modes can significantly reduce the order of the original problem. Regular polynomials are also used with some degree of success. Substituting the linear transformation and pre-multiplying the equations by the transpose of the assumed-mode matrix yields:

$$U^T M U \ddot{\eta} + U^T C U \dot{\eta} + U^T K U \eta = U^T Q \qquad (7.6\text{-}39)$$

This approximation, by utilizing the assumed modes, dose not decouple the equations, but reduces the order of the equations.

7.7 Examples

A number of rotor-bearing systems are analyzed and presented by using **DyRoBeS** in this chapter. The primary use of the computer software is for large flexible rotor-bearing systems. Thus, some complex and practical systems are presented to illustrate the ability of this program to deal with more complicated configuration. The examples presented here serve several purposes:

1. To demonstrate the use of **DyRoBeS-Rotor** and **DyRoBeS-BePerf**
2. To compare results with analytical solutions or previously published data
3. To demonstrate the consistency of results obtained from various calculations
4. To demonstrate the comprehensiveness of the analytical capabilities
5. To demonstrate the easy-to-use graphical presentation of results
6. To explain some fundamentals of rotordynamics

Most of the features available in this program are demonstrated in these examples, it is however not possible to include every combinations. The features that are not illustrated in these examples should be relatively easy to implement.

Example 7.1: A Uniform Shaft Supported by Fluid Film Bearings

This example as shown in Figure 7.7-1 is taken from Lund (1974). The uniform shaft with a length of 50 in., a diameter of 4 in., a Young's modulus of 3.0E07 psi, and a weight density of 0.283 Lb/in^3, is supported at the ends by two identical fluid film bearings. For the purpose of comparison with the original publication, the shaft shear deformation, rotatory inertia, and gyroscopic effects are not considered in this example. These effects have already been discussed in a previous example (Example 5.1), and have demonstrated a minimal influence on the critical speed analysis. The two fluid film bearings are plain cylindrical bearings with a journal diameter of 4 in., a bearing radial clearance of 0.002 in., a bearing axial length of 1 in., and an oil viscosity of 6.9 centiPoise (1.0E-06 Reyns). Both linear and nonlinear analyses are performed in this example. The results are discussed in details.

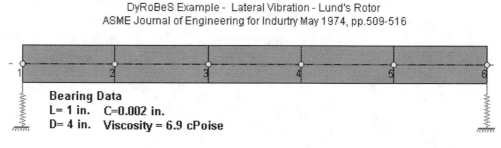

Figure 7.7-1 Example 7.1

For linear analysis, we need to know the linearized bearing coefficients (stiffness and damping) about the static equilibrium position. To calculate the bearing performance, the bearing static load is required as an input for the bearing analysis. Therefore, the first step is to calculate the bearing static loads. There are no external loads applied to the rotor. Therefore, the bearing static loads are due to the shaft weight only. Since this is a simple and symmetrical shaft, the static load on each bearing is equal to half the shaft weight. However, for complicated systems with more than 2 bearings, the static loads calculation can be very involving. Although the determination of bearing loads and static deflection is not "dynamics", it is included in the ***DyRoBeS-Rotor***. To perform the static load calculation in this example, gravity constants must be given as shown in Figure 7.7-2.

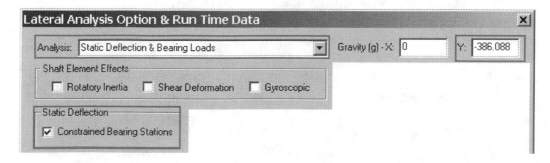

Figure 7.7-2 Analysis input for the static deflection and bearing load calculation

The negative gravity constant indicates that the gravity force is vertically down. Since we do not have the bearing coefficients yet, we need to put some dummy stiffness in the bearing station, or model the bearing as a nonlinear bearing and "Check" the constrained bearing stations in the analysis input, to perform the static load calculation. The results of static deflection and bearing loads are presented in Figure 7.7-3. As expected in this example, the load on each bearing is exactly half of the shaft weight.

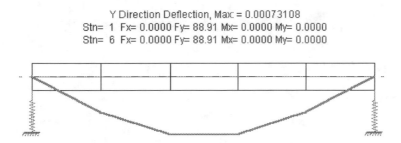

Figure 7.7-3 Static deflection and bearing loads

Once we know the bearing static loads, we are ready to calculate the bearing static performance (minimum film thickness, power loss, etc.) and dynamic coefficients by using ***DyRoBeS-BePerf***. For comparison purposes, the turbulent effect is neglected and the Reynolds boundary condition is used for the bearing calculation. The bearing geometry is shown in Figure 7.7-4, with exaggerated clearance in the figure for illustrative purposes. The data input for the bearing analysis is shown in Figure 7.7-5. The Standard coordinate system is used in the bearing analysis.

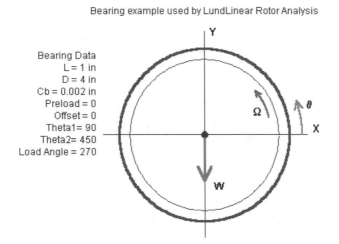

Figure 7.7-4 Bearing geometry

The journal equilibrium locus for a speed range of 1,500 to 10,000 rpm is presented in Figure 7.7-6. Since only the dynamics of the rotor is studied in this example, the bearing static performance, such as pressure, film thickness, and frictional power loss, are not presented here. Only the linearized bearing stiffness and damping are presented in

Figures 7.7-7 and 7.7-8. Note that the cross-coupled dampings (C_{xy} and C_{yx}) and cross-coupled stiffness (K_{yx}) are negative in this case by using the Standard coordinate system.

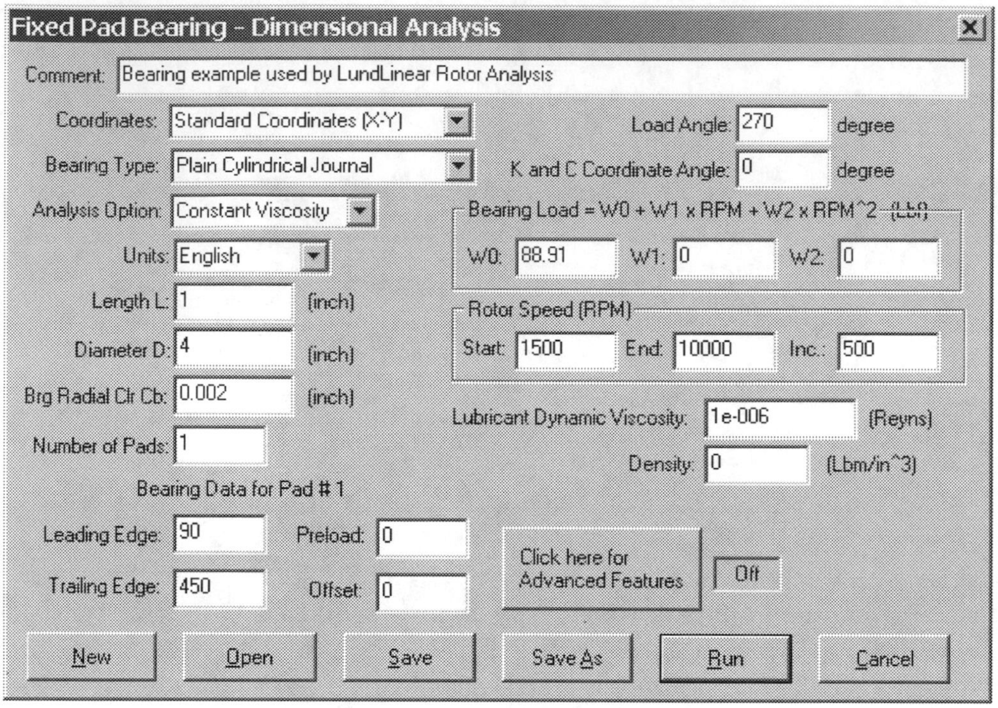

Figure 7.7-5 Bearing data input

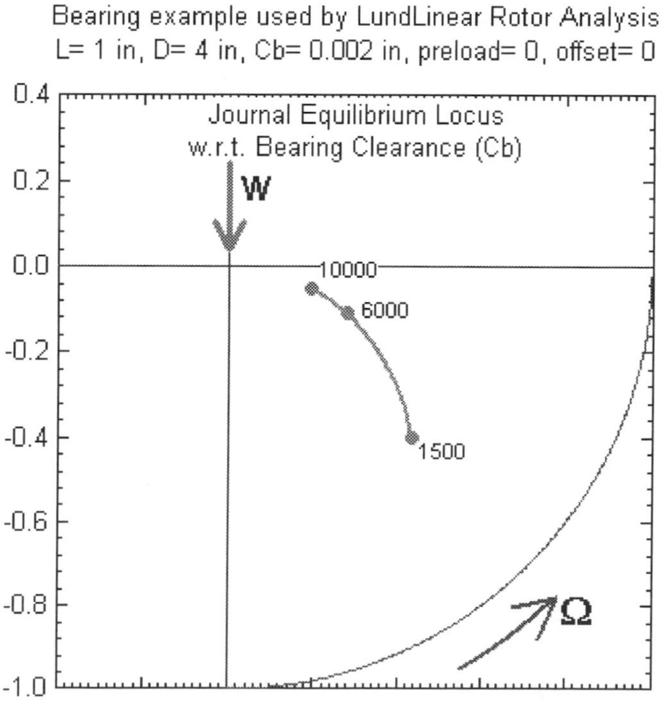

Figure 7.7-6 Journal static equilibrium locus

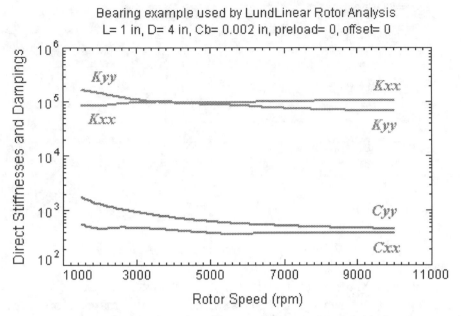

Figure 7.7-7 Bearing direct stiffnesses and dampings

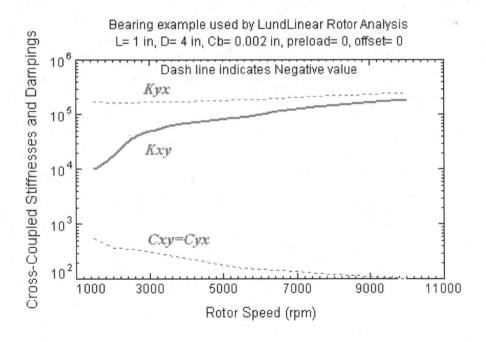

Figure 7.7-8 Bearing cross-coupled stiffnesses and dampings

The linearized bearing coefficients are computed by using **DyRoBeS-BePerf** and are saved into a file. This bearing data file is then imported by the **DyRoBeS-Rotor** to be used in the linear rotordynamics analyses, as shown in Figure 7.7-9. These linearized bearing coefficients are speed dependent.

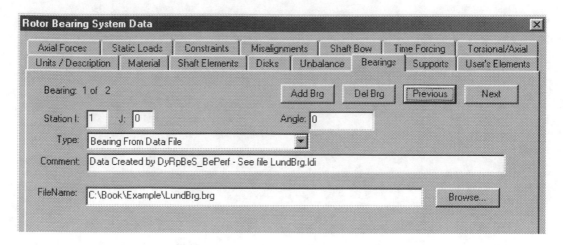

Figure 7.7-9 Bearing input in *DyRoBeS-Rotor*

The Critical Speed Map is commonly used to estimate the location of the critical speeds by overlapping the bearing dynamic stiffness on the map. However, extreme caution must be taken while interpreting the map, due to the undamped and isotropic assumptions made in the calculation. Figure 7.7-10 shows that the horizontal critical speed occurs at around 6000 rpm and that the vertical critical speed occurs at around 7000 rpm. Since the fluid film bearings have significant damping effect on the system response, as explained in Chapter 3, the two separate critical speeds may not be observed in the actual rotor unbalance response. Also, due to the high damping effect, the peak response due to unbalance will occur at a speed higher than the predicted critical speeds, as explained in Chapter 3. The first two rigid bearing critical speeds are found to be 7627 rpm (127 cps) and 30550 rpm (509 cps) and they are in agreement with the previous published values.

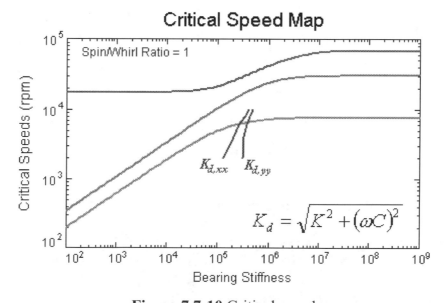

Figure 7.7-10 Critical speed map

The Whirl Speed Map from the whirl speed and stability analysis is presented in Figure 7.7-11. Since the bearing coefficients are speed dependent, **DyRoBeS-Rotor** uses spline function to interpolate these coefficients for the various rotor speeds of the analysis. The first four damped natural frequencies are shown in the whirl speed map. Note that "F" denotes the forward modes and "B" denotes the backward modes. The higher natural frequencies are very high and therefore not shown on the graph. The synchronous excitation line is also overlapped in the map, and the intersections between the damped natural frequency curves and the excitation line are referred to as damped critical speeds. It shows that the third mode is synchronous at about 7,500 rpm, and the first and second modes do not intersect at the excitation line. This means that the third mode can be excited by mass unbalance, and the first and second modes will not be excited by mass unbalance. The speed at 7,500 rpm is recognized as the first critical speed of the system.

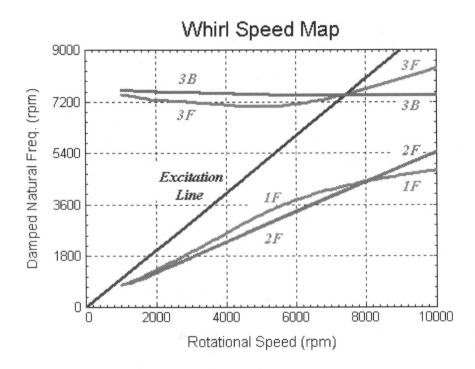

Figure 7.7-11 Whirl speed map

Due to the non-symmetric properties of the bearing coefficients and the gyroscopic effect (which is not included in this example), each vibration mode is split into two modes known as forward and backward precessional modes. Further, due to the existence of damping, some modes are overdamped and become non-vibratory real modes with zero natural frequencies, such as the first and second backward modes in this example. The real modes will not be shown in the Whirl Speed Map. It also shows that the damped natural frequency curves can intersect with each other and switch their order. For example, the first mode at 6,000 rpm becomes the second mode at 9,000 rpm, since the first and second modes intersect at around 8,000 rpm. The third forward mode and backward mode switch order at around 7,400 rpm. The damping factors for the first four modes are plotted in the Stability Map, as shown in Figure 7.7-12. Negative damping factor (or logarithmic decrement) indicates system instability in the linear sense. It shows that the damping factor of the first forward mode approaches zero as the rotor speed approaches 9,120 rpm, and then becomes negative as the rotor speed increases. This speed is known as the *instability threshold.* The rotor behavior beyond instability threshold cannot be predicated by linear analysis and nonlinear simulation should be used when rotor speed is near and above the instability threshold. Since the Whirl Speed Map and Stability Map are very complex in nature, caution must be taken when connecting the modes. Mode shapes should always be present when preparing these maps.

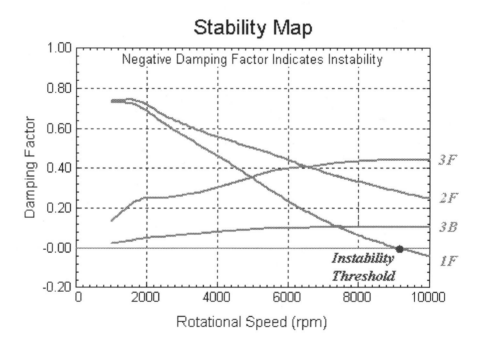

Figure 7.7-12 Stability map

The mode shapes for the first four modes at 2,000 and 6,000 rpm are shown in Figure 7.7-13. At 2,000 rpm, the first two modes are typically "so-called" rigid body modes with almost zero strain energy in the shaft. Since this is a symmetric rotor system, the first mode is a classical conical mode and the second mode is a classical translatory (cylindrical) mode. They are all forward modes. Their corresponding backward modes are overdamped, **real** non-vibratory modes, and not shown on the Whirl Speed Map. The third and fourth modes are bending modes with little motion at the bearing locations. The third mode is a forward precession and the fourth mode is a backward precession. It is evident from the stability map that as speed increases, the Logarithmic Decrements (or Damping Factors) of the first two forward modes decrease rapidly, and the third and fourth modes increase. At 6,000 rpm, it shows that the second mode (translatory mode) and the third mode (bending mode) influence each other as they approach. The translatory mode has some bending in the shaft and the bending mode has some motions at the bearing locations. They are not pure rigid body or pure bending modes any more.

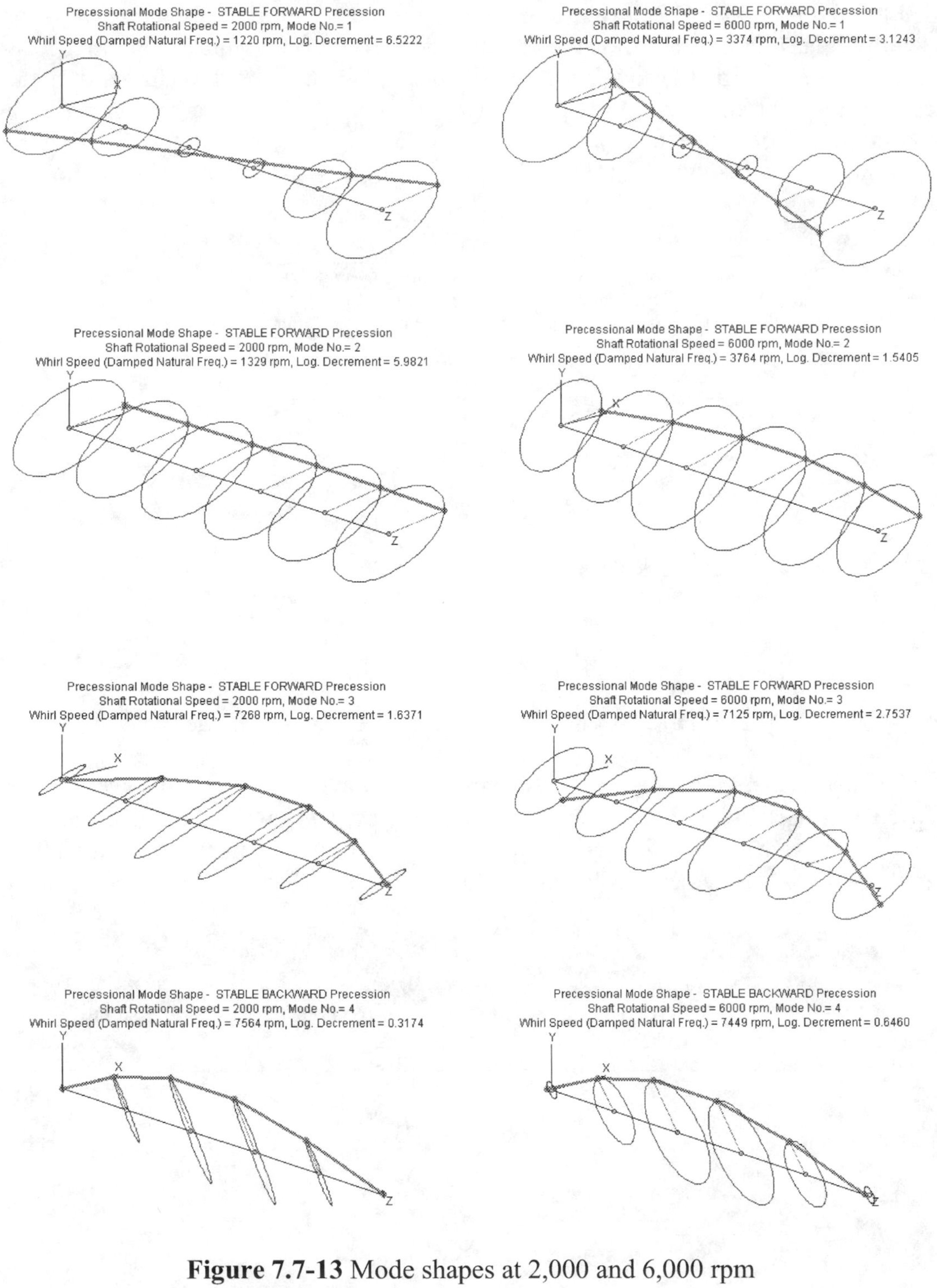

Precessional Mode Shape - STABLE FORWARD Precession
Shaft Rotational Speed = 2000 rpm, Mode No.= 1
Whirl Speed (Damped Natural Freq.) = 1220 rpm, Log. Decrement = 6.5222

Precessional Mode Shape - STABLE FORWARD Precession
Shaft Rotational Speed = 6000 rpm, Mode No.= 1
Whirl Speed (Damped Natural Freq.) = 3374 rpm, Log. Decrement = 3.1243

Precessional Mode Shape - STABLE FORWARD Precession
Shaft Rotational Speed = 2000 rpm, Mode No.= 2
Whirl Speed (Damped Natural Freq.) = 1329 rpm, Log. Decrement = 5.9821

Precessional Mode Shape - STABLE FORWARD Precession
Shaft Rotational Speed = 6000 rpm, Mode No.= 2
Whirl Speed (Damped Natural Freq.) = 3764 rpm, Log. Decrement = 1.5405

Precessional Mode Shape - STABLE FORWARD Precession
Shaft Rotational Speed = 2000 rpm, Mode No.= 3
Whirl Speed (Damped Natural Freq.) = 7268 rpm, Log. Decrement = 1.6371

Precessional Mode Shape - STABLE FORWARD Precession
Shaft Rotational Speed = 6000 rpm, Mode No.= 3
Whirl Speed (Damped Natural Freq.) = 7125 rpm, Log. Decrement = 2.7537

Precessional Mode Shape - STABLE BACKWARD Precession
Shaft Rotational Speed = 2000 rpm, Mode No.= 4
Whirl Speed (Damped Natural Freq.) = 7564 rpm, Log. Decrement = 0.3174

Precessional Mode Shape - STABLE BACKWARD Precession
Shaft Rotational Speed = 6000 rpm, Mode No.= 4
Whirl Speed (Damped Natural Freq.) = 7449 rpm, Log. Decrement = 0.6460

Figure 7.7-13 Mode shapes at 2,000 and 6,000 rpm

As the speed increases, the frequencies of the third and fourth modes switch order at around 7,400 rpm, and the frequencies of the first and second modes switch order at around 8,000 rpm. When the rotor speed reaches 9,200 rpm, the first forward mode (translatory mode) becomes unstable in the linear sense. The whirling frequency at the onset of instability is 4,665 rpm, which is very close to one-half of the rotor speed. The unstable mode shape at 9,200 rpm is shown in Figure 7.7-14. Note that at the onset of instability, the rotor whirls at a natural frequency of 4,665 rpm which is the first forward mode, and not at its first critical speed of 7,500 rpm which is the third forward mode. It is commonly referred to as "Oil Whirl" since the rotor whirls with a predominated rigid body motion. The linear theory is no longer valid when the rotor speed is beyond the onset of instability and nonlinear theory should be used.

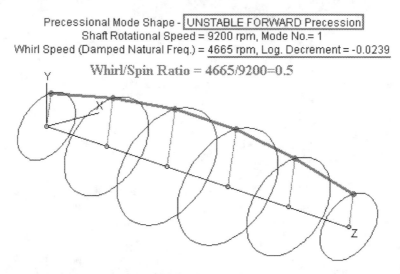

Precessional Mode Shape - UNSTABLE FORWARD Precession
Shaft Rotational Speed = 9200 rpm, Mode No.= 1
Whirl Speed (Damped Natural Freq.) = 4665 rpm, Log. Decrement = -0.0239
Whirl/Spin Ratio = 4665/9200=0.5

Figure 7.7-14 Unstable forward precessional mode

The steady state unbalance response can be calculated with these linearized bearing coefficients. Let us assume the unbalance forces are uniformly distributed along the shaft. The unbalance force is assumed to be 0.2 oz-in per shaft element. That is, 0.1 oz-in at each end of the element as shown in Figure 7.7-15.

Rotor Bearing System Data

| Axial Forces | Static Loads | Constraints | Misalignments | Shaft Bow | Time Forcing | Torsional/Axial |
| Units / Description | Material | Shaft Elements | Disks | Unbalance | Bearings | Supports | User's Elements |

	Ele	Sub	Left Unb.	Left Ang.	Right Unb.	Right Ang.	Comments
1	1	1	0.1	0	0.1	0	0.2 oz-in / element
2	2	1	0.1	0	0.1	0	
3	3	1	0.1	0	0.1	0	
4	4	1	0.1	0	0.1	0	
5	5	1	0.1	0	0.1	0	
6							

Figure 7.7-15 Unbalance input

The steady state unbalance responses at station 1 (bearing station) and station 3 are plotted in Figures 7.7-16, 7-7-17. They show that the amplitudes of motion are different in both directions, especially around the critical speeds range, due to the asymmetric properties of the bearings. From the x displacement, it shows two separate critical speeds, however, from the *y* displacement, there is only one peak response. Thus, the observed critical speeds may be different as observed from the X or Y vibration probes. It also shows that the peak response occurs at different speeds, at different finite element stations. This phenomenon is better observed by using the 3D shaft response *Animation* provided in **DyRoBeS-Rotor**, as illustrated in Figure 7.7-18. The *Animation* feature allows readers to visualize the rotor motion for startup/shutdown or at constant speed.

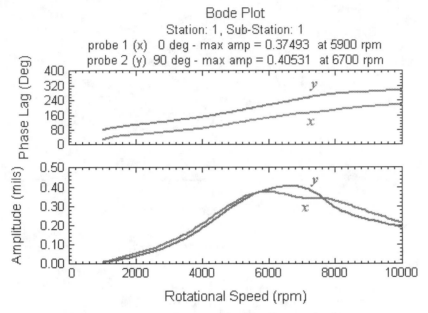

Figure 7.7-16 Bode plot for station 1

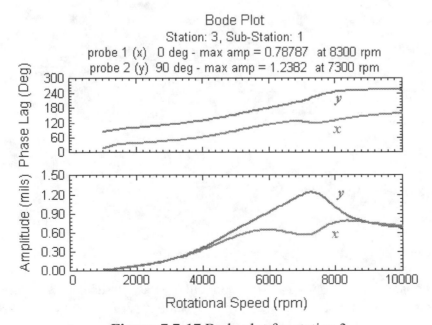

Figure 7.7-17 Bode plot for station 3

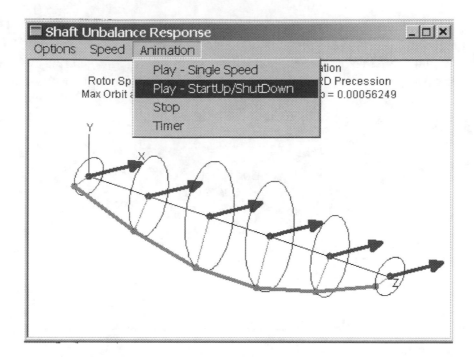

Figure 7.7-18 Shaft 3-dimensional response plot

When the rotor speed is beyond the instability threshold, the linear theory is no longer valid and nonlinear theory should be used in this region.   Now let us examine the dynamic behavior of the rotor at the instability regime by using the nonlinear analysis. The bearings are now modeled as nonlinear plain cylindrical bearings, as shown in Figure 7.7-19.

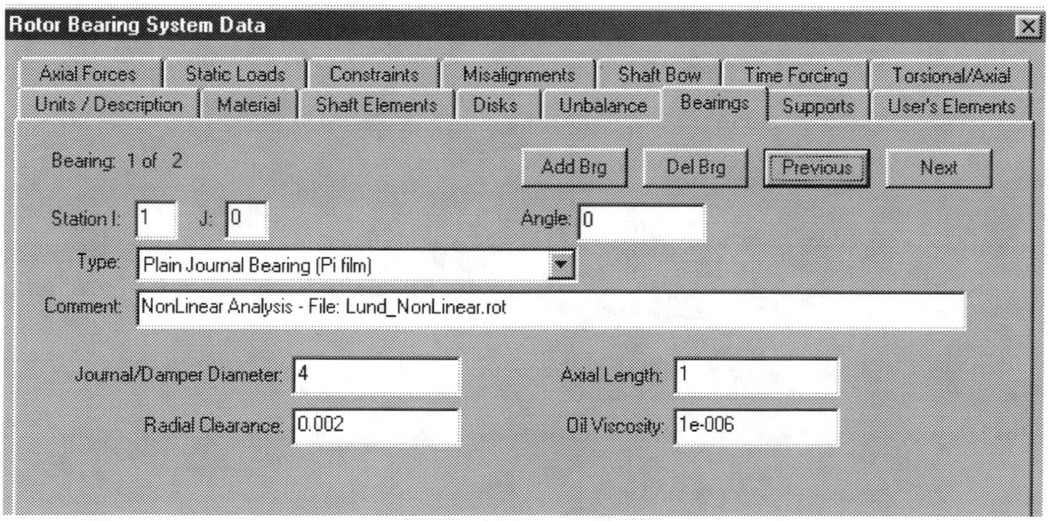

Figure 7.7-19 Nonlinear bearing data input in *DyRoBeS-Rotor*

The time transient analysis starts at 0 second and ends at 1 second, with an increment of 0.0001 seconds at a rotor speed of 9,200 rpm, as shown below:

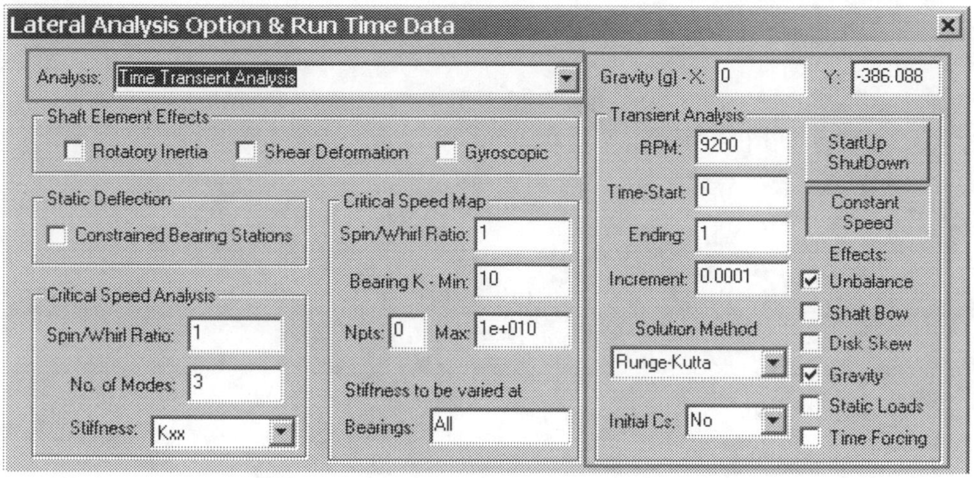

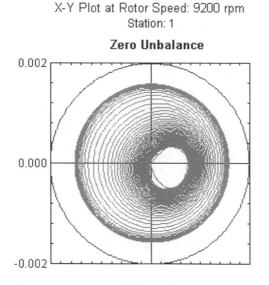

Figure 7.7-20 Time transient analysis input

Four different unbalance forces, 0, 0.2, 0.3, and 0.4 oz-in per element, are considered in the time transient analysis. Let us examine the zero unbalance force first. For zero unbalance force, the rotor whirls in a limit cycle, at very close to half of the rotor speed (whirl/spin ratio =0.5), as predicted by the linear theory. Figure 7.7-21 shows the rotor whirling orbit at the bearing station from 0 to 1 second. This self-excited motion (sub-synchronous vibration) is verified by using the FFT (Fast Fourier Transform) to convert the time-domain signals to the discrete frequency domain data.

Figure 7.7-21 Rotor whirling orbit at the bearing station with zero unbalance

Figure 7.7-22 is the FFT spectrum for Figure 7.7-21, which shows the whirling frequency at 4687 rpm with a delta frequency of 73 rpm. Since the response frequency spectrum is commonly used in examining the calculated results from the time transient analysis. Therefore, a brief description of the relationship between the parameters used in the time transient analysis and the FFT spectrum is given below.

Fast Fourier Transform (FFT)

A series of discrete data in time domain calculated by using the time transient analysis is transformed to the discrete frequency components by using FFT. The size of the time increment Δt determines the number of data points and sampling frequency. The number of data points calculated by using time transient analysis is:

$$n = \frac{(t_f - t_i)}{\Delta t} + 1 = \frac{\left(\text{Ending time} - \text{Statring time}\right)}{\text{Increment}} + 1 \qquad (7.7\text{-}1)$$

The Sampling Frequency (SF) is:

$$SF = \frac{1}{\Delta t} \qquad (7.7\text{-}2)$$

However, the FFT algorithm requires the number of FFT data points (N) to be a power of 2. Therefore, the number of data points used in FFT (N) is less than or equal to the number of data points calculated by time transient analysis (n). That is: $N \leq n$ and N must be a power of 2. The FFT calculates the discrete periodic waveforms (frequency harmonics) which sum up to make the original time domain signal. The frequency of the harmonic components in the FFT depends on the sampling frequency (SF) and the number of FFT data points (N). The range of frequencies in the FFT is from 0 (DC signal) to $SF/2$ (the Nyquist frequency, or maximum frequency in the spectrum). The number of frequency points in the FFT spectrum is: $N/2+1$. The harmonics ranges from 0 to $N/2$. The frequency for a given harmonic index is:

$$f_i = \left(\frac{i}{N}\right) \cdot SF \qquad , i = 0, 1, 2, \ldots, N/2 \qquad (7.7\text{-}3)$$

The frequency interval (delta frequency) between frequency harmonics is:

$$\Delta f = \frac{SF}{N} = \frac{1}{N \cdot \Delta t} \qquad (7.7\text{-}4)$$

The maximum frequency (Nyquist frequency) in the FFT spectrum is $SF/2$. If we would like to cover the frequency range up to mX of the rotor speed in the FFT spectrum, then we need the sample frequency to be at least $2m$X of the rotor speed ($SF = 2$ times mX rps). This means that the time step Δt used in the transient analysis has to be equal to or less than $1/SF = 1/(2m$X rps). Typically, smaller time step is used in the analysis for numerical convergence and results accuracy.

In this example, the time increment used in the analysis is 0.0001 seconds, the sampling frequency is $SF=1/0.0001 = 10000$ Hz. For a 1 second period, the number of sampled data is 10001 points. However, for FFT to be efficient, the data points used in the FFT must be a power of 2. Therefore, for 10001 data points, only 8192 points are used in the FFT (N=8192). The frequency range (The Nyquist frequency) is $SF/2$ = 5000 HZ (300,000 cpm), which is 32.6 times the synchronous speed. This is definitely enough for the range of the frequency of interest. The time step, which determined the sampling frequency, is normally decided by the numerical accuracy and convergence. The frequency interval (delta frequency) is $\Delta f = SF/N$ = 10000 Hz/8192 = 1.22 Hz = 73 rpm.

Notice from Figure 7.7-22 that the amplitude at the whirl frequency is about 0.001, which is smaller than the limit cycle orbit. This is due to the time interval for the transient response from 0 to 1 second (N=8192, Δf = 73 rpm). It takes about 0.6 seconds for the rotor to reach its steady state limit cycle. Figure 7.7-23 shows the FFT spectrum from 0 to 2 seconds (N=16384, Δf = 37 rpm). The delta frequency is now 37 rpm and the amplitude is about 0.015, which is very close to the limit cycle orbit.

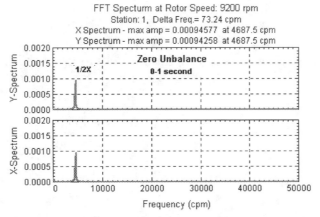

Figure 7.7-22 FFT spectrum for 1 second

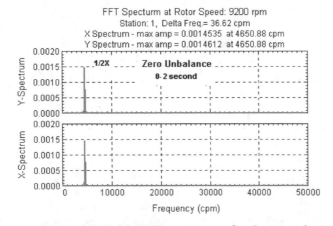

Figure 7.7-23 FFT spectrum for 2 seconds

The complete rotor whirling orbit is shown in Figure 7.7-24. It shows that the rotor whirls at its forward precessional mode, as predicated by the linear theory. For the linear theory, the Z axis is the shaft static equilibrium position. For the nonlinear transient analysis, the Z axis is the bearing geometric center line.

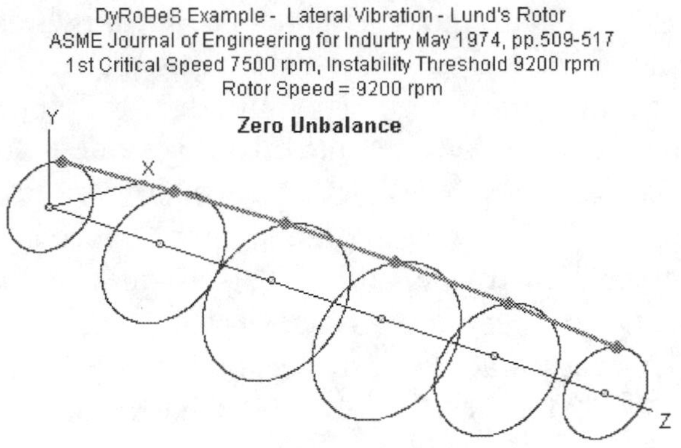
DyRoBeS Example - Lateral Vibration - Lund's Rotor
ASME Journal of Engineering for Indurtry May 1974, pp.509-517
1st Critical Speed 7500 rpm, Instability Threshold 9200 rpm
Rotor Speed = 9200 rpm
Zero Unbalance
Y
X
Z

Figure 7.7-24 Shaft steady state response at 9200 rpm – zero unbalance

Figure 7.7-25 shows the rotor steady state response (limit cycle) for unbalance of 0.2 oz-in/element. Figure 7.7-26 is the FFT spectrum at the bearing station, and it shows that the motion is predominated by the sub-synchronous component and that the synchronous vibration is small.

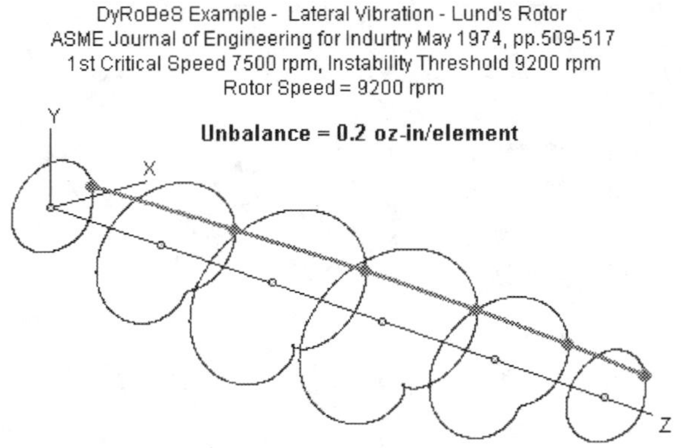
DyRoBeS Example - Lateral Vibration - Lund's Rotor
ASME Journal of Engineering for Indurtry May 1974, pp.509-517
1st Critical Speed 7500 rpm, Instability Threshold 9200 rpm
Rotor Speed = 9200 rpm
Unbalance = 0.2 oz-in/element
Y
X
Z

Figure 7.7-25 Shaft steady state response at 9200 rpm – 0.2 oz-in unbalance

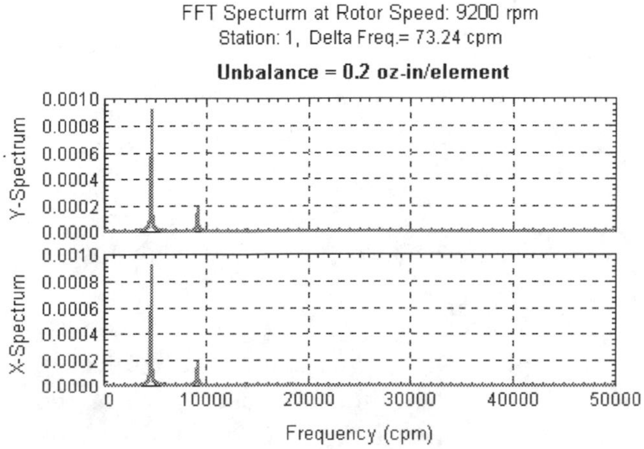
FFT Specturm at Rotor Speed: 9200 rpm
Station: 1, Delta Freq.= 73.24 cpm
Unbalance = 0.2 oz-in/element
Y-Spectrum
X-Spectrum
0.0010
0.0008
0.0006
0.0004
0.0002
0.0000
0
10000
20000
30000
40000
50000
Frequency (cpm)

Figure 7.7-26 FFT at bearing station at 9200 rpm – 0.2 oz-in unbalance

Figures 7.7-27 and 7.7-28 show the rotor response for unbalance of 0.3 oz-in per element. Figures 7.7-29 and 7.7-30 show the rotor response for unbalance of 0.4 oz-in per element. It shows that as the unbalance increases, the synchronous response (forced response) increases and the sub-synchronous response (self-excitation) decreases. They also show that the rotor response shape changes from mainly rigid body mode, as shown in Figure 7.7-24, where sub-synchronous dominates, to bending mode, as shown in Figure 7.7-29, where synchronous vibration dominates.

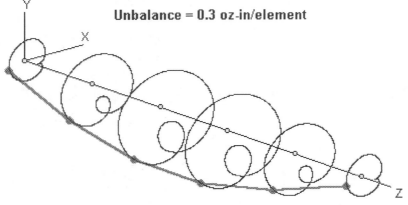

Figure 7.7-27 Shaft steady state response at 9200 rpm – 0.3 oz-in unbalance

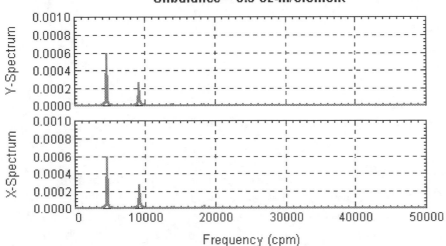

Figure 7.7-28 FFT at bearing station at 9200 rpm – 0.3 oz-in unbalance

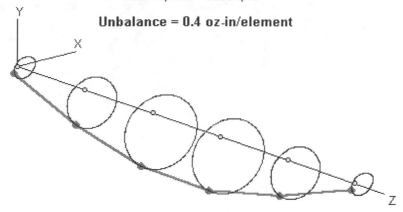

Figure 7.7-29 Shaft steady state response at 9200 rpm – 0.4 oz-in unbalance

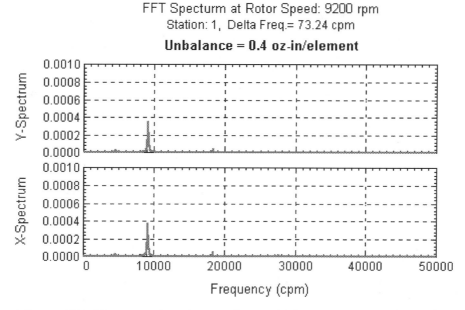

Figure 7.7-30 FFT at bearing station at 9200 rpm – 0.4 oz-in unbalance

Figures 7.7-31 and 7.7-32 show the limit cycle motions at the bearing station (station 1) and at station 3 for various unbalance forces at 9200 rpm. They show that as the unbalance increases from zero to 0.4 oz-in, the amplitude of rotor orbit decreases, and the vibration component changes from mainly sub-synchronous to synchronous vibration. However, as unbalance increases from 0.4 oz-in to a higher value, the rotor response amplitude increases too, as illustrated in Figures 7.7-33 and 7.7-34. It becomes mainly the synchronous vibration caused by the large unbalance excitation. Figures 7.7-33 and 7.7-34 show the limit cycle motions at the bearing station (station 1) and at station 3 for various unbalance forces at 9200 rpm.

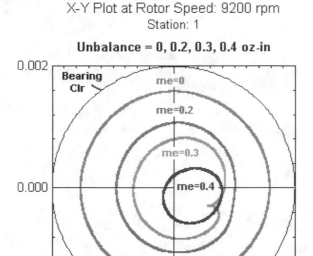

Figure 7.7-31 Rotor limit cycle motions at bearing station for various unbalance forces

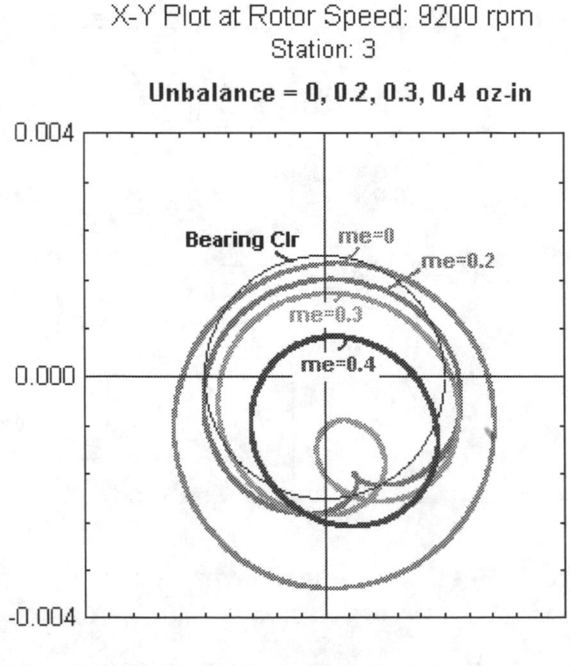

Figure 7.7-32 Rotor limit cycle motions at station 3 for various unbalance forces

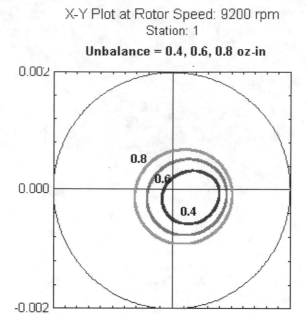

Figure 7.7-33 Rotor limit cycle motions at bearing station
for unbalance forces greater than 0.4 oz-in

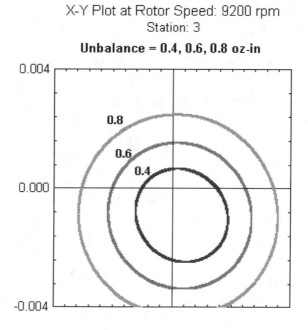

Figure 7.7-34 Rotor limit cycle motions at station 3
for unbalance forces greater than 0.4 oz-in

Example 7.2: Transient Response Through Critical Speeds

The transient response with acceleration was illustrated by a simple 2 DOF system in Chapter 3. An industrial flexible rotor system taken from Childs (1972) is presented here to demonstrate the ability of dealing with a more complicated configuration. A schematic of the system is shown below:

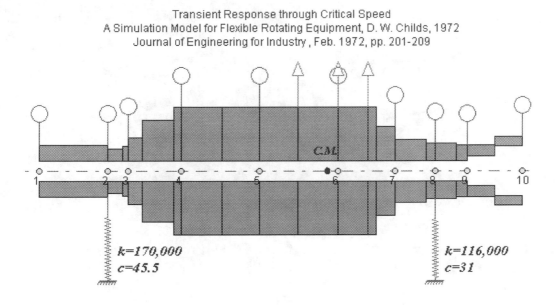

Figure 7.7-35 Rotor model for example 7.2

The rotor system is supported by two isotropic bearings, with bearing properties shown in the Figure 7.7-35. The transient simulation is performed at 5,000 rpm to 18,000 rpm in 0.07 seconds, as presented in the reference paper. Before we perform the nonlinear transient response analysis, it is beneficial to examine the results from the linear analysis first. The first three undamped forward synchronous critical speeds are calculated by using the Critical Speed Analysis and found to be 6751, 11487, and 46695 rpm. Their mode shapes and associated potential energy distributions are shown in Figure 7.7-36. It shows that the first two modes are classic rigid body modes with most of the potential energy in the real bearing (station 8) for mode 1, and front bearing (station 2) for mode 2. Note that the two rigid body modes do not necessarily have to be one translational and one conical. The third mode is the first bending mode with strain energy dominated by the elastic deformation of the rotor. The rotor transient response going through the first two rigid body modes is examined in this example.

The Whirl Speed Map, as shown in Figure 7.7-37, can be obtained from the complete Whirl Speed/Stability Analysis. The damped critical speeds obtained from the Whirl Speed Map are in good agreement with the undamped critical speeds obtained from the Critical Speed Analysis. Note that the Whirl Speed Map also can be used to identify the backward critical speeds.

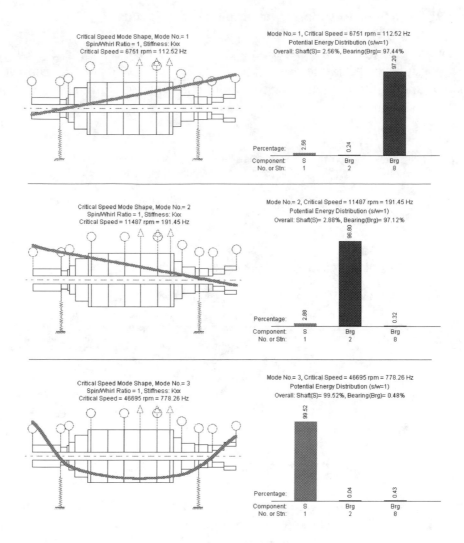

Figure 7.7-36 Critical speed mode shapes and potential energies

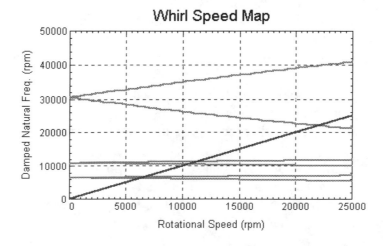

Figure 7.7-37 Whirl speed map for example 7.2

The steady state unbalance response is calculated for a rotor speed from 500 rpm to 25,000 rpm with an increment of 100 rpm. The steady state responses in Bode plots at the front bearing (station 2), rear bearing (station 8), and mid span (station 5) are shown in Figure 7.7-38.

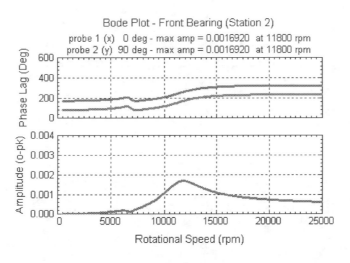

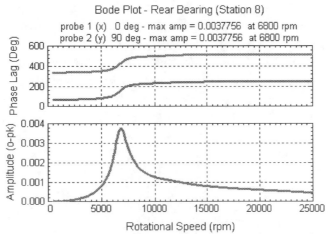

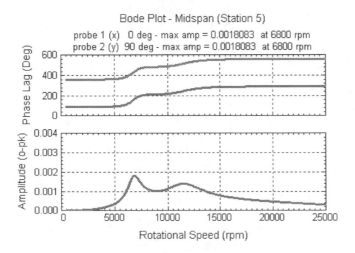

Figure 7.7-38 Bode plots

Let us plot the amplitudes in the same graph as shown in Figure 7.7-39. It shows that the left end of the rotor is mainly excited by the second mode and the right end of the rotor is mainly excited by the first mode. The mid span is excited by both modes. These results are in agreement with the mode shapes shown previously.

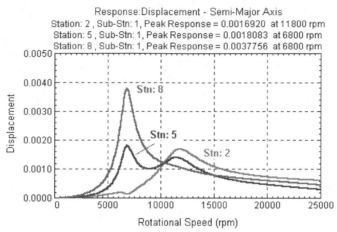

Figure 7.7-39 Steady state unbalance responses at stations 2, 5, and 8

The transient responses through critical speeds are determined by numerical integration of the equations of motion. The transient responses vs. time at both bearings (stations 2 and 8) are shown in Figure 7.7-40. The transient simulation is performed at the rotor speed from 5,000 rpm to 18,000 rpm in 0.07 seconds with a linear acceleration.

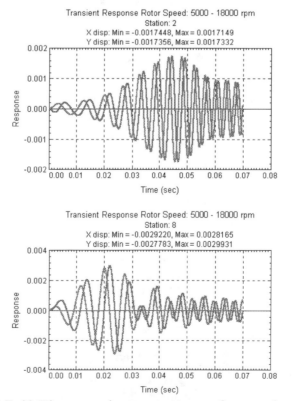

Figure 7.7-40 Time transient responses at front and rear bearings

The results are in agreement with the previously published data. It is interesting to show the responses vs. speed. The transient responses through critical speeds at stations 2, 5 and 8 are shown in Figure 7.7-41.

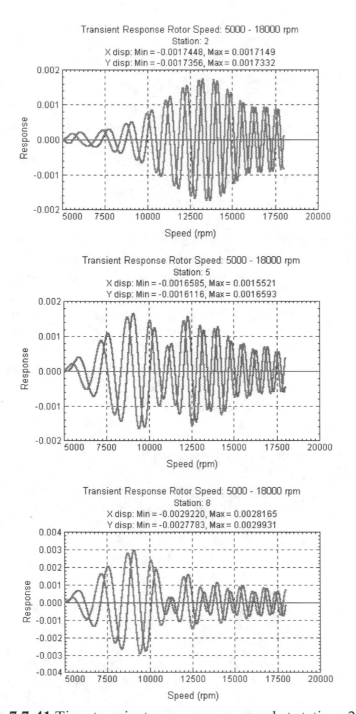

Figure 7.7-41 Time transient responses vs. speed at stations 2, 5 and 8

To compare with the steady state response, the amplitudes of the transient response at stations 2, 5 and 8 are plotted against speed in Figure 7.7-42.

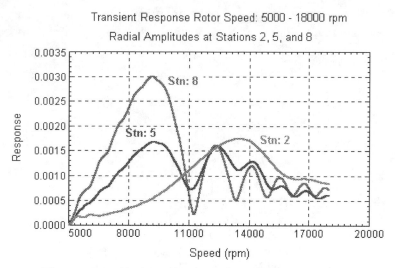

Figure 7.7-42 Amplitudes of the transient response

The maximum transient amplitude at station 8 is smaller than the amplitude of the steady state response. This has been demonstrated in Chapter 3 that the amplitude of transient response with linear acceleration is smaller than the amplitude of steady state response. The maximum transient amplitudes at stations 5 and 8 are nearly the same as those obtained from the steady state response. The amplitude fluctuation after the critical speeds is due to coexistence of the natural vibration excited at the resonance region, and the forced vibration excited by the unbalance. These two motions alternately reinforce each other and cancel each other, giving the appearance of a *beating phenomenon*. The beating phenomenon decreases as the natural motion is damped out with time. Rotor motion during the startup can be observed and animated by the 3D transient response plot, as shown in Figure 7.7-43.

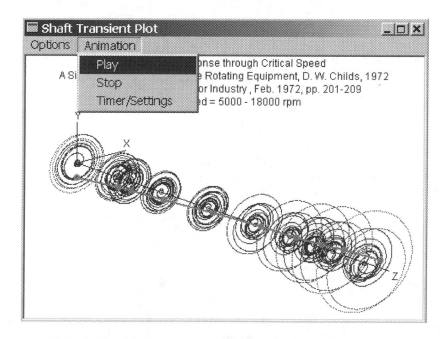

Figure 7.7-43 Animation of shaft transient response during startup

Example 7.3: Dual Rotor System

The rotor system used in this example is a dual-shaft system as shown in Figure 7.7-44. Details on rotor configuration are included for reference purposes. The system consists of two shafts interconnected by an intershaft bearing between station 4 of shaft 1 and station 10 of shaft 2. Shaft (1,2) is modeled as a (6,4) station, (5,3) element, and (24,16) degrees-of-freedom rotating assembly with stations, as indicated in the system configuration plot. The speed ratio of shaft 2 to shaft 1 is 1.5. That is: $\Omega_2 = 1.5\Omega_1$. The system parameters are listed below for reference.

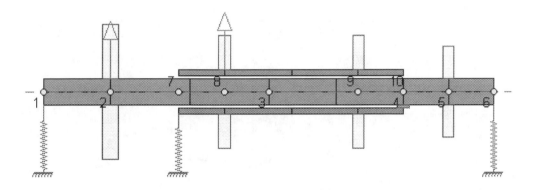

Figure 7.7-44 Dual rotor system

```
************************* Material Properties ****************************
          Property      Mass        Elastic        Shear
             no        Density      Modulus        Modulus

              1       .77702E-03   .30000E+08    .11540E+08
*************************************************************************
*************************** Shaft Elements ******************************
      Sub   Left      ------ Mass ------    --- Stiffness ----
  Ele Ele   End        Inner    Outer       Inner    Outer     Material
  no   no   Loc  Length Diameter Diameter   Diameter Diameter    no

   1
       1    .000  3.0000  .0000   1.2000     .0000   1.2000       1
   2
       1   3.000  3.5000  .0000   1.2000     .0000   1.2000       1
       2   6.500  3.5000  .0000   1.2000     .0000   1.2000       1
   3
       1  10.000  3.0000  .0000   1.2000     .0000   1.2000       1
       2  13.000  3.0000  .0000   1.2000     .0000   1.2000       1
   4
       1  16.000  2.0000  .0000   1.2000     .0000   1.2000       1
   5
       1  18.000  2.0000  .0000   1.2000     .0000   1.2000       1
  -------
   7
       1   6.000  2.0000  1.5000  2.0000    1.5000   2.0000       1
   8
       1   8.000  3.0000  1.5000  2.0000    1.5000   2.0000       1
       2  11.000  3.0000  1.5000  2.0000    1.5000   2.0000       1
   9
       1  14.000  2.0000  1.5000  2.0000    1.5000   2.0000       1
*************************************************************************
```

```
*************************** Rigid/Flexible Disks ***************************
 Stn                   Diametral       Polar      Skew    Skew
 no        Mass        Inertia        Inertia      X       Y      Offset

  2     .28000E-01    .12000         .24000      .0000   .0000    .0000
  5     .24000E-01    .90000E-01     .18000      .0000   .0000    .0000
  8     .19000E-01    .65000E-01     .13000      .0000   .0000    .0000
  9     .13000E-01    .43000E-01     .86000E-01  .0000   .0000    .0000


****************************************************************************

****************** Rotor Equivalent Rigid Body Properties ******************
 Rotor Left End              C.M.              Diametral    Polar   Speed
 no    Location  Length   Location    Mass     Inertia    Inertia   Ratio

  1     .000     20.000    9.943    .69576E-01   3.705    .423164   1.0000
  2    6.000     10.000   10.578    .42680E-01   .4816    .224344   1.5000

 --- Sum ---     20.000   10.184    .11226       4.197    .647507


****************************************************************************

*************************** Bearing Coefficients ***************************
 StnI, J   Angle   rpm    --------------- Coefficients -----------------

  1   0    .00                     (Linear Bearing)

 Kxx Kxy Kyx Kyy   150000.     .000000      .000000    150000.
 Cxx Cxy Cyx Cyy   20.0000     .000000      .000000    20.0000
 Krr Krs Ksr Kss   .000000     .000000      .000000    .000000
 Crr Crs Csr Css   .000000     .000000      .000000    .000000

  4  10    .00                     (Linear Bearing)

 Kxx Kxy Kyx Kyy   50000.0     .000000      .000000    50000.0
 Cxx Cxy Cyx Cyy   15.0000     .000000      .000000    15.0000
 Krr Krs Ksr Kss   .000000     .000000      .000000    .000000
 Crr Crs Csr Css   .000000     .000000      .000000    .000000

  6   0    .00                     (Linear Bearing)

 Kxx Kxy Kyx Kyy   100000.     .000000      .000000    100000.
 Cxx Cxy Cyx Cyy   20.0000     .000000      .000000    20.0000
 Krr Krs Ksr Kss   .000000     .000000      .000000    .000000
 Crr Crs Csr Css   .000000     .000000      .000000    .000000

  7   0    .00                     (Linear Bearing)

 Kxx Kxy Kyx Kyy   100000.     .000000      .000000    100000.
 Cxx Cxy Cyx Cyy   10.0000     .000000      .000000    10.0000
 Krr Krs Ksr Kss   .000000     .000000      .000000    .000000
 Crr Crs Csr Css   .000000     .000000      .000000    .000000


****************************************************************************

*********************** Unbalance (M x Ecc) ***********************
    Ele   SubEle   ----- Left End -----      ----- Right End ----
    no     no      Magnitude    Angle        Magnitude    Angle
     2      1      .26600E-04    .00          .00000       .00
     8      1      .26600E-04    .00          .00000       .00
****************************************************************************
****************************************************************************
```

Since this is an isotropic system with constant bearing properties, the critical speeds can be determined by using the Critical Speed Analysis. The forward synchronous critical speeds are determined by use of Spin/Whirl Ratio = 1 (Spin(1)/Whirl = 1 and Spin(2)/Whirl = 1.5) for the shaft 1 unbalance excitation and by use of Spin/Whirl Ratio = 0.6667 (Spin(1)/Whirl = 0.6667 and Spin(2)/Whirl = 1) for the shaft 2 unbalance excitation. The undamped forward synchronous critical speeds determined by using the critical speed analysis are summarized below. The critical speeds in rpm are given for shaft 1 and shaft 2 rotational speeds. The associated mode shapes are shown in Figure 7.7-45. As expected, the mode shapes are very similar for every critical speed due to shaft 1 and 2 excitations in this example.

Spin(1,2)/Whirl Ratio	1.0, 1.5 (shaft 1 excitation)	0.6667, 1 (shaft 2 excitation)
Mode 1	8250/12375	5233/7849
Mode 2	15292/22938	10062/15093
Mode 3	21836/32754	14489/21733

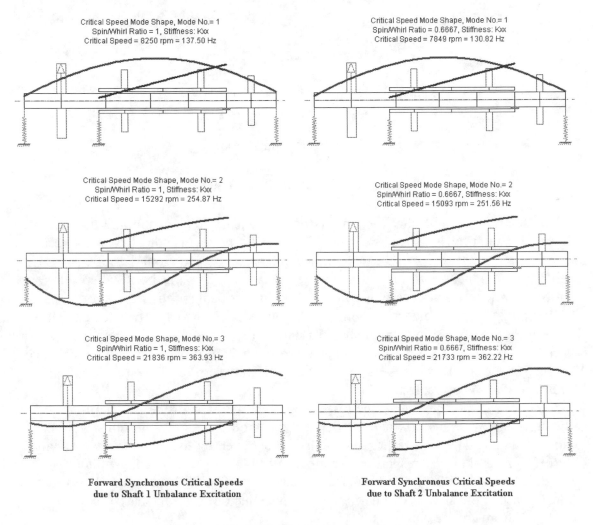

Forward Synchronous Critical Speeds
due to Shaft 1 Unbalance Excitation

Forward Synchronous Critical Speeds
due to Shaft 2 Unbalance Excitation

Figure 7.7-45 Critical speed mode shapes

The Whirl Speed Map is shown in Figure 7.7-46. The damped critical speeds determined from the Whirl Speed Map are in agreement with the critical speeds determined from the Critical Speed Analysis. Note that the critical speeds due to shaft 1 and 2 excitations are the same precessional modes, $1f$, $2f$, and $3f$ at different rotor speeds. Therefore, the mode shapes for ($1f_1,1f_2$), ($2f_1,2f_2$), and ($3f_1,3f_2$) are very similar.

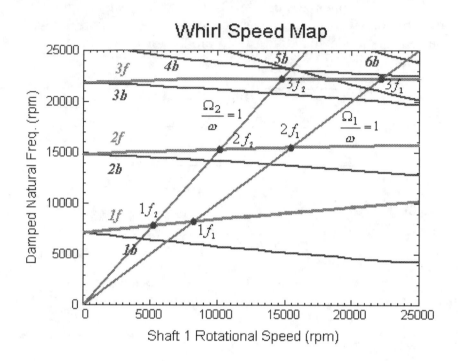

Figure 7.7-46 Whirl speed map

The steady state unbalance response (Bode and Polar Plots) at station 2 due to shaft 1 unbalance excitation is shown in Figure 7.7-47. It shows that the first peak response occurs around 8300 rpm and the second peak response occurs around 15500 rpm. This result is in agreement with the Critical Speed Analysis. The third critical speed is well damped, therefore, no peak response around the third critical speed is observed.

The steady state unbalance response at station 2 due to shaft 2 unbalance excitation is shown in Figure 7.7-48. Note that shaft 1 rotational speed is shown in the graph. Again, one can see that the first peak response occurs around 5250 rpm (shaft 1 rotational speed) and the second peak response occurs around 10000 rpm. This result is again in agreement with the critical speed analysis.

The total steady state unbalance response will be the sum of the two harmonics: One is synchronous with shaft 1 rotational speed and the other is synchronous with shaft 2 rotational speed. The total response can also be obtained by using the transient analysis. Figure 7.7-49 is the total response at stations 2 and 7 for shaft speeds of 8000 (shaft 1) and 12000 rpm (shaft 2) due to both shaft excitations. The FFT spectrum shows that the total response is the sum of two harmonic components; one is from shaft 1 excitation and the other is from shaft 2 excitation.

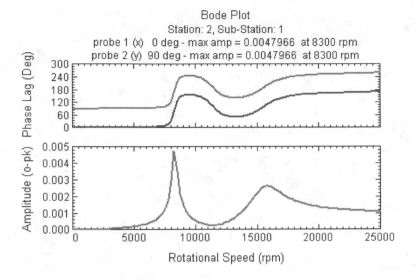

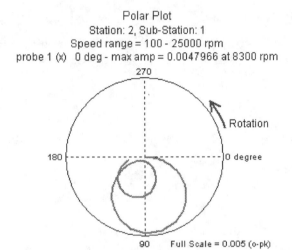

Figure 7.7-47 Steady state response at station 2 due to shaft 1 unbalance

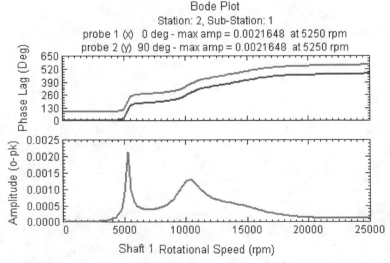

Figure 7.7-48 Steady state response at station 2 due to shaft 2 unbalance

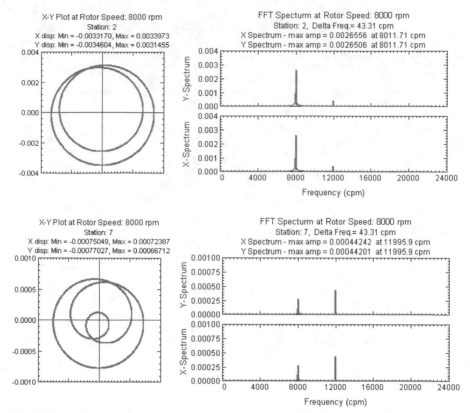

Figure 7.7-49 Steady state responses at stations 2 and 7 due to shafts 1 and 2 unbalances

Non-Linear Bearing

Next, let us consider the bearing at station 6 as a nonlinear bearing. The nonlinear bearing forces at station 6 are given by (Nelson et. al. 1989):

$$F_x = -\left[\left(k_1 r + k_3 r^3\right)\left(\frac{x}{r}\right)\right] - c\dot{x}$$

$$F_y = -\left[\left(k_1 r + k_3 r^3\right)\left(\frac{y}{r}\right)\right] - c\dot{y}$$

where

$$r = \sqrt{x^2 + y^2}$$
$$k_1 = 5.0E04 \quad \text{Lbf/in}$$
$$k_3 = 5.0E10 \quad \text{Lbf/in}^3$$
$$c = 20 \quad \quad \text{Lbf-s/in}$$

For the purposes of comparison to the previous publication, the unbalance force at station 2 is retained in this case and the unbalance force at station 8 is removed. In addition to the unbalance force at station 2, a side load at the nonlinear bearing location (station 6) is applied as illustrated in Figure 7.7-50.

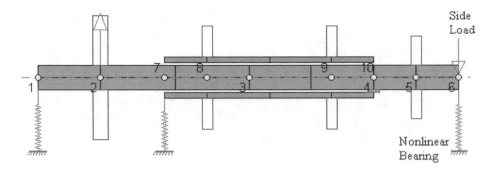

Figure 7.7-50 Dual rotor with nonlinear bearing and side load

The time transient analysis is used to analyze this system. The steady state response orbits at stations 2, 6, 7, and 9 at zero side load are shown in Figure 7.7-51. Since this nonlinear bearing is a general isotropic bearing, the steady state displacement orbits are circular for zero side load. This result can also be obtained by using the Steady State Synchronous Response Analysis with general nonlinear isotropic bearings. Readers are encouraged to experiment with this two different analysis options.

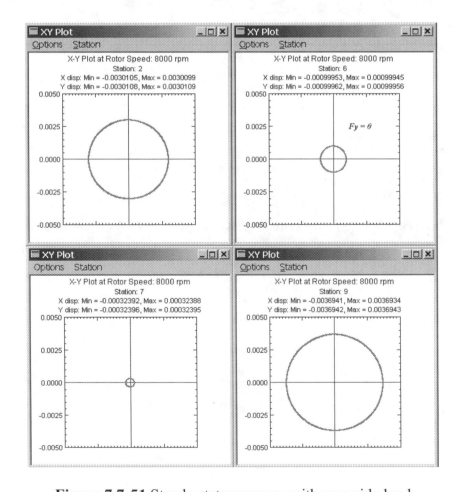

Figure 7.7-51 Steady state response with zero side load

With non-zero side load, the orbits are no longer centered circular. Therefore, the transient analysis must be employed in order to study the side load effect. The steady state response orbits at stations 2, 6, 7 and 9 for a side load of 100 Lbf in the negative Y direction, are shown in Figure 7.7-52. These results are in good agreement with the previous publication by using Trigonometric Collocation Method.

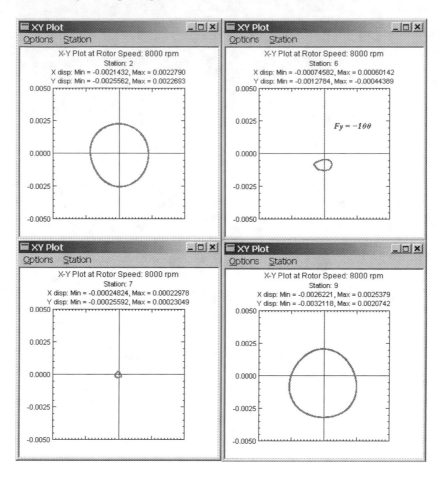

Figure 7.7-52 Steady state response with Fy = - 100 Lbf side load

Torsional Vibration 8

8.1 General Introduction

Torsional vibration can cause serious failures in rotating geared trains if the complete system is not properly evaluated and designed. Torsional failures of couplings, shafts, and gears can destroy the entire equipment. Since the complete geared train consists of many rotating components from different manufacturers, in contrast to lateral vibrations that have been carefully analyzed by the component manufacturers, the torsional vibration for the complete geared train has very often been ignored until catastrophic failures occur. A typical compressor system, as shown in Figure 8.1-1, consists of a motor, couplings, spacer, gear, and high-speed compressors. Very often the equipment users specify and acquire each rotating component without analyzing the torsional vibrations. Figure 8.1-2 shows the catastrophic failure on couplings, which also damaged the motor shaft and compressor after the equipment user changed the motor without analyzing the torsional vibration. Both free and forced torsional analyses are discussed in this Chapter. The most common analyses for torsional vibration are:

- Natural frequencies and normal modes calculation
- Steady state forced response analysis
- Startup transient analysis
- Short circuit transient analysis

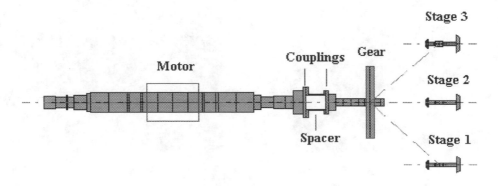

Figure 8.1-1 A geared train for torsional vibration

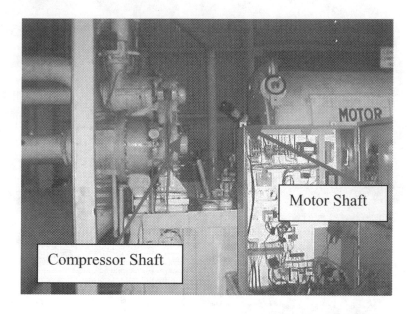

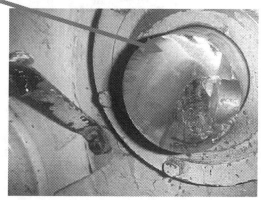

Figure 8.1-2 Failure due to torsional resonance

8.2 Element Equation

For pure torsional vibration, the motion of each finite element station is described by a rotational displacement (θ) about the spinning axis. There are two degrees-of-freedom for a typical shaft element (one for each end), as shown in Figure 8.2-1. The rotational displacement at section s is denoted by $\theta(s,t)$. The equation of motion for a typical element is derived from the energy expressions.

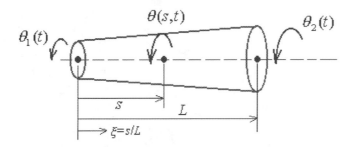

Figure 8.2-1 Torsional degrees-of-freedom for a typical element

The kinetic energy and potential energy for a typical element with homogeneous material properties are:

$$T = \frac{1}{2}\int_0^L \rho I_p \dot{\theta}^2 \, ds = \frac{1}{2}\int_0^1 \rho I_p L \dot{\theta}^2 \, d\xi \tag{8.2-1}$$

and

$$V = \frac{1}{2}\int_0^L GJ\left(\frac{\partial \theta}{\partial s}\right)^2 ds = \frac{1}{2}\int_0^1 \frac{GJ}{L}\left(\frac{\partial \theta}{\partial \xi}\right)^2 d\xi \tag{8.2-2}$$

Where ρ is the mass density and G is the shear modulus. I_p is the polar **area** moment of inertia about the centroidal axis, and J is a geometric property of the cross-section. For a circular cross-section, $J = I_p$. For other cross-sections, J is less than I_p and readers can refer texts on advanced strength of materials or elasticity to obtain expressions for J (Young, 1989, Ker Wilson, 1956). The local coordinate s is measured from 0 to L and the local non-dimensional coordinate $\xi=s/L$ is measured from 0 to 1. The rotational displacement within the element can be approximately expressed in terms of the two end displacements and shape functions as:

$$\theta(s,t) = N_1(s)\,\theta_1(t) + N_2(s)\,\theta_2(t) \tag{8.2-3}$$

The linear shape functions can be obtained from the two end boundary conditions:

$$\theta(0,t) = \theta_1(t)$$
$$\theta(L,t) = \theta_2(t) \tag{8.2-4}$$

The linear shape functions are:

$$N_1 = 1 - \frac{s}{L} = 1 - \xi$$

$$N_2 = \frac{s}{L} = \xi$$

(8.2-5)

Substitution of the displacement approximation, Eq. (8.2-3), into the energy expressions leads to:

$$T = \frac{1}{2} \sum_{i=1}^{2} \sum_{j=1}^{2} m_{ij} \dot{\theta}_i \dot{\theta}_j$$

(8.2-6)

and

$$V = \frac{1}{2} \sum_{i=1}^{2} \sum_{j=1}^{2} k_{ij} \theta_i \theta_j$$

(8.2-7)

The stiffness and mass coefficients are:

$$k_{ij} = \int_0^1 \frac{GJ}{L} N_i' N_j' \, d\xi$$

(8.2-8)

$$m_{ij} = \int_0^1 \rho L I_p N_i N_j \, d\xi$$

(8.2-9)

The generalized force (torque) is determined from the virtual work of the external forces:

$$\delta W = \int_0^L \tau(x,t) \, \delta\theta \, dx = \sum_{i=1}^{2} T_i \, \delta\theta_i$$

(8.2-10)

The generalized torque is:

$$T_i = \int_0^L \tau(x,t) N_i \, dx$$

(8.2-11)

Note that the symbol T is also used for the torque, which should not be confused with the kinetic energy.

For a conical (tapered) element with a circular cross-section ($J=I_p$), as illustrated in Figure 8.2-1, the internal polar area moment of inertia is expressed in terms of the left end properties:

$$I_p = \frac{\pi}{2}\left(r_o^4 - r_i^4\right) = I_{p,L}\left[1 + \delta_1\xi + \delta_2\xi^2 + \delta_3\xi^3 + \delta_4\xi^4\right] \qquad (8.2\text{-}12)$$

where $I_{p,L}$ is the polar area moment of inertia of the left cross-section of the element.

$$I_{p,L} = \frac{\pi}{2}\left(r_{o,L}^4 - r_{i,L}^4\right) \qquad (8.2\text{-}13)$$

The coefficients δ_i are given in appendix A and are not repeated here. By substituting the shape function and integrating the energy expressions, we obtain:

$$K = \begin{bmatrix} k_{11} & k_{12} \\ k_{21} & k_{22} \end{bmatrix} = \begin{bmatrix} k & -k \\ -k & k \end{bmatrix} \qquad (8.2\text{-}14)$$

$$k = \frac{GI_{p,L}}{L}\left(1 + \frac{\delta_1}{2} + \frac{\delta_2}{3} + \frac{\delta_3}{4} + \frac{\delta_4}{5}\right) \qquad (8.2\text{-}15)$$

and

$$M = \begin{bmatrix} m_{11} & m_{12} \\ m_{21} & m_{22} \end{bmatrix} \qquad (8.2\text{-}16)$$

$$m_{11} = \rho L I_{p,L}\left(\frac{1}{3} + \frac{\delta_1}{12} + \frac{\delta_2}{30} + \frac{\delta_3}{60} + \frac{\delta_4}{105}\right) \qquad (8.2\text{-}17)$$

$$m_{22} = \rho L I_{p,L}\left(\frac{1}{3} + \frac{\delta_1}{4} + \frac{\delta_2}{5} + \frac{\delta_3}{6} + \frac{\delta_4}{7}\right) \qquad (8.2\text{-}18)$$

$$m_{12} = m_{21} = \rho L I_{p,L}\left(\frac{1}{6} + \frac{\delta_1}{12} + \frac{\delta_2}{20} + \frac{\delta_3}{30} + \frac{\delta_4}{42}\right) \qquad (8.2\text{-}19)$$

For a cylindrical element, where the left and right area properties are the same ($\delta_i = 0$), the matrices degenerate into more familiar expressions:

$$K = \frac{GI_p}{L}\begin{bmatrix} 1 & -1 \\ -1 & 1 \end{bmatrix} \qquad (8.2\text{-}20)$$

$$M = \frac{\rho L I_p}{6}\begin{bmatrix} 2 & 1 \\ 1 & 2 \end{bmatrix} \qquad (8.2\text{-}21)$$

The mass matrices derived by utilizing the shape functions are referred to as *consistent mass matrices*, since the same shape functions are used to derive both mass and stiffness matrices. For the cylindrical element in the torsional vibration, it is sometimes assumed that the element inertia is lumped at the element's two end points with half of the inertia at each end. This diagonal mass matrix is referred to as *lumped mass matrix* for the cylindrical element:

$$M = \frac{\rho L I_p}{2} \begin{bmatrix} 1 & 0 \\ 0 & 1 \end{bmatrix} \tag{8.2-22}$$

8.3 Two-inertia System

The simplest torsional model in rotating machinery applications includes a driver and driven units (two-inertia system) as shown in Figure 8.3-1.

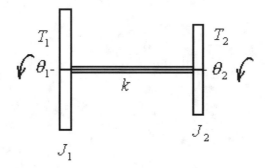

Figure 8.3-1 Two-inertia torsional model

θ_1 and θ_2 are the independent rotational coordinates for inertia J_1 and J_2. The two inertias are connected by an elastic spring of k. The two inertias are also subject to two external torques T_1 and T_2. The equations of motion can be obtained from the Lagrange's equations or Newton's law of motion, which in matrix form are:

$$\begin{bmatrix} J_1 & 0 \\ 0 & J_2 \end{bmatrix} \begin{Bmatrix} \ddot{\theta}_1 \\ \ddot{\theta}_2 \end{Bmatrix} + \begin{bmatrix} k & -k \\ -k & k \end{bmatrix} \begin{Bmatrix} \theta_1 \\ \theta_2 \end{Bmatrix} = \begin{Bmatrix} T_1 \\ T_2 \end{Bmatrix} \tag{8.3-1}$$

For the free vibration (eigenvalue) problem, assuming harmonic motion:

$$\begin{Bmatrix} \theta_1 \\ \theta_2 \end{Bmatrix} = \begin{Bmatrix} A_1 \\ A_2 \end{Bmatrix} e^{j\omega t} \tag{8.3-2}$$

Substituting these normal mode solutions into the homogeneous governing equations gives:

$$\begin{bmatrix} k - \omega^2 J_1 & -k \\ -k & k - \omega^2 J_2 \end{bmatrix} \begin{Bmatrix} A_1 \\ A_2 \end{Bmatrix} = \begin{Bmatrix} 0 \\ 0 \end{Bmatrix} \tag{8.3-3}$$

The above equations are satisfied for any A_1 and A_2 if the determinant is zero, which leads to the characteristic equation:

$$J_1 J_2 \omega^4 - k(J_1 + J_2)\omega^2 = 0 \qquad\qquad (8.3\text{-}4)$$

Thus, the two natural frequencies and normal modes of the system are found to be:

$$\omega_0 = 0 \qquad\qquad \text{and} \qquad \left(\frac{A_1}{A_2}\right)_0 = 1 \qquad\qquad (8.3\text{-}5)$$

$$\omega_1 = \sqrt{\frac{k(J_1 + J_2)}{J_1 J_2}} \qquad \text{and} \qquad \left(\frac{A_1}{A_2}\right)_1 = \frac{-J_2}{J_1} \qquad (8.3\text{-}6)$$

The normal mode with zero natural frequency is known as the *rigid-body mode* without any elastic deformation in the system, and all the inertias move in the same direction with the same amplitudes, as illustrated in Eq. (8.3-5). In rotating machinery, this rigid body mode is an important characteristic and can be used to verify the accuracy of the computational tools. In the first elastic mode with non-zero natural frequency, the inertias move in opposite directions and the shaft is actively involved with a node point, which remains stationary. Figure 8.3-2 shows the mode shapes for a two-inertia system with $J_1 = 1$, $J_2 = 3$, and $k = 1200$.

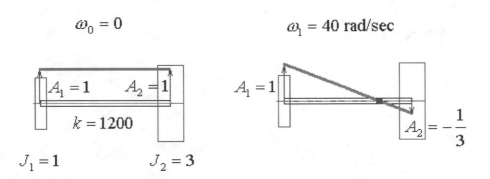

Figure 8.3-2 Mode shapes for the two-inertia system

8.4 Geared System

For a geared system, the torsional stiffness at the gear mesh is much stiffer than other torsional flexibilities of the system and thus can be considered a rigid connection. Consider the simplest geared system as shown in Figure 8.4-1, where the speed ratio of shaft 2 to shaft 1 is n, that is:

$$n = \frac{\Omega_2}{\Omega_1} \qquad\qquad (8.4\text{-}1)$$

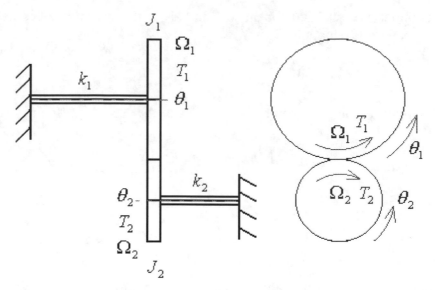

Figure 8.4-1 The simplest geared system

In this illustration, the direction of angular coordinates is assumed to be the same, regardless of the direction of the shaft rotation. The direction of the external torques is the same as the direction of the shaft rotation. Also, for illustrative purposes, the system is constrained and rigid body mode does not exist in this example.

The kinetic energy and potential energy of the system are:

$$T = \frac{1}{2}J_1\dot{\theta}_1^2 + \frac{1}{2}J_2\dot{\theta}_2^2 \tag{8.4-2}$$

$$V = \frac{1}{2}k_1\theta_1^2 + \frac{1}{2}k_2\theta_2^2 \tag{8.4-3}$$

The virtual work done by the external torques is:

$$\delta W = T_1\delta\theta_1 + (-T_2)\delta\theta_2 \tag{8.4-4}$$

Note that the negative sign for T_2 exists because the torque is specified to be in the direction of the shaft rotation in this case and it is in the opposite direction of the angular displacement θ_2. There are two angular coordinates specified in the system, however, they are related by the gear ratio. The total independent generalized coordinate is only one and this is a single degree-of-freedom system. Thus, the geared system can be reduced to an equivalent single shaft system. Consider shaft 1 as the reference shaft and substitute the geometric constraint $(\theta_2 = -n\theta_1)$ into the energy and work expressions. This yields a single DOF system:

$$T = \frac{1}{2}J_1\dot{\theta}_1^2 + \frac{1}{2}n^2J_2\dot{\theta}_1^2 = \frac{1}{2}\left(J_1 + n^2J_2\right)\dot{\theta}_1^2 \tag{8.4-5}$$

$$V = \frac{1}{2}k_1\theta_1^2 + \frac{1}{2}n^2k_2\theta_1^2 = \frac{1}{2}\left(k_1 + n^2k_2\right)\theta_1^2 \tag{8.4-6}$$

and

$$\delta W = T_1\delta\theta_1 + (-T_2)\delta\theta_2 = \left(T_1 + nT_2\right)\delta\theta_1 \tag{8.4-7}$$

Thus, the equivalent inertia and stiffness of shaft 2 referred to shaft 1 are n^2J_2 and n^2k_2. The equivalent torque is nT_2. By applying Lagrange's equation, gives the equation of motion:

$$\left(J_1 + n^2J_2\right)\ddot{\theta}_1 + \left(k_1 + n^2k_2\right)\theta_1 = \left(T_1 + nT_2\right) \tag{8.4-8}$$

The natural frequency is determined from the homogeneous equation:

$$\omega = \sqrt{\frac{\left(k_1 + n^2k_2\right)}{\left(J_1 + n^2J_2\right)}} \tag{8.4-9}$$

The equivalent single shaft system with shaft 1 as the reference shaft is shown below.

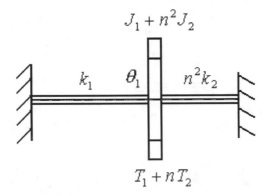

Figure 8.4-2 The equivalent one degree-of-freedom system

Note that if shaft 2 is chosen as the reference shaft, the same procedure as described above yields:

$$\left(\frac{1}{n^2}J_1 + J_2\right)\ddot{\theta}_2 + \left(\frac{1}{n^2}k_1 + k_2\right)\theta_2 = \left(\frac{-1}{n}T_1 - T_2\right) \tag{8.4-10a}$$

or

$$\left(J_1 + n^2J_2\right)\ddot{\theta}_2 + \left(k_1 + n^2k_2\right)\theta_2 = \left(-nT_1 - n^2T_2\right) \tag{8.4-10b}$$

By substituting θ_2 with $-n\theta_1$, the above equation, derived by using shaft 2 as the reference shaft, is identical to the equation derived by using shaft 1 as the reference shaft.

Therefore, the choice of the reference shaft is irrelevant to the analysis results. In **DyRoBeS**, shaft 1 is always used as the reference shaft for convenience.

If the angular coordinates for a geared system are chosen such that the displacements are in the same direction of the shaft rotation, as shown in Figure 8.4-3, then the geometric relationship becomes $\theta_2 = n\theta_1$. Thus, the equivalent inertia and stiffness of shaft 2 referred to shaft 1 are $n^2 J_2$ and $n^2 k_2$. The virtual work done by the external torques is:

$$\delta W = T_1 \delta\theta_1 + T_2 \delta\theta_2 = (T_1 + nT_2)\delta\theta_1 \tag{8.4-11}$$

Note that the torque is in the same direction as the angular displacement now. The equivalent torque is nT_2. Therefore, the results are identical to the results presented before.

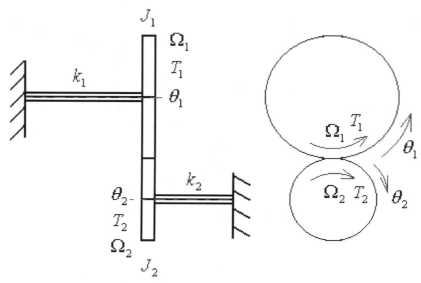

Figure 8.4-3 Angular coordinates in the same direction as the rotational speeds

Consider a practical and simple gear train system as shown in Figure 8.4-4. The system consists of a motor (J_1), a gear (J_2), a pinion (J_3), and a compressor (J_4). The speed ratio (compressor speed/motor speed) is n. There are four lumped inertias for each component and there is an angular displacement associated with each component.

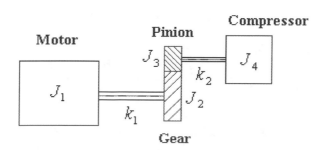

Figure 8.4-4 Simple geared system

The kinetic energy and potential energy of the system are:

$$T = \frac{1}{2}\left(J_1\dot{\theta}_1^2 + J_2\dot{\theta}_2^2 + J_3\dot{\theta}_3^2 + J_4\dot{\theta}_4^2\right) \tag{8.4-12}$$

$$V = \frac{1}{2}\left[k_1\left(\theta_2 - \theta_1\right)^2 + k_2\left(\theta_4 - \theta_3\right)^2\right] \tag{8.4-13}$$

There are two external torques (driving torque acting on the motor and resistant torque acting on the compressor) and the virtual work done by these two external torques is:

$$\delta W = T_m\delta\theta_1 + T_c\delta\theta_4 \tag{8.4-14}$$

Note that the angular displacements are chosen so that the displacements are in the same direction of the shaft rotation in this case. Although the system has four physical angular coordinates (one for each inertia), there is a constraint due to gear (rigid) connection. Therefore, this is a three-degrees-of-freedom system (3=4-1) with $\theta_3 = n\theta_2$. The three independent generalized coordinates (q_1, q_2, q_3) can be defined by the following relationship with the four physical coordinates $(\theta_1, \theta_2, \theta_3, \theta_4)$:

$$q_1 = \theta_1, \quad q_2 = \theta_2, \quad q_3 = \theta_4/n \tag{8.4-15a}$$

or

$$\theta_1 = q_1, \quad \theta_2 = q_2, \quad \theta_3 = nq_2, \quad \theta_4 = nq_3 \tag{8.4-15b}$$

Substituting the above relationship into the energy expressions yields:

$$T = \frac{1}{2}\left(J_1\dot{q}_1^2 + J_2\dot{q}_2^2 + J_3n^2\dot{q}_2^2 + J_4n^2\dot{q}_3^2\right) \tag{8.4-16}$$

$$V = \frac{1}{2}\left[k_1\left(q_2 - q_1\right)^2 + k_2n^2\left(q_3 - q_2\right)^2\right] \tag{8.4-17}$$

and

$$\delta W = T_m\delta q_1 + (T_c n)\delta q_3 \tag{8.4-18}$$

The equations of motion for the complete gear train can be derived from the Lagrange's equations as follows:

$$\begin{bmatrix} J_1 & 0 & 0 \\ 0 & J_2 + n^2J_3 & 0 \\ 0 & 0 & n^2J_4 \end{bmatrix}\begin{Bmatrix} \ddot{q}_1 \\ \ddot{q}_2 \\ \ddot{q}_3 \end{Bmatrix} + \begin{bmatrix} k_1 & -k_1 & 0 \\ -k_1 & k_1 + n^2k_2 & -n^2k_2 \\ 0 & -n^2k_2 & n^2k_2 \end{bmatrix}\begin{Bmatrix} q_1 \\ q_2 \\ q_3 \end{Bmatrix} = \begin{Bmatrix} T_m \\ 0 \\ nT_c \end{Bmatrix} \tag{8.4-19}$$

A numerical example taken from Wachel and Szenasi (1993) for Figure 8.4-4 is presented below. The parameters are:

$n = 2$ (compressor rpm/motor rpm)

$J_1 = 15000$ in-Lb-s^2 (Motor inertia)
$J_2 = 600$ in-Lb-s^2 (Gear inertia)
$J_3 = 150$ in-Lb-s^2 (Pinion inertia)
$J_4 = 2500$ in-Lb-s^2 (Compressor inertia)

$k_1 = 2E09$ in-Lb/rad (Motor shaft stiffness)
$k_2 = 5E07$ in-Lb/rad (Compressor shaft stiffness)

The natural frequencies calculated by using **DyRoBeS** are listed below. They are in agreement with the published data.

```
***************   Torsional Frequency Analysis   ******************

    **********   Shaft  1  Speed Ratio =  1.000   **********
    **********   Shaft  2  Speed Ratio =  2.000   **********

      Mode          RPM            R/S             HZ

        0         .000000      *** Rigid Body Mode ***
        1         1647.14        172.488         27.4523
        2         13358.5        1398.90         222.641
```

8.5 Couplings

The driven unit is connected to the driver by a coupling or couplings with a spacer. The coupling is probably the most important component in the elimination of torsional vibration problems, and is often the only mechanical component that can be changed in the final design stage. There are many types of couplings used in rotating machinery. Some require regular maintenance and some require less maintenance. The choice of coupling depends on the cost, installation and maintenance requirements, along with torsional vibration analysis. Three types of couplings commonly used in turbomachinery are shown in Figure 8.5-1. Gear type coupling is the most cost effective, and is widely used in many industrial applications. However, it requires regular maintenance (lubrication). Disc coupling does not require lubrication as the disc packs are pre-assembled and can be easily retrofitted in the field. However, the disc pack is very soft in the axial direction so axial vibration should be checked when disc type coupling is used for some applications. The resilient (Holset) type coupling with rubber blocks is suitable to provide resilience and damping for severe applications, such as synchronous motor or engine drive applications. The rubber elements provide low stiffness and good damping characteristics compared to the gear and disc type couplings. The disadvantages of the rubber couplings include cost, large weight, inertia and space requirements, and difficulty balancing.

Gear and disc couplings are usually modeled as linear mass-spring components. However, the rubber type coupling is modeled as a non-linear component due to its highly non-linear torque-deflection curve. The linearized coupling dynamic stiffness and damping for a rubber type coupling at any instant can be obtained as:

$$k_c = \frac{\partial T}{\partial \theta} \tag{8.5-1}$$

$$C_c = \frac{k_c}{D_m \omega} \tag{8.5-2}$$

where D_m is the dynamic magnifier of the coupling for a specified rubber hardness. The coupling damping varies directly with torsional stiffness and inversely with frequency and dynamic magnifier.

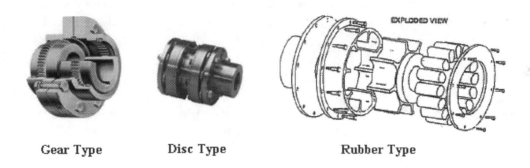

Gear Type   Disc Type   Rubber Type

Figure 8.5-1 Couplings

8.6 System Equations of Motion

The system equations of motion that describe the behavior of the entire system are obtained by assembling the elemental matrices and vectors. The element coordinates must be transformed to the generalized global coordinates and the displacement continuity must be enforced in the assembly process. For a geared system, the equations of motion of each branch should be referenced to shaft 1 (the reference shaft) by multiplying the inertia, damping, and stiffness matrices by n^2 and multiplying the load vector by n, where n is the speed ratio of the branch rotor to the reference rotor. The following table gives the conversion between the actual and equivalent systems for a geared train:

Parameters	Actual	Equivalent
Displacement	q	q/n
Inertia	J	Jn^2
Stiffness	k	kn^2
Damping	c	cn^2
Torque	T	Tn

The assembled system equations of motion are as follows:

$$M\ddot{q} + C\dot{q} + Kq = T(q,t) \qquad (8.6\text{-}1)$$

where q is the generalized coordinates referenced to shaft 1 (the reference shaft) and T is the generalized torque vector. For a linear system, the torque is a function of time only. For a nonlinear system, the torque can also be function of displacement. M and K are real symmetric matrices in general. Moreover, M derived from kinetic energy is positive definite and K derived from potential energy is positive semidefinite in rotating machinery applications. In general, the damping matrix C in torsional vibration cannot be constructed from the element level analogous to the assembly of the element inertia and the stiffness matrices. For torsional vibration, the damping is usually specified in the system level by the critical damping factors for a number of modes of interest. The torsional critical damping factors for systems with dry type couplings have been reported in the range of 1-5 percent (0.01-0.05). Typically, 2 percent (0.02) of the critical damping factor is used in the forced response calculation for conservative design. For systems with resilient rubber couplings, the critical damping factor of the coupling modes (where the modal displacements are dominated by the couplings) can go up to 6-10 percent. Again, 6 percent (0.06) is normally used for conservative design. If the damping is specified by the critical damping factors, the proportional damping matrix is generated by the procedures described below. For systems with nonlinear flexible coupling, the linearized coupling stiffness is assembled into the global stiffness matrix K for the use of the following procedure.

From the orthogonality relationship, we have:

$$\Phi^T M \Phi = \hat{M} = diag(M_r)$$
$$\Phi^T K \Phi = \hat{K} = diag(K_r) = diag(\omega_r^2 M_r) \qquad (8.6\text{-}2)$$
$$\Phi^T C \Phi = \hat{C} = diag(C_r) = diag(2\xi_r \omega_r M_r)$$

where Φ is the modal matrix obtained from the undamped eigenvalue problem:

$$\left(K - \omega_i^2 M\right)\varphi_i = 0 \qquad (8.6\text{-}3)$$

$$\Phi = \left[\varphi_1 \varphi_2 \varphi_3 ... \varphi_N\right] \qquad (8.6\text{-}4)$$

From the modal mass and damping equations, we have:

$$\Phi^{-1} = \hat{M}^{-1}\Phi^T M \qquad (8.6\text{-}5)$$

$$C = \Phi^{-T}\hat{C}\Phi^{-1} \qquad (8.6\text{-}6)$$

Note that modal mass and damping matrices, $\hat{M}$ and $\hat{C}$, are diagonal matrices. The physical damping matrix can be expressed as:

$$C = \sum_{i=1}^{\hat{N}<<N} \left(\frac{2\xi_i \omega_i}{M_i} \right) (M\varphi_i)(M\varphi_i)^T \tag{8.6-7}$$

where the number of retained modes ($\hat{N}$) is far less than the total number of modes (N) in practice. The damping in the torsional system has little influence on the natural frequencies, since it is generally very small. The undamped natural frequencies are usually calculated and plotted in the frequency interference diagram (Campbell Diagram). The damping matrix is required only when the forced response calculation is performed.

8.7 Natural Frequencies and Modes

For pure torsional vibration, the natural frequencies are independent of the rotational speed. This is in contrast to the lateral vibration, where the natural frequencies are generally a function of the shaft rotational speed. The system torsional undamped and damped natural frequencies can be determined from the linearized homogeneous equation of motion. For undamped natural frequencies, the real eigenvalue problem is:

$$\left(K - \omega_i^2 M \right) \varphi_i = 0 \qquad i=1,2,3,\dots N_{\text{dof}} \tag{8.7-1}$$

For damped natural frequencies, the complex eigenvalue problem is:

$$\begin{bmatrix} -M^{-1}C & -M^{-1}K \\ I & 0 \end{bmatrix} y = \lambda \, y \tag{8.7-2}$$

In rotating machinery applications, the system torsional stiffness matrix is positive semidefinite. The complete geared rotating system is unrestrained and rigid body motion is permitted in the system. Therefore, one (and only one) zero natural frequency will be found in the frequency calculation. The associated mode is referred to as a *rigid body mode*. The general motion of a torsional rotating system consists of a combination of rigid body motion and elastic motion.

Since damping has an imperceptible effect on the torsional natural frequencies, the undamped natural frequencies are usually calculated and plotted in the frequency interference diagram, also known as *Campbell diagram*, as illustrated in Figure 8.7-1. This is the most important design analysis in torsional vibration. The various excitation frequency lines are superimposed in the diagram. The intersections between the natural frequencies and the excitation frequency lines are the interference points where resonance can occur. In general, the interference points for the steady state operating conditions should be at least 10% away from the operating speed. However, it may not always be possible to avoid the interference points. In these cases, the generalized modal torque or forced response for these interference points need to be examined to determine the degree of concern. The generalized modal torque for the modal analysis is given by:

$$\Gamma = \varphi^T T \tag{8.7-3}$$

It shows from Eq. (8.7-3) that if the excitation is at or near the node point, the overall modal torque is still small and the corresponding interference will not cause any vibration concerns. Steady state excitation frequencies for various applications are summarized in Corbo and Malanoski (1996). Most AC motors and generators produce fluctuations at line frequency and twice line frequency by a number of different phenomena. For the reciprocating engines, the excitation frequencies are dependent upon the number of cylinders, number of strokes, and firing orders. In general, the excitation frequencies are harmonics of the fundamental frequency, such as 1X, 2X, nX, etc.

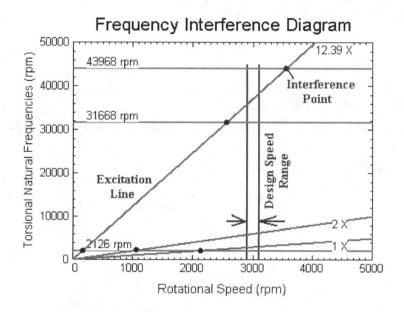

Figure 8.7-1 Torsional frequency interference diagram

For the synchronous motors and variable frequency drivers, the transient interference points can occur during startup where resonance can be damaging to the system. For the synchronous motors, the excitation frequency of the pulsating torque is equal to twice the slip frequency and decreases linearly from twice line frequency at zero motor speed to zero frequency at synchronous speed. The slip frequency can be written as:

$$f_s = f_L \left(\frac{N_{syn} - N_m}{N_{syn}} \right) \quad \text{Hz} \tag{8.7-4}$$

where f_L is the line frequency (50 or 60 Hz), N_m is the motor speed (rpm) at a given time during the startup, and N_{syn} is the synchronous speed (rpm) calculated by the following expression:

$$N_{syn} = \frac{120 \cdot f_L}{\text{No of Poles}} \quad \text{rpm} \tag{8.7-5}$$

The excitation frequency for the synchronous motor pulsating torque during startup is equal to 2X the slip frequency:

$$\omega_{exc} = 2\pi \cdot 2f_s = 4\pi \cdot f_s \qquad\qquad (8.7\text{-}6)$$

All natural frequencies less than 2X line frequency will be excited by the pulsating torque during synchronous motor startup. The torsional resonant (critical) speeds are obtained by setting the excitation frequency to the system natural frequency. The resonance speeds are expressed as follows:

$$N_{cr} = N_{syn}\left(1 - \frac{\omega_i}{4\pi\, f_L}\right) \qquad \text{where } \omega_i < 4\pi\, f_L \qquad i = 1,2,\dots \qquad (8.7\text{-}7)$$

As the natural frequency decreases, transient critical speed increases and approaches the synchronous speed. If all the system natural frequencies are above 2X line frequency, there will be no resonant speeds during the startup. Figure 8.7-2 shows the Frequency Interference Diagram for a compressor system driven by a 50 Hz 4 poles synchronous motor. It shows that the complete geared train has to go through two torsional critical speeds before reaching the synchronous speed. The synchronous speed is 1500 rpm. Two natural frequencies below the 2X line frequency are: 2775 cpm and 681 cpm. From Eq. (8.7-7), the two transient critical speeds due to the oscillating torque are 806 rpm and 1330 rpm.

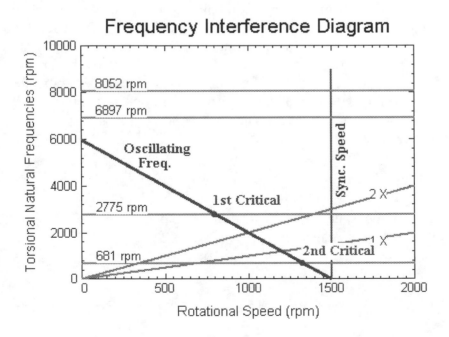

Figure 8.7-2 Frequency interference diagram for a synchronous motor

8.8 Steady State Forced Response

At steady state conditions, the general rotational motion of rotating machinery consists of a combination of rigid body motion and elastic deformation.

$$q = q_r + q_e \tag{8.8-1}$$

The steady state excitation (driving and load) torque includes a constant torque and a harmonic excitation torque.

$$T = T_0 + T_e \cos(\omega_{exc} + \theta_e) = T_0 + T_{ec} \cos(\omega_{exc}) + T_{es} \sin(\omega_{exc}) \tag{8.8-2}$$

Since the rigid body motion does not contribute to the element vibratory torque and stress, separating the rigid body motion and elastic deformation yields two sets of equations:

$$M\ddot{q}_r + C\dot{q}_r + Kq_r = T_0(t) \tag{8.8-3}$$

for rigid body motion, and

$$M\ddot{q}_e + C\dot{q}_e + Kq_e = T_{ec} \cos(\omega_{exc}t) + T_{es} \sin(\omega_{exc}) \tag{8.8-4}$$

for elastic motion.

Depending on the type of machinery and applications, the excitation frequency and magnitude can be constant or a function of rotor rotational speed. The magnitudes of the excitation torques are usually determined from the test data. Some values have been recommended by Wachel and Szenasi (1993) and extended by Corbo and Malanoski (1996). In general, one percent (1%) of the transmitted torque is recommended for 1X excitation and half percent (0.5%) of the transmitted torque is recommended for 2X excitation. For propellers, 15% is recommended.

For steady state elastic motion, the solution is:

$$q_e = q_{ec} \cos(\omega_{exc}t) + q_{es} \sin(\omega_{exc}t) \tag{8.8-5}$$

On differentiation of the solution and substitution into the equations of motion, the following set of linear algebra equations apply:

$$\begin{bmatrix} K - \omega_{exc}^2 M & \omega_{exc}C \\ -\omega_{exc}C & K - \omega_{exc}^2 M \end{bmatrix} \begin{Bmatrix} q_{ec} \\ q_{es} \end{Bmatrix} = \begin{Bmatrix} T_{ec} \\ T_{es} \end{Bmatrix} \tag{8.8-6}$$

Once the steady state solution is established, the vibratory torque for element i can be obtained:

$$T_i = k_i(q_{e,i+1} - q_{e,i}) + c_i(\dot{q}_{e,i+1} - \dot{q}_{e,i}) = |T_i|\cos(\omega_{exc}t + \theta_i) \tag{8.8-7}$$

The steady state stress then can be calculated. For circular cross sections, the maximum stress occurs at the outer surface:

$$\tau_{max} = \frac{|T| \cdot d_o / 2}{J}, \qquad J = \frac{\pi}{32}\left(d_o^4 - d_i^4\right) \tag{8.8-8}$$

d_o and d_i are the outer and inner diameters. For non-circular cross sections, the torsional stresses generally have to be determined by using finite element programs with the calculated torques from the above torsional analysis. The equation of motion for the rigid body motion, Eq. (8.8-3), can be numerically integrated to obtain the acceleration time and speed profile.

8.9 Transient Motion

Two types of transient motions are considered in the torsional vibration analysis. One is the startup transient motion and the other is due to the short circuit torque excitation.

8.9.1 Startup Transient Motion

In many applications, such as a synchronous motor driven system, the rotor torsional motion needs to be studied during startup and through critical speeds. The startup time for the rotating equipment to reach the full design speed is also an important design parameter for the equipment manufacturers. The general expression for the equations of motion of a torsional system is given as follows:

$$M\ddot{q} + C\dot{q} + Kq = T(t, q, \dot{q}) \tag{8.9-1}$$

In general, the driving torque at any instant during startup includes a constant average torque and an oscillating torque:

$$T_d = T_{avg} + T_{osc} \sin(\omega_{exc} t) \tag{8.9-2}$$

where the magnitudes of the torque are usually a function of the rotational speed. Figure 8.9-1 shows the typical torque curves for a synchronous motor at a reduced voltage start. The excitation frequency of the pulsating torque for a synchronous motor during startup is equal to the twice the slip frequency and is a linear function of the rotor speed:

$$\omega_{exc}(rad/\sec) = 2\pi \cdot 2f_s = 4\pi\, f_L\left(\frac{N_{syn} - N_m}{N_{syn}}\right) = 4\pi\, f_L - \frac{4\pi\, f_L}{N_{syn}}\left(N_m\right) \tag{8.9-3}$$

The load torque is the resistance of the system. Figure 8.9-1 also shows a typical compressor total load curve for throttled condition during startup. The load curves for different inlet valve positions or loading conditions can be obtained from the equipment manufacturers. The coupling torsional stiffness at different loading conditions is commonly listed in the coupling catalogue. Various numerical integration schemes, such as Newmark-β, Wilson-θ, Runge-Kutta, and Gear's methods, as described in the lateral vibration, can be employed to solve this equation of motion. Note that the total motion is integrated in this case.

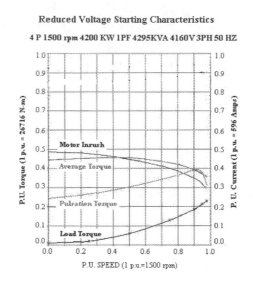

Figure 8.9-1 Torque curves for a synchronous motor driven compressor

8.9.2 Short Circuit Transient Motion

Short circuit occurs very often during the initial phase of plant development, particularly in the process plant. The general expression for the air gap fault torque is commonly normalized with respect to the rated torque:

$$T = T_{rated} \left[T_0\, e^{-a_0 t} + T_1\, e^{-a_1 t} \sin\left(\omega t + \phi_1\right) + T_2\, e^{-a_2 t} \sin\left(2\omega t + \phi_2\right) \right] \qquad (8.9\text{-}4)$$

where

T_{rated}, T_0, T_1, T_2	are rated torque and torque components.
a_0, a_1, a_2	are time constants.
ϕ_1, ϕ_2	are phase angles.

The rated torque and horsepower are related by:

$$T_{rated}\left(Lb_f - in\right) = \frac{12 \times 33000 \times hp}{2\pi \times rpm} = \frac{63025 \times hp}{rpm} \qquad (8.9\text{-}5)$$

The following table summarizes the typical short circuit torques for a 4200 kW, 4160V, 3PH, 4 Poles, 50 Hz **synchronous** motor. The Line-to-Line short circuit is a fault between two of the phase circuits while the motor is running. It produces an excitation torque which has 1X and 2X the line frequency components. The three-phase short circuit would occur when all three phases are shorted together, and produces an excitation torque with the fundamental (1X) frequency component only. Figures 8.9-2 and 8.9-3 show the short circuit excitation torques for a Line-to-Line fault and three-phase fault.

4200 kW, 4 Poles, 50 Hz **Synchronous** Motor		
	L-L Fault	3 Phase Fault
T_0	-1.29861	-1.12280
a_0	8.10062	11.15171
T_1	-3.56847	-2.95105
a_1	6.55393	6.95576
ϕ_1	0	0
T_2	1.44020	0
a_2	1.16981	0
ϕ_2	0	0

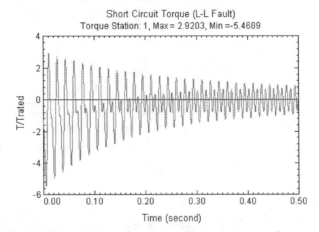

Figure 8.9-2 Line to Line short circuit excitation torque for a synchronous motor

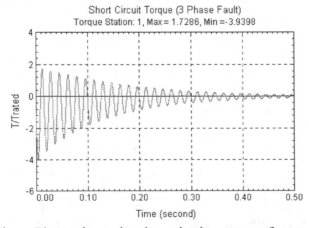

Figure 8.9-3 Three Phase short circuit excitation torque for a synchronous motor

The following table summarizes the short circuit torques for a typical 2000 HP, 2 poles, 60 Hz **induction** motor.

2000 HP, 2 Poles, 60 Hz **Induction** Motor		
	L-L Fault	3 Phase Fault
T_0	0	0
a_0	0	0
T_1	6.07	5.94
a_1	38.5	57.3
ϕ_1	0	0
T_2	-3.04	0
a_2	37.6	0
ϕ_2	0	0

From the above two examples, the short circuits often produce torques which are several times the rated torque of the motor. The Line-to-Line short circuit produces a larger torque than that of a 3-phase short circuit. When the AC motor is switched on, large transient torque occurs at line frequency and then quickly dies out. The general torque expression for the short circuit torque can also be used for the startup torque.

8.10 Examples

Two numerical examples are presented in this session. One is a marine steam turbine propulsion system. The steady state forced response is discussed in this example. The second example is a three-stages compressor system driven by a synchronous motor. Startup transient motion is analyzed in this example.

Example 8.1 – Marine Steam Turbine Propulsion System

A marine steam turbine propulsion system presented in Ehrich (1992) is used in the first example. The system configuration is shown in Figure 8.10-1. The propeller is driven by two steam turbines through double speed reduction gearing. The system consists of three branches with 5 shafts. Each shaft has two (2) finite element stations and the system has a total of ten (10) finite element stations. The gear connections are assumed to be rigid. Therefore, there are four constraints from the gear meshes in the system. The total independent generalized coordinates (degrees-of-freedom) in the system are six (6); i.e., 6=10-4. The design speeds are given below:

	Speed (rpm)	Speed ratio (n)
Propeller	85.0	1.0000
High-Pressure Intermediate Shaft	799.8	9.4094
High-Pressure Turbine	6650.1	78.2365
Low-Pressure Intermediate Shaft	799.8	9.4094
Low-Pressure Turbine	3403.6	40.0424

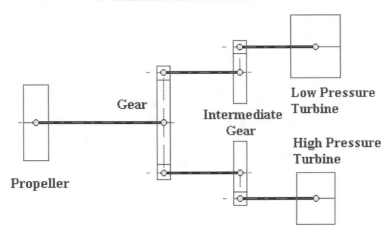

Figure 8.10-1 Marine steam turbine propulsion system

The system is modeled with lumped inertias and elastic torsional stiffness for each shaft. The values of polar moment of inertia, stiffness, and damping for the actual and equivalent systems are listed below.

Inertia (Lb-in-s^2)	Actual	Equivalent (n^2J)
Propeller	2.454E06	2.454E06
Gear	0.826E06	0.826E06
High-Pressure Intermediate Gear	2.723E04	2.411E06
High-Pressure Turbine	2.612E02	1.599E06
Low-Pressure Intermediate Gear	1.283E04	1.136E06
Low-Pressure Turbine	1.509E04	24.20E06

Stiffness (Lb-in/rad)	Actual	Equivalent (n^2k)
Propeller shaft	826E06	826E06
High-Pressure Intermediate Shaft	24.17E06	2140E06
High-Pressure Turbine Shaft	14.26E06	87301E06
Low-Pressure Intermediate Shaft	203.94E06	18056E06
Low-Pressure Turbine Shaft	30.51E06	48925E06

Damping (Lb-in-s/rad)	Actual	Equivalent (n^2c)
Propeller	3.86E06	3.86E06
High-Pressure Turbine	42.738	0.2616E06
Low-Pressure Turbine	108.77	0.1744E06

In addition to the direct dampings from the propeller and turbines, a critical damping factor of 0.00398 (logarithmic decrement of 0.025) is used to estimate the combination of damping effects from elastic hysteresis of the shaft, sliding fits, bearings, gear element, and others, as discussed in the previous publication. The steady state constant torque varies with the square of the propeller speed:

$$T_p = T_o \left(\frac{\Omega_p}{\Omega_o} \right)^2 \tag{8.10-1}$$

where T_o (22.24E06 Lb-in) is the propeller torque at rated speed and Ω_o (8.901 rad/sec = 85 rpm) is the rated speed. It is assumed that the magnitude of the steady state exciting (oscillating) torque at propeller T_s is equal to 10 percent of the steady-state constant torque. This excitation frequency from the propeller is at the fundamental blade passing frequency. The number of blades is five (5) in this example. The excitation frequency is five times the propeller speed. Therefore, the steady state exciting torque is:

$$T_e = 0.1 \cdot T_p \cdot \cos(5\Omega_p)$$

$$= 28070 \cdot \Omega_p^2 \cdot \cos(5\Omega_p) = 28070 \cdot \left(\frac{2\pi}{60} \right)^2 N_p^2 \cos(5\Omega_p) \tag{8.10-2}$$

The undamped natural frequencies and modes are obtained from the eigenvalue problem. The frequency interference diagram is shown in Figure 8.10-2. The first three natural frequencies are found to be 177.7, 220.2, and 1282.6 rpm. Since the excitation frequency is five times the propeller speed, a 5X excitation line is drawn in the interference diagram. Two critical speeds (interference points) are found below the propeller rated speed and they are: 35.5 rpm for the first mode and 44.0 rpm for the second mode. They are equal to the natural frequencies divided by the number of blades. The mode shapes in the generalized equivalent coordinates for the first two modes are plotted in Figures 8.10-3 and 8.10-4.

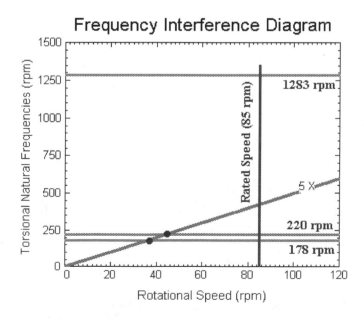

Figure 8.10-2 Frequency interference diagram

Torsional Vibration Mode Shape
Mode No.= 1, Frequency = 178 cpm = 2.96 Hz

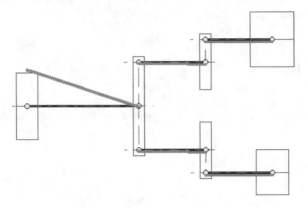

Figure 8.10-3 The first mode shape

Torsional Vibration Mode Shape
Mode No.= 2, Frequency = 220 cpm = 3.67 Hz

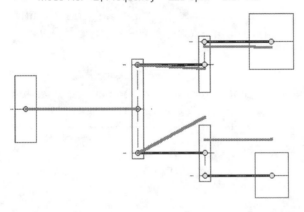

Figure 8.10-4 The second mode shape

At the first critical speed, the maximum amplitude of vibration occurs at the propeller while relatively small amplitudes occur at the gear and turbine branches. It is expected that the resonance response at the propeller and vibratory torque at the associated shaft can be very high due to the excitation from the propeller. At the second critical speed, all motion occurs at the turbine branches and the propeller branch has zero amplitude. This is a consequence of the exact equality of the turbine branch frequencies. Since the modal excitation torque is the torque times the modal displacement, there will be no resonance response from the second natural mode in the turbine branches due to propeller excitation. This system is referred to as a *nodal driver*, since the gear is a node point in the torsional vibration. The above natural frequencies and modes, calculated by using the matrix Q-R algorithm from **DyRoBeS**, are in agreement with the results from the Holzer method in reference.

In order to evaluate the forced response, the damping matrix is formulated with the specified modal damping (critical damping factor) at the first mode and the direct damping at the propeller and turbines:

$$C = \left(\frac{2\xi_1\omega_1}{M_1}\right)(M\varphi_1)(M\varphi_1)^T + C_d \tag{8.10-3}$$

Once the damping matrix is established, the steady state solution can be obtained using:

$$q_e = q_{ec}\cos(\omega_{exc}t) + q_{es}\sin(\omega_{exc}t) \tag{8.10-4}$$

It should be noted that in this example the input data (inertia, stiffness, damping, torque, and speed ratio) are actual values. However, the solution procedures are performed in the equivalent system and the program automatically converts the actual values to the equivalent values. Once the response for the equivalent system is solved, the actual response can be obtained by the reverse conversion. The steady state vibratory torque in the propeller shaft is shown in Figure 8.10-5. As expected, the maximum vibratory torque occurs at the first critical speed (35.5 rpm), with an amplitude of 4.166E06 Lb-in. There is no resonance as the speed passes through the second critical speed. The result is in agreement with those results obtained from the Energy Balance Method in reference.

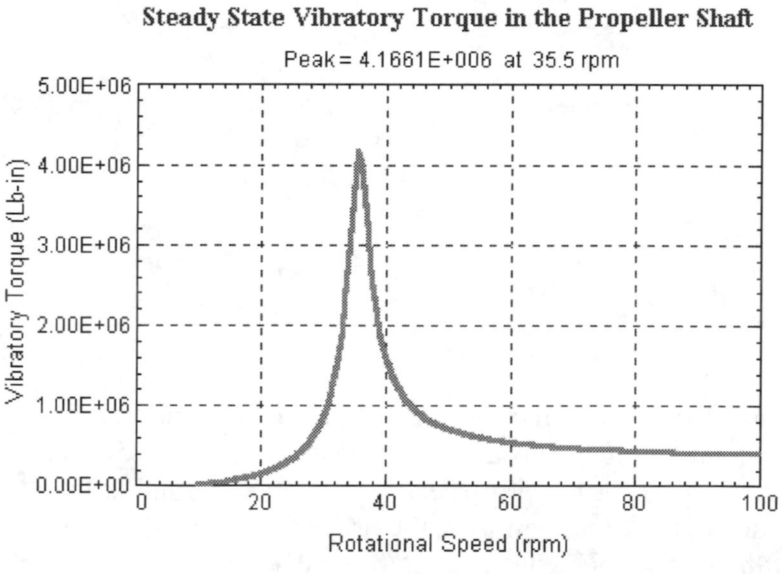

Steady State Vibratory Torque in the Propeller Shaft
Peak = 4.1661E+006 at 35.5 rpm

Figure 8.10-5 Steady state vibratory torque at propeller shaft

This complete system can also be modeled as a three-branches and 3-rotors system. The intermediate gears and turbines are combined into a single rotor system as presented in the reference and shown in Figure 8.10-6. Again, the complete system has eight (8) finite element stations and two (2) constraints due to the gear connections. The equivalent system has a total of six (6) degrees-of-freedom.

Marine Steam Turbine Propulsion System
3 Branches and 3 Rotors

Figure 8.10-6 Alternative model for the marine steam turbine propulsion system

The results from this 3-branches and 3-rotors system are identical to the results obtained from the previous 3-branches and 5-rotors system. The first two mode shapes in the generalized equivalent coordinates are plotted in Figures 8.10-7 and 8.10-8 for reference purposes.

Torsional Vibration Mode Shape
Mode No.= 1, Frequency = 178 cpm = 2.96 Hz

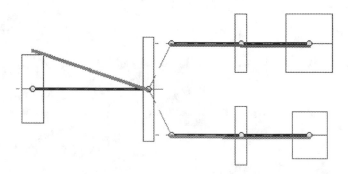

Figure 8.10-7 The first mode shape

Torsional Vibration Mode Shape
Mode No.= 2, Frequency = 220 cpm = 3.67 Hz

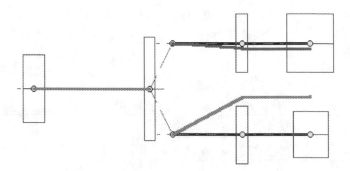

8.10-8 The second mode shape

Example 8.2 – A Three-Stage Compressor Driven by a Synchronous Motor

This example is a three-stages compressor driven by a synchronous motor. ***DyRoBeS*** is capable of modeling this system with distributed inertia and stiffness properties. However, the lumped inertias and stiffnesses are provided by their manufacturers, respectively. Therefore, a lumped model is used and a schematic of the system configuration is shown in Figure 8.10-9. A 50 Hz, 4200 kW, 4-poles, synchronous motor is the prime driver. The synchronous speed is 1500 rpm. The speed ratio is shown in the following table:

Unit	Speed Ratio
Motor – Bull Gear	1
Stage 1	8.43
Stage 2	12.04
Stage 3	16.85

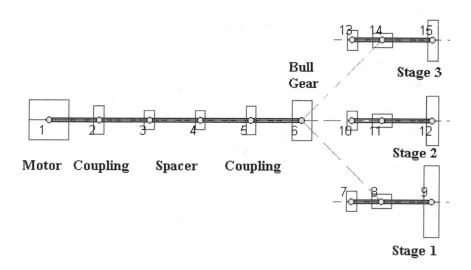

Figure 8.10-9 A synchronous motor driven 3-stages compressor

The inertia and stiffness have been converted into the effective inertia and stiffness by multiplying the square of the speed ratio. The effective inertia and stiffness are given in the following tables:

Station No.	Effective Inertia (kg-m2)	Comment
1	158.3	Motor
2	2.872	Coupling Outer Hub
3	16.27	Coupling Inner Hub + 1/2 Spacer

4	16.35	Coupling Inner Hub + 1/2 Spacer
5	2.872	Coupling Outer Hub
6	165.6	Bull Gear
7	0.4034	Stage 1 Collar
8	3.749	Stage 1 Pinion
9	248.4	Stage 1 Impeller
10	0.8233	Stage 2 Collar
11	2.067	Stage 2 Pinion
12	77.5	Stage 2 Impeller
13	1.614	Stage 3 Collar
14	1.68	Stage 3 Pinion
15	28.29	Stage 3 Impeller

Station I	Station J	Effective Stiffness (Nm/rad)	Comment
1	2	1.4510E+07	Motor
2	3	1.4789E+06	Coupling
3	4	4.8807E+08	Spacer
4	5	1.4789E+06	Coupling
5	6	7.5850E+06	Bull Gear
7	8	8.9020E+07	Stage 1 Collar end
8	9	8.8220E+07	Stage 1 Impeller end
10	11	1.7120E+08	Stage 2 Collar end
11	12	8.1850E+07	Stage 2 Impeller end
13	14	1.6040E+08	Stage 3 Collar end
14	15	1.7180E+07	Stage 3 Impeller end

There are three geometric constraints:

Gear Mesh Connection	Comment
6-8	Bull Gear – Stage 1
6-11	Bull Gear – Stage 2
6-14	Bull Gear – Stage 3

The steady state Frequency Interference Diagram is shown in previous Figure 8.7-2. It shows that the system has to go through two torsional critical speeds during startup before it reaches the synchronous speed. The mode shapes for the two critical speeds are shown in Figure 8.10-10 And 8.10-11. It should be noted that the first critical speed is mode number 2 and the second critical speed is mode number 1, as illustrated in Figure 8.7-2.

Mode Shape - First Critical Speed

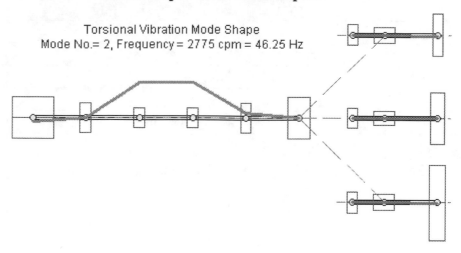

Torsional Vibration Mode Shape
Mode No.= 2, Frequency = 2775 cpm = 46.25 Hz

Figure 8.10-10 The first critical speed mode shape

Mode Shape - Second Critical Speed

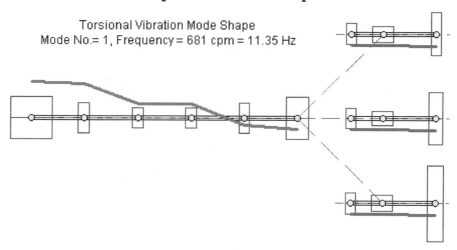

Torsional Vibration Mode Shape
Mode No.= 1, Frequency = 681 cpm = 11.35 Hz

Figure 8.10-11 The second critical speed mode shape

The two couplings are rubber type couplings and their dynamic characteristics are nonlinear. However, they are oversized and the torque-deflection curve in the operating range is very close to linear. Only the linearized stiffness values were available in this case. In view of the mode shapes, significant motions occur at the couplings. Therefore, 6% of modal damping factors are assumed for these two modes for the forced response analysis. The driving torques and load torques are shown in Figure 8.10-12. Since the loads are quite evenly distributed among three stages, the three stages load curves are very close. The motor is started with reduced voltage.

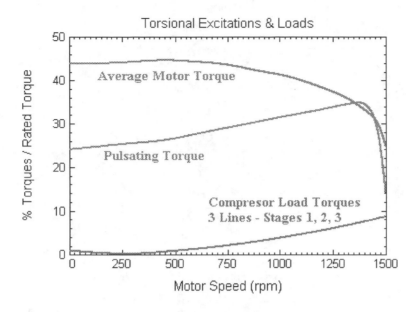

Figure 8.10-12 Torque curves

The startup transient analysis is performed in this example. The startup time for the system to reach the synchronous speed is around 16 seconds, as shown in Figure 8.10-13.

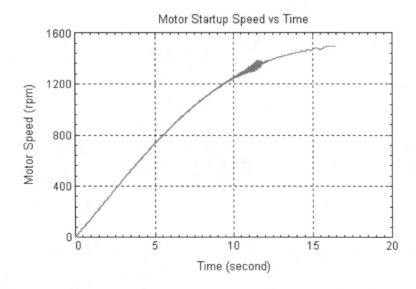

Figure 8.10-13 Motor speed vs. time

The transient vibratory torque, with respect to the steady state torque at the bull gear shaft, is shown in Figure 8.10-14. It shows that the transient vibratory torque caused by the motor pulsating torque during the startup, can be more than 2 times the steady state torque.

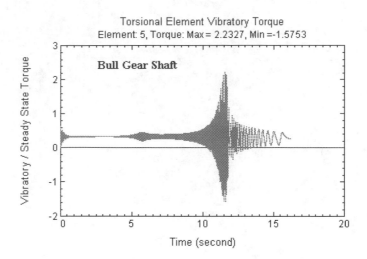

Figure 8.10-14 Bull gear vibratory torque

The vibratory torque at the stage 1 impeller shaft is shown in Figure 8.10-15. Again, the vibratory torque can be several times larger than the steady state torque. It should be noted that if the effective inertia and stiffness are used in the analysis, then the actual torque is obtained by dividing the effective torque by the speed ratio, as discussed previously.

$$T_{actual} = \frac{T_{effective}}{n},\quad \text{where } n \text{ is the speed ratio.}$$

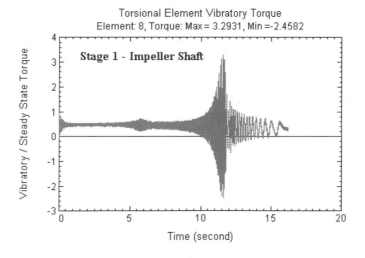

Figure 8.10-15 Stage impeller shaft vibratory torque

Balancing 9

9.1 General Introduction

For a high-speed rotor system, even a small imbalance may cause a large synchronous vibration and oscillatory force to be transmitted to the bearings and the supporting structure. The rotors always have some amount of residual unbalance no matter how well they are balanced. Thus it is necessary to remove the unbalance of a rotor to as large an extent as possible for the smooth operation. A comprehensive review on balancing classification, principals, and methods was presented by Gunter and Jackson (1992). Another complete review on the fundamentals of balancing was well documented by Schenck Trebel Corporation (1983). There are many methods in balancing a rotor system. Some methods, such as modal balancing, require knowledge in the rotor-bearing system in order to build a good mathematical model and simulate the force-response relationship. Some methods, such as influence coefficient method, do not require the rotor-bearing mathematical model of the system, but trial weights are required to obtain the influence coefficients. Readers are encouraged to refer the reference books for more details.

DyRoBeS provides many useful tools for the rotordynamics analysis, one of them s the balancing calculation based on the influence coefficient method (Tessarzik et. al., 1972; Lund and Tonnesen, 1972). The concept of influence coefficient method is based on the linear system theory, which assumes a cause-effect relationship for the components of a system. The input (force) - output (response) relation represents the cause and effect relationship of the system. The application of the influence coefficient method in the balancing of a rotor system is illustrated in the following block diagram assuming the system is linear:

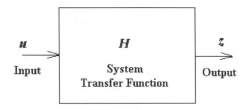

Figure 9.9-1 Force-Response relationship

In lateral vibration, the input (unbalance force) and output (steady state response) are vectors. A vector can be represented by a complex number with amplitude and angle. The linear relationship between the input and output is:

$$z = H\,u \tag{9.1-1}$$

Note that the input u, output z, and transfer function H are all complex numbers in this application. A linear system is a system which satisfies the principal of superposition. When the system is subject to an input (force) u_1, it provides an output (response) z_1. When the system is subject to an input u_2, it provides a corresponding output z_2. For a linear system, it is necessary that the excitation $(C_1u_1 + C_2u_2)$ result in a response $(C_1z_1 + C_2z_2)$ for all pairs of inputs (u_1, u_2) and all pairs of constants (C_1, C_2). The relationships are summarized below:

$$z_1 = H\,u_1 \tag{9.1-2}$$
$$z_2 = H\,u_2 \tag{9.1-3}$$
$$\left(C_1z_1 + C_2z_2\right) = H\left(C_1u_1 + C_2u_2\right) \tag{9.1-4}$$

9.2 Influence Coefficient Method

Once we know the concept of linear system, we are ready to discuss the most widely used balancing technique in practice: Influence Coefficient Method. Assuming there is n number of balancing planes where the corrections can be made and m number of probes (or measurement planes) where the amplitude and angle can be measured. For linear systems, the relationship between the force and response is:

$$z_{(mx1)} = H_{(mxn)}\,u_{(nx1)} \tag{9.2-1}$$

where

$$z_{(mx1)} = \left\{ \begin{array}{c} z_1 \\ z_2 \\ \vdots \\ z_{m-1} \\ z_m \end{array} \right\}, \qquad u_{(nx1)} = \left\{ \begin{array}{c} u_1 \\ u_2 \\ \vdots \\ u_{n-1} \\ u_n \end{array} \right\} \tag{9.2-2}$$

It is assumed that all the unbalance forces are located in these n balancing planes, and the responses at the measured m probes are critical to the smooth operation of the machine. The goal is to minimize/eliminate the responses at the measurement planes by correcting the unbalance forces at the balancing planes. However, if the balancing planes are inappropriately selected, such as at or near the nodal points, then it will not be feasible to balance the rotor system due to the insensitive response to the trial weights and physical limits on the corrections. If the measurement planes are inappropriately selected, then small responses at these planes do not ensure the smooth operation of the machine. Therefore, the selection of these balancing and measurement planes is the key to the successful balancing of a rotor system and the smooth operation of the machine. Knowledge in the rotor dynamic behavior and mode shapes are certainly beneficial to the successful use of the influence coefficient method.

The elements in the influence coefficient matrix (or transfer function matrix) are evaluated by adding trial weights in each balance planes, one at a time, and the rotor response data are taken in all the m measurements. The procedure for the determination of the unbalance and correction is summarized below:

1. Measure the initial response due to the unbalance vector $\boldsymbol{u}$ which is to be determined and corrected in this procedure:

$$z_0 = \boldsymbol{H}\,\boldsymbol{u} \tag{9.2-3}$$

Note that z_0 is the response due to the unbalance only. Any runout (or slow roll vector) should be compensated in the calculation.

2. Apply trial weight in each balance plane, one at a time, and measure the response at the same condition, such as speed and load, as the initial response to determine the influence coefficient. Let $\boldsymbol{T}_j$ be the additional force vector due to the trial weight added in plane j, the force-response relation yields:

$$z_j = \boldsymbol{H}\left(\boldsymbol{u} + \boldsymbol{T}_j\right) \tag{9.2-4}$$

where

$$\boldsymbol{T}_j = \left\{ \begin{array}{c} 0 \\ \vdots \\ T_j \\ \vdots \\ 0 \end{array} \right\}_{(n\times 1)} \tag{9.2-5}$$

From the principal of superposition, the net response due to the trail weight alone is:

$$\left(z_j - z_0\right) = \boldsymbol{H}\,\boldsymbol{T}_j \tag{9.2-6}$$

or

$$\left\{ \begin{array}{c} z_{1,j} - z_{1,0} \\ \vdots \\ z_{i,j} - z_{i,0} \\ \vdots \\ \vdots \\ z_{m,j} - z_{m,0} \end{array} \right\}_{(m\times 1)} = \begin{bmatrix} H_{11} & \cdots & H_{1j} & \cdots & H_{1n} \\ \vdots & \vdots & \vdots & \vdots & \vdots \\ H_{i1} & \cdots & H_{ij} & \cdots & H_{in} \\ \vdots & \vdots & \vdots & \vdots & \vdots \\ \vdots & \vdots & \vdots & \vdots & \vdots \\ H_{m1} & \cdots & H_{mj} & \cdots & H_{mn} \end{bmatrix}_{(m\times n)} \left\{ \begin{array}{c} 0 \\ \vdots \\ T_j \\ \vdots \\ 0 \end{array} \right\}_{(n\times 1)} \tag{9.2-7}$$

Therefore, the influence coefficients in the jth column are determined by the trial weight at the jth plane:

$$H_{ij} = \frac{z_{i,j} - z_{i,0}}{T_j}, \qquad i = 1, 2, ..., m \tag{9.2-8}$$

Repeat this step for every trail weight from 1 to n planes; the complete influence coefficient matrix is then determined. This above equation can also be extended to different operating conditions, such as speeds and loads. Assuming N is the number of conditions, then the force-response relationship becomes:

$$\begin{Bmatrix} z^{(1)}_{(mx1)} \\ \vdots \\ z^{(k)}_{(mx1)} \\ \vdots \\ z^{(N)}_{(mx1)} \end{Bmatrix}_{(Nmx1)} = \begin{bmatrix} H^{(1)}_{(mxn)} \\ \vdots \\ H^{(k)}_{(mxn)} \\ \vdots \\ H^{(N)}_{(mxn)} \end{bmatrix}_{(Nmxn)} u_{(nx1)} \tag{9.2-9}$$

The influence coefficients can be obtained by using the following flowchart:

for $k=1,...,N$ (number of conditions)

   for $j=1,..., n$ (number of balancing planes)

      for $i=1,...,m$ (number of measurements)

$$H^k_{ij} = \frac{z^k_{i,j} - z^k_{i,0}}{T_j} \tag{9.2-10}$$

      end (i)

   end (j)

end (k)

3. In order to solve Eq. (9.2-9), the order of (Nxm) must be greater than or equal to (n)

$$(N \times m) \geq n \tag{9.2-11}$$

Once the unbalance vector u is determined from Eq. (9.2-9) by using the least square method, the correction u_c can be made to minimize the residual vibration.

$$u_c = -u \tag{9.2-12}$$

9.3 Examples

Several examples are presented here to illustrate the use of *DyRoBeS* balancing calculation. Simple examples are mainly for the validation purposes and the detailed calculation is also presented.

Example 9.1: Single Plane Balancing

This example is taken from Gunter and Jackson (1992). The rotor weighs 1000 Lb and is operated at 6000 rpm. The initial steady-state response has an amplitude of 3.0 mils at a phase angle of 270°. A trial weight of 0.174 oz (m) is placed at a radius of 9 inches (e) at a position of 30° from the timing mark against rotor rotation. The amplitude of the trial unbalance (me) is 1.566 oz-in (use 1.57 for comparison purposes). The magnitude of the unbalance force caused by the trial weight at 6000 rpm is:

$$U_T = me\Omega^2 = \frac{1.57}{16 \times 386}\left(\frac{6000}{60} \cdot 2\pi\right)^2 \cong 100 \quad Lb$$

This is about 10 percent of the rotor weight. The steady-state response with the trial weight on the rotor has an amplitude of 2 mils at 170°.
The balancing calculation is summarized below:
The initial response:

$$z_0 = 3.0\angle 270° = 0 - j\,3$$

The trial weight vector:

$$T_j = 1.57\angle 30° = 1.35966 + j\,0.78500$$

The response after the trial weight is applied (response due to trial weight and the original unbalance):

$$z_j = 2.0\angle 170° = -1.96962 + j\,0.34730$$

The influence coefficient:

$$H = \frac{z_j - z_0}{T_j} = -0.02044 + j\,2.47366 = 2.47375\angle 90.5°$$

The original unbalance force:

$$u = H^{-1}z_0 = -1.21269 + j\,0.01002 = 1.21273\angle 179.5°$$

The balancing correction:

$$\boldsymbol{u}_c = -\boldsymbol{u} = 1.21269 - j\,0.01002 = 1.21273\angle359.5°$$

That is, a correction balance of 1.21 oz-in (0.134 oz at 9 in radius) should be placed at the angle of 359.5° against the rotation to minimize the steady-state response. If the trial weight is left in the rotor after the trial run, then the combination of the trim plus the trial is equal to the total required balance. The trim balance is:

$$\boldsymbol{u}_{trim} = \boldsymbol{u}_c - \boldsymbol{T}_j = -0.14697 - j\,0.79502 = 0.80849\angle259.5°$$

This is, a trim balance of 0.81 oz-in should be placed at an angle of 259.5° against rotation if the trial weight is left-in after the trial run. The force vector diagram can be shown in Figure 9.3-1:

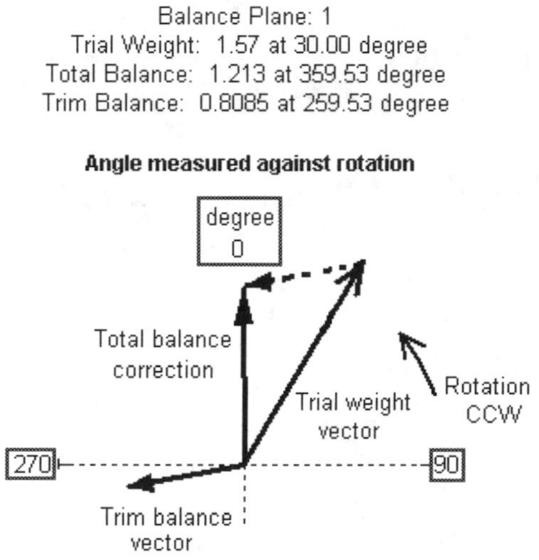

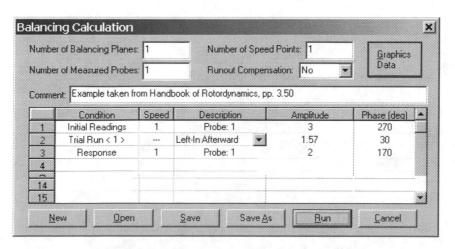

Figure 9.3-1 Force vector diagram

The input data for this example is shown in Figure 9.3-2:

Figure 9.3-2 Data input for example 9.1

DyRoBeS allows you to leave the trial weight in the rotor or remove it after the trial run. For graphic presentation, it also allows you to define the direction of shaft rotation, angle of the trial weight with respect to the direction of shaft rotation, and position of the zero degree mark, under the *Graphic Data* tab in the input screen. The calculated results can be viewed in graphic formats or text output. Figure 9.3-3 shows the balance correction vectors. Figure 9.3-4 shows the influence coefficient.

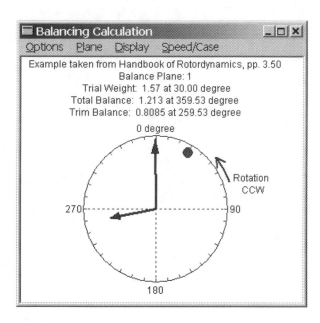

Figure 9.3-3 Balance vectors and trial weight

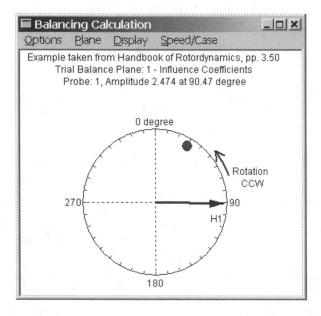

Figure 9.3-4 Influence coefficient and trial weight

The text output is also included for reference.

```
Example taken from Handbook of Rotordynamics, pp. 3.50

*****   Number of Speeds or Cases   :   1
*****   Number of Balancing Planes  :   1
*****   Number of Measurement Probes:   1

*****   NO Runout Compensation

======  Initial Response (Without Trails)  ======
   Speed     Probe        Amplitude      Phase Angle
     1         1            3.0000          270.00

**************  Trial Unbalance Run: 1  **************
   Plane       Amplitude      Phase Angle    Afterward
     1           1.5700         30.000        Left-In

---------  Response to Trial Unbalance  ---------
   Speed     Probe        Amplitude      Phase Angle
     1         1            2.0000          170.00


<<<<<<<<   Total Balance Correction   >>>>>>>>
   Correction Required to Balance the Rotor
   Plane No.       Amplitude      Phase Angle
     1              1.2127          359.53


<<<<<<<<<   Trim Balance Correction   >>>>>>>>>>
   Correction Required if Trial Weight Left-in
   Trim Balance = Total Balance - Trial Weight
   Plane No.       Amplitude      Phase Angle
     1              0.80849         259.53

*********   The Influence Coefficients   *********
                              Influence Coef.
   Trial-Run  Speed  Probe   Amplitude     Phase
      1         1      1       2.4737        90.
```

Example 9.2: Two-Plane Balancing with Runout Compensation

This example is again taken from Gunter and Jackson (1992). The pertinent parameters are listed in Figure 9.3-5. Runout compensation is demonstrated in the example. The *runout*, sometimes called *slow-roll vector*, is taken at a very low speed of 300 rpm in this case. The unbalance force has minimal influence at this low speed. This run-out is not caused by the unbalance force, therefore, should be compensated (subtracted) in the balancing calculation. The run-out readings taken during the coastdown at a speed of 300 rpm are: $0.5\angle 272°$ and $0.4\angle 123°$ at probe 1 and 2, respectively. The initial response readings at design speed are: $1.8\angle 148°$ and $3.6\angle 115°$ at probe 1 and 2, respectively. The first trial unbalance has a magnitude of 4.9 at $120°$. The response readings at both probes with the first trial weight mounted are: $1.1\angle178°$ and $2.0\angle98°$. The first trial weight is removed after the trial run. The second trial unbalance has a magnitude of 4.9 at $220°$. The response readings at both probes with the second trial weight on are: $2.1\angle98°$ and $3.7\angle102°$. The second trial weight is left in the rotor after the trial run.

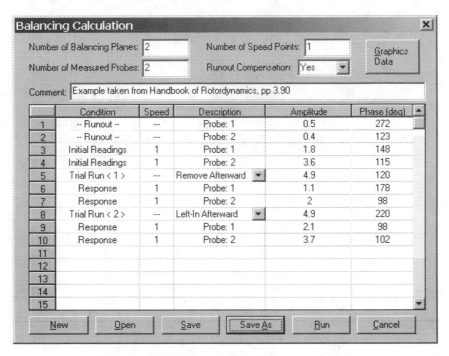

Figure 9.3-5 Data input for example 9.2

The calculation is summarized below:
The initial response due to the unbalance is the total response subtracts the run-out:

$$z_0 = \begin{Bmatrix} z_1 \\ z_2 \end{Bmatrix}_0 = \begin{Bmatrix} 1.8\angle148° - 0.5\angle272° \\ 3.6\angle115° - 0.4\angle123° \end{Bmatrix} = \begin{Bmatrix} 2.12051\angle136.73° \\ 3.20438\angle114.00° \end{Bmatrix}$$

The first trial weight vector:

$$T_1 = 4.9\angle120°$$

The compensated response after the first trial weight is applied:

$$z_1 = \begin{Bmatrix} z_1 \\ z_2 \end{Bmatrix}_1 = \begin{Bmatrix} 1.1\angle178° - 0.5\angle272° \\ 2.0\angle98° - 0.4\angle123° \end{Bmatrix} = \begin{Bmatrix} 1.23965\angle154.27° \\ 1.64618\angle92.11° \end{Bmatrix}$$

The first trial weight is removed before the second trial runs.
The second trial weight vector:

$$T_2 = 4.9\angle220°$$

The compensated response after the second trial weight is applied:

$$z_2 = \left\{ \begin{matrix} z_1 \\ z_2 \end{matrix} \right\}_2 = \left\{ \begin{matrix} 2.1\angle98° - 0.5\angle272° \\ 3.7\angle102° - 0.4\angle123° \end{matrix} \right\} = \left\{ \begin{matrix} 2.59779\angle96.85° \\ 3.32965\angle99.53° \end{matrix} \right\}$$

The four influence coefficients are calculated using Eq. (9.2-10):

$$H = \begin{bmatrix} 0.20617\angle175.01° & 0.34092\angle182.37° \\ 0.36446\angle194.11° & 0.16986\angle165.36° \end{bmatrix}$$

Since this is a square matrix, the inverse of this influence coefficient matrix can be easily calculated. The unbalance, which produces the initial response, can be found by:

$$u = H^{-1}z_0$$

The total balance correction is:

$$u_c = -u = \left\{ \begin{matrix} 7.48729\angle84.96° \\ 5.32091\angle179.73° \end{matrix} \right\}$$

Since the second trial weight is left in the rotor, the trim balance is:

$$u_{2,trim} = u_{2,c} - T_2 = 3.54022\angle116.28°$$

The locations of trial, total, and trim balance vectors on both planes are shown in Figure 9.3-6. The result for the balancing plane 1 is viewed from the equipment side and the direction of shaft rotation is counterclockwise. The result for the balancing plane 2 is viewed from the driver side and the direction of rotation is clockwise. Since the second trial weight is left in the rotor, therefore a trim balance vector is also shown in this plane.

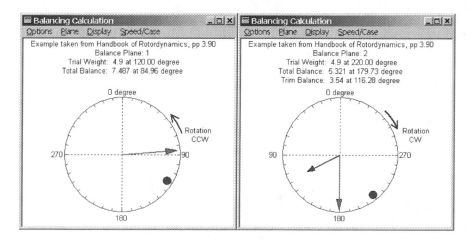

Figure 9.3-6 Balance vectors

The text output is listed below:

```
Example taken from Handbook of Rotordynamics, pp 3.90

*****   Number of Speeds or Cases   :   1
*****   Number of Balancing Planes  :   2
*****   Number of Measurement Probes:   2

*****   Runout (slow-roll vectors)  *****
    Probe        Amplitude        Phase Angle
      1           0.50000           272.00
      2           0.40000           123.00


======   Initial Response (Without Trails)  ======
    Speed   Probe      Amplitude       Phase Angle
      1       1          1.8000          148.00
      1       2          3.6000          115.00


**************   Trial Unbalance Run: 1   **************
    Plane       Amplitude       Phase Angle    Afterward
      1          4.9000          120.00          Remove


---------   Response to Trial Unbalance   ---------
    Speed   Probe      Amplitude       Phase Angle
      1       1          1.1000          178.00
      1       2          2.0000          98.000


**************   Trial Unbalance Run: 2   **************
    Plane       Amplitude       Phase Angle    Afterward
      2          4.9000          220.00          Left-In


---------   Response to Trial Unbalance   ---------
    Speed   Probe      Amplitude       Phase Angle
      1       1          2.1000          98.000
      1       2          3.7000          102.00

<<<<<<<<   Total Balance Correction   >>>>>>>>
  Correction Required to Balance the Rotor
    Plane No.      Amplitude       Phase Angle
      1             7.4873           84.957
      2             5.3209           179.73

<<<<<<<<<   Trim Balance Correction   >>>>>>>>>>
  Correction Required if Trial Weight Left-in
  Trim Balance = Total Balance - Trial Weight
    Plane No.      Amplitude       Phase Angle
      2             3.5402           116.28


*********   The Influence Coefficients   *********
                              Influence Coef.
  Trial-Run   Speed   Probe   Amplitude     Phase
      1         1       1      0.20617       175.
      1         1       2      0.36446       194.
      2         1       1      0.34092       182.
      2         1       2      0.16986       165.
```

Example 9.3: Two-Plane Balancing at Six Speeds

This example is taken from Gunter (1987). The rotor system to be balanced is a 70 MW gas turbine. It is a two-plane balancing using two vibration probes and six speed cases. The six speed sets include the various turbine and generator critical speeds. In this example, the hand calculation is not practical and computer program is preferable for this type of calculation. The text input and output with graphics are listed below.

```
Example taken from ROTORBAL - 70 MW Gas Turbine

*****   Number of Speeds or Cases   :   6
*****   Number of Balancing Planes  :   2
*****   Number of Measurement Probes:   2


*****   NO Runout Compensation

======   Initial Response (Without Trails)  ======
  Speed     Probe        Amplitude        Phase Angle
    1         1            1.7000            339.00
    1         2            4.6000            54.000
    2         1            2.8000            226.00
    2         2            6.7000            10.000
    3         1            3.9000            145.00
    3         2            3.7000            333.00
    4         1            4.5000            103.00
    4         2            4.7000            302.00
    5         1            5.4000            74.000
    5         2            6.5000            113.00
    6         1            1.9800            98.000
    6         2            5.7000            114.00

**************   Trial Unbalance Run: 1  *************
  Plane         Amplitude        Phase Angle     Afterward
    1            20.000            359.00         Left-In

---------   Response to Trial Unbalance  ---------
  Speed     Probe        Amplitude        Phase Angle
    1         1            2.6000            313.00
    1         2            5.9000            7.0000
    2         1            3.9000            232.00
    2         2            4.4000            4.0000
    3         1            4.2000            160.00
    3         2            2.8000            340.00
    4         1            3.5000            120.00
    4         2            5.4000            325.00
    5         1            4.1000            73.000
    5         2            3.7000            97.000
    6         1            1.5000            141.00
    6         2            3.1000            99.000

**************   Trial Unbalance Run: 2  *************
  Plane         Amplitude        Phase Angle     Afterward
    2            10.000            270.00         Left-In

---------   Response to Trial Unbalance  ---------
```

Speed	Probe	Amplitude	Phase Angle
1	1	1.8000	319.00
1	2	4.3000	15.000
2	1	3.1000	244.00
2	2	3.4000	9.0000
3	1	3.2000	107.00
3	2	1.9000	329.00
4	1	2.4000	122.00
4	2	4.4000	330.00
5	1	2.4000	61.000
5	2	3.5000	101.00
6	1	1.0200	170.00
6	2	3.1100	104.00

<<<<<<<<   Total Balance Correction   >>>>>>>>
Correction Required to Balance the Rotor

Plane No.	Amplitude	Phase Angle
1	26.867	347.83
2	20.639	267.30

<<<<<<<<<   Trim Balance Correction   >>>>>>>>>>
Correction Required if Trial Weight Left-in
Trim Balance = Total Balance - Trial Weight

Plane No.	Amplitude	Phase Angle
1	8.2159	319.71
2	10.660	264.77

********* The Influence Coefficients *********

Trial-Run	Speed	Probe	Influence Coef. Amplitude	Phase
1	1	1	0.65281E-01	279.
1	1	2	0.21766	317.
1	2	1	0.57655E-01	248.
1	2	2	0.11846	202.
1	3	1	0.54915E-01	228.
1	3	2	0.49103E-01	134.
1	4	1	0.77078E-01	242.
1	4	2	0.10636	26.
1	5	1	0.65130E-01	258.
1	5	2	0.15575	313.
1	6	1	0.67568E-01	230.
1	6	2	0.14110	312.
2	1	1	0.83143E-01	210.
2	1	2	0.17475	257.
2	2	1	0.10809	105.
2	2	2	0.10554	258.
2	3	1	0.34210	118.
2	3	2	0.10027	271.
2	4	1	0.11046	26.
2	4	2	0.10867	214.
2	5	1	0.18221	359.
2	5	2	0.32108E-01	318.
2	6	1	0.78362E-01	12.
2	6	2	0.27106E-01	279.

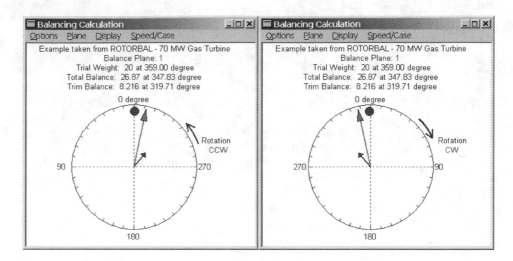

Figure 9.3-7 Balance vectors

Appendix A: Mass Properties

1. Mass properties of a homogeneous hollow cylinder

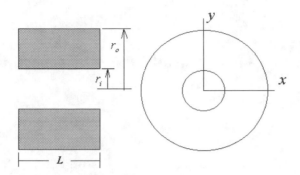

Figure A-1 Cylindrical element

Basic parameters

r_o = outer radius, d_o = outer diameter

r_i = inner radius, d_i = inner diameter

L = element length, ρ = mass density

Properties of cross-section

Area

$$A = \pi \left(r_o^{\,2} - r_i^{\,2}\right) = \frac{\pi}{4}\left(d_o^{\,2} - d_i^{\,2}\right) \qquad (A\text{-}1)$$

Area moment of inertia

$$I = I_x = I_y = \frac{\pi}{4}\left(r_o^{\,4} - r_i^{\,4}\right) = \frac{\pi}{64}\left(d_o^{\,4} - d_i^{\,4}\right) \qquad (A\text{-}2)$$

$$I_z = I_x + I_y = 2I \qquad (A\text{-}3)$$

Properties of solid

Center of Mass $(0,0,L/2)$ $\qquad\qquad\qquad\qquad\qquad$ (A-4)

Mass $m = \rho\,A\,L$ $\qquad\qquad\qquad\qquad\qquad\qquad$ (A-5)

Mass moment of inertia

$$I_d = I_{xx} = I_{yy} = \frac{m}{12}\left(3r_o^{\,2} + 3r_i^{\,2} + L^2\right) = \frac{m}{48}\left(3d_o^{\,2} + 3d_i^{\,2} + 4L^2\right) \tag{A-6}$$

$$I_p = I_{zz} = \frac{m}{2}\left(r_o^{\,2} + r_i^{\,2}\right) = \frac{m}{8}\left(d_o^{\,2} + d_i^{\,2}\right) \tag{A-7}$$

2. Mass properties of a homogeneous linearly tapered (conical) element

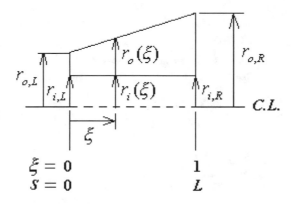

Figure A-2 Tapered element

Each end of the tapered element is associated with an inner and outer radii, r_i and r_o, with the subscripts L and R referring to the Left and Right ends of the element, respectively. The cross section is circular. The local position coordinate s is from 0 (left end) to L (right end). The non-dimensional coordinate ξ is from 0 (left end) to 1 (right end). The inner and outer radii of any inner cross section are:

$$r_i(\xi) = r_{i,L}(1-\xi) + r_{i,R}(\xi) = r_{i,L}\left[1 + (\Delta_i - 1)\xi\right] \tag{A-8}$$

$$r_o(\xi) = r_{o,L}(1-\xi) + r_{o,R}(\xi) = r_{o,L}\left[1 + (\Delta_o - 1)\xi\right] \tag{A-9}$$

where

$$\xi = \frac{s}{L}, \qquad \Delta_o = \frac{r_{o,R}}{r_{o,L}}, \qquad \Delta_i = \frac{r_{i,R}}{r_{i,L}} \tag{A-10}$$

Properties of cross-section

The cross-sectional area, area moment of inertia, and polar moment of inertia of the cross-section are:

$$A(\xi) = \pi\left(r_o{}^2 - r_i{}^2\right) = A_L\left(1 + \alpha_1\xi + \alpha_2\xi^2\right) \tag{A-11}$$

$$I(\xi) = I_x(\xi) = I_y(\xi) = \pi\left(r_o{}^4 - r_i{}^4\right)/4 = I_L\left(1 + \delta_1\xi + \delta_2\xi^2 + \delta_3\xi^3 + \delta_4\xi^4\right) \tag{A-12}$$

$$I_p(\xi) = I_z(\xi) = 2I(\xi) \tag{A-13}$$

where the left end area properties are:

$$A_L = \pi\left(r_{o,L}^2 - r_{i,L}^2\right) \tag{A-14}$$

$$I_L = \pi\left(r_{o,L}^4 - r_{i,L}^4\right)/4 \tag{A-15}$$

And the coefficients are:

$$\alpha_1 = 2\left[r_{o,L}^2\left(\Delta_o - 1\right) - r_{i,L}^2\left(\Delta_i - 1\right)\right]/\left(r_{o,L}^2 - r_{i,L}^2\right) \tag{A-16}$$

$$\alpha_2 = \left[r_{o,L}^2\left(\Delta_o - 1\right)^2 - r_{i,L}^2\left(\Delta_i - 1\right)^2\right]/\left(r_{o,L}^2 - r_{i,L}^2\right) \tag{A-17}$$

$$\delta_1 = 4\left[r_{o,L}^4\left(\Delta_o - 1\right) - r_{i,L}^4\left(\Delta_i - 1\right)\right]/\left(r_{o,L}^4 - r_{i,L}^4\right) \tag{A-18}$$

$$\delta_2 = 6\left[r_{o,L}^4\left(\Delta_o - 1\right)^2 - r_{i,L}^4\left(\Delta_i - 1\right)^2\right]/\left(r_{o,L}^4 - r_{i,L}^4\right) \tag{A-19}$$

$$\delta_3 = 4\left[r_{o,L}^4\left(\Delta_o - 1\right)^3 - r_{i,L}^4\left(\Delta_i - 1\right)^3\right]/\left(r_{o,L}^4 - r_{i,L}^4\right) \tag{A-20}$$

$$\delta_4 = \left[r_{o,L}^4\left(\Delta_o - 1\right)^4 - r_{i,L}^4\left(\Delta_i - 1\right)^4\right]/\left(r_{o,L}^4 - r_{i,L}^4\right) \tag{A-21}$$

Properties of solid

The conical element mass

$$m = \int_0^L \rho\, A\, ds = \int_0^1 \rho\, A(\xi)\, L\, d\xi = \rho A_L L\left(1 + \frac{\alpha_1}{2} + \frac{\alpha_2}{3}\right) \tag{A-22}$$

The mass center of gravity from left end

$$s_c = \frac{\int_0^L \rho\, A\, s\, ds}{m} = \frac{\rho A_L L^2 \left(\dfrac{1}{2} + \dfrac{\alpha_1}{3} + \dfrac{\alpha_2}{4} \right)}{\rho A_L L \left(1 + \dfrac{\alpha_1}{2} + \dfrac{\alpha_2}{3} \right)} = \frac{L\left(6 + 4\alpha_1 + 3\alpha_2 \right)}{\left(12 + 6\alpha_1 + 4\alpha_2 \right)} \tag{A-23}$$

The mass transverse moment of inertia with respect to center of gravity axes

$$I_d = \rho I_L L \left(1 + \frac{\delta_1}{2} + \frac{\delta_2}{3} + \frac{\delta_3}{4} + \frac{\delta_4}{5} \right) + \rho A_L L^3 \left(\frac{1}{3} + \frac{\alpha_1}{4} + \frac{\alpha_2}{5} \right) - \left(m \cdot s_c^{\,2} \right) \tag{A-24}$$

The mass polar moment of inertia

$$I_p = \rho (2I_L) L \left(1 + \frac{\delta_1}{2} + \frac{\delta_2}{3} + \frac{\delta_3}{4} + \frac{\delta_4}{5} \right) \tag{A-25}$$

Note that for the cylindrical element, $\Delta_i = \Delta_o = 1$, all the coefficients (α_i and δ_i) are zero and the equations are identical to the cylindrical element equations presented previously.

Appendix B: Elemental Matrices

The shape factor κ for a hollow circular cross-sectional beam is given by (Cowper, 1966):

$$\kappa = \frac{6(1+\nu)(1+\Re^2)^2}{(7+6\nu)(1+\Re^2)^2 + (20+12\nu)\Re^2} \tag{B-1}$$

Where ν is the Poisson ratio and $\Re = \dfrac{r_i}{r_o}$ is the ratio of inner to outer radii. The most common cases used in the rotordynamics analysis are:

For a solid shaft: $\kappa = \dfrac{6(1+\nu)}{(7+6\nu)}$ (for ν=0.3, κ=0.886)

For a thin-walled tube: $\kappa = \dfrac{2(1+\nu)}{(4+3\nu)}$ (for ν=0.3, κ=0.531)

The transverse shear effect parameter is defined as:

$$\Phi = \frac{12EI}{kAGL^2} \tag{B-2}$$

where

    E = Young's modulus
    G = shear modulus
    I = area moment of inertia
    A = cross-section area
    L = element length

Since elemental mass and stiffness matrices in single plane are required for the analysis of isotropic systems, therefore they are listed in a single plane of motion. The shaft element is assumed to be isotropic and axis-symmetric about the axis of rotation. The material is assumed to be homogeneous. The elemental matrices for complete two planes of motion can be easily obtained by extending the matrices of single plane of motion. The gyroscopic matrix couples two planes of motion, thus a complete (8x8) matrix is given below.

1. **Translational mass matrix:**

$$
\boldsymbol{M}_{T,XZ}^{e} = \begin{bmatrix}
m_{11} & & \text{Sym} & \\
m_{41} & m_{44} & & \\
m_{51} & m_{54} & m_{55} & \\
m_{81} & m_{84} & m_{85} & m_{88}
\end{bmatrix}
\tag{B-3}
$$

where

$$
m_{11} = \frac{\rho A L}{420(1+\Phi)^2}\left(156 + 294 \cdot \Phi + 140 \cdot \Phi^2\right)
$$

$$
m_{41} = \frac{\rho A L}{420(1+\Phi)^2}\left(22 + 38.5 \cdot \Phi + 17.5 \cdot \Phi^2\right)L
$$

$$
m_{51} = \frac{\rho A L}{420(1+\Phi)^2}\left(54 + 126 \cdot \Phi + 70 \cdot \Phi^2\right)
$$

$$
m_{81} = \frac{\rho A L}{420(1+\Phi)^2}\left(-13 - 31.5 \cdot \Phi - 17.5 \cdot \Phi^2\right)L
$$

$$
m_{44} = \frac{\rho A L}{420(1+\Phi)^2}\left(4 + 7 \cdot \Phi + 3.5 \cdot \Phi^2\right)L^2
$$

$$
m_{84} = \frac{\rho A L}{420(1+\Phi)^2}\left(-3 - 7 \cdot \Phi - 3.5 \cdot \Phi^2\right)L^2
$$

$$
m_{54} = -m_{81}, \quad m_{55} = m_{11}, \quad m_{85} = -m_{41}, \quad m_{88} = m_{44}
$$

2. **Rotatory mass matrix:**

$$
\boldsymbol{M}_{R,XZ}^{e} = \begin{bmatrix}
n_{11} & & \text{Sym} & \\
n_{41} & n_{44} & & \\
n_{51} & n_{54} & n_{55} & \\
n_{81} & n_{84} & n_{85} & n_{88}
\end{bmatrix}
\tag{B-4}
$$

where

$$
n_{11} = \frac{\rho I}{30L(1+\Phi)^2}\left(36\right)
$$

$$
n_{41} = \frac{\rho I}{30L(1+\Phi)^2}\left(3 - 15 \cdot \Phi\right)L
$$

$$
n_{44} = \frac{\rho I}{30L(1+\Phi)^2}\left(4 + 5 \cdot \Phi + 10 \cdot \Phi^2\right)L^2
$$

$$
n_{84} = \frac{\rho I}{30L(1+\Phi)^2}\left(-1 - 5 \cdot \Phi + 5 \cdot \Phi^2\right)L^2
$$

$$
n_{51} = -n_{11}, \quad n_{81} = n_{41}, \quad n_{54} = -n_{41}, \quad n_{55} = n_{11}, \quad n_{85} = -n_{41}, \quad n_{88} = n_{44}
$$

3. Stiffness matrix due to shaft bending and shear effects

$$
K^e_{XZ} = \frac{EI}{L^3(1+\Phi)}
\begin{bmatrix}
12 & & & \text{Sym} \\
6L & (4+\Phi)L^2 & & \\
-12 & -6L & 12 & \\
6L & (2-\Phi)L^2 & -6L & (4+\Phi)L^2
\end{bmatrix}
\tag{B-5}
$$

4. Stiffness matrix due to axial load (P)

$$
K^e_{XZ,a} =
\begin{bmatrix}
k_{11} & & \text{Sym} & \\
k_{41} & k_{44} & & \\
k_{51} & k_{54} & k_{55} & \\
k_{81} & k_{84} & k_{85} & k_{88}
\end{bmatrix}
\tag{B-6}
$$

where

$$
k_{11} = \frac{P}{30L(1+\Phi)^2}\left(36 + 60\Phi + 30\Phi^2\right)
$$

$$
k_{41} = \frac{P}{30L(1+\Phi)^2}(3L)
$$

$$
k_{44} = \frac{P}{30L(1+\Phi)^2}\left(4 + 5\Phi + 2.5\Phi^2\right)L^2
$$

$$
k_{84} = \frac{P}{30L(1+\Phi)^2}\left(-1 - 5\Phi - 2.5\Phi^2\right)L^2
$$

$$
k_{51} = -k_{11}, \quad k_{81} = k_{41}, \quad k_{54} = -k_{41}, \quad k_{55} = k_{11}, \quad k_{85} = -k_{41}, \quad k_{88} = k_{44}
$$

5. Gyroscopic matrix

$$
G^e =
\begin{bmatrix}
0 & & & & & & & \\
g_{21} & 0 & & \text{Skew} & \text{Sym.} & & & \\
g_{31} & 0 & 0 & & & & & \\
0 & g_{42} & g_{43} & 0 & & & & \\
0 & g_{52} & g_{53} & 0 & 0 & & & \\
g_{61} & 0 & 0 & g_{64} & g_{65} & 0 & & \\
g_{71} & 0 & 0 & g_{74} & g_{75} & 0 & 0 & \\
0 & g_{82} & g_{83} & 0 & 0 & g_{86} & g_{87} & 0
\end{bmatrix}
\tag{B-7}
$$

where

$$g_{21} = -2n_{11} = \frac{-\rho I}{15L(1+\Phi)^2} \quad (36)$$

$$g_{31} = 2n_{41} = \frac{\rho I}{15L(1+\Phi)^2}(3-15\cdot\Phi)L$$

$$g_{43} = -2n_{44} = \frac{-\rho I}{15L(1+\Phi)^2}\left(4+5\cdot\Phi+10\cdot\Phi^2\right)L^2$$

$$g_{83} = -2n_{84} = \frac{-\rho I}{15L(1+\Phi)^2}\left(-1-5\cdot\Phi+5\cdot\Phi^2\right)L^2$$

$$g_{61} = -g_{21}, \quad g_{71} = g_{31}, \quad g_{42} = g_{31}, \quad g_{52} = g_{21}, \quad g_{82} = g_{31},$$

$$g_{53} = g_{31}, \quad g_{64} = g_{31}, \quad g_{74} = -g_{83}, \quad g_{65} = g_{21}, \quad g_{75} = -g_{31},$$

$$g_{86} = -g_{31}, \quad g_{87} = g_{43}$$

Note that the coefficients in the gyroscopic matrix are close related to the coefficients in the rotatory mass matrix.

6. Gravitational force vector

Gravity force vectors in (X-Z) and (Y-Z) planes are different due to the gravitational constants, g_x and g_y in the X and Y directions, respectively.

$$Q_{g,XZ} = \rho A g_x \begin{Bmatrix} \left(\dfrac{L}{2}\right) \\ \left(\dfrac{L^2}{12}\right) \\ \left(\dfrac{L}{2}\right) \\ \left(\dfrac{-L^2}{12}\right) \end{Bmatrix}, \qquad Q_{g,YZ} = \rho A g_y \begin{Bmatrix} \left(\dfrac{L}{2}\right) \\ \left(\dfrac{-L^2}{12}\right) \\ \left(\dfrac{L}{2}\right) \\ \left(\dfrac{L^2}{12}\right) \end{Bmatrix} \qquad \text{(B-8)}$$

7. From one plane of motion to two planes of motion

For the complete (8x8) mass and stiffness matrices, the (4x4) symmetric matrix in the (X-Z) plane is given as:

$$A_{XZ} = \begin{bmatrix} a_{11} & & \text{Sym} & \\ a_{41} & a_{44} & & \\ a_{51} & a_{54} & a_{55} & \\ a_{81} & a_{84} & a_{85} & a_{88} \end{bmatrix}$$

The complete (8x8) matrix for both planes are given in the following format:

$$
A = \begin{bmatrix}
\boldsymbol{a}_{11} & & & & & & & \\
0 & a_{22} & & & & \textit{Sym.} & & \\
0 & a_{32} & a_{33} & & & & & \\
\boldsymbol{a}_{41} & 0 & 0 & \boldsymbol{a}_{44} & & & & \\
\boldsymbol{a}_{51} & 0 & 0 & \boldsymbol{a}_{54} & \boldsymbol{a}_{55} & & & \\
0 & a_{62} & a_{63} & 0 & 0 & a_{66} & & \\
0 & a_{72} & a_{73} & 0 & 0 & a_{76} & a_{77} & \\
\boldsymbol{a}_{81} & 0 & 0 & \boldsymbol{a}_{84} & \boldsymbol{a}_{85} & 0 & 0 & \boldsymbol{a}_{88}
\end{bmatrix}
\tag{B-9}
$$

where the coefficients in the (Y-Z) plane are given by the coefficients in the (X-Z):

$$a_{22} = a_{11}, \quad a_{33} = a_{44}, \quad a_{66} = a_{55}, \quad a_{77} = a_{88},$$

$$a_{32} = -a_{41}, \quad a_{62} = a_{51}, \quad a_{72} = -a_{81},$$

$$a_{63} = -a_{54}, \quad a_{73} = a_{84}, \quad a_{76} = -a_{85},$$

Appendix C: Lubricant Properties

Lubricant plays an important role in the fluid film bearing analysis. Dynamic (absolute) viscosity is the single most important property among all the chemical and physical properties of a lubricant when determining the lubrication application of a given lubricant. Without knowing the dynamic viscosity of the fluid (μ), the Reynolds equation could not even be solved. The viscosity is usually measured by allowing the fluid to flow through a long capillary under its own head. This direct measured viscosity is called the *kinematic viscosity* (ν). The kinematic viscosities of a lubricant in centiStoke (cSt) frequently reported by the manufacturers are measured at 40 and 100°C (previously 100 and 210°F) at atmosphere pressure and low-shear rates. The higher the viscosity, the more slowly the oil flows under a given set of conditions. The viscosity classification system developed by a joint ASLE-ASTM group, which has been adopted by the International Standards Organization (ISO), is shown in Table C-1. Figure C-1 summarized by Texaco Lubricants Company shows the viscosity grade comparisons for various systems.

Viscosity system for industrial fluid lubricants

ISO viscosity grade numbers (ISO VG)	Midpoint viscosity (cSt at 40°C) Mid	Viscosity ranges (cSt at 40°C) min	max
2	2.2	1.98	2.42
3	3.2	2.88	3.52
5	4.6	4.14	5.06
7	6.8	6.12	7.48
10	10	9.00	11.00
15	15	13.50	16.50
22	22	19.80	24.20
32	32	28.80	35.20
46	46	41.40	50.60
68	68	61.20	74.80
100	100	90.00	110.00
150	150	135.00	165.00
220	220	198.00	242.00
320	320	288.00	352.00
460	460	414.00	506.00
680	680	612.00	748.00
1000	1000	900.00	1100.00
1500	1500	1350.00	1650.00

Table C-1 Viscosity Grade

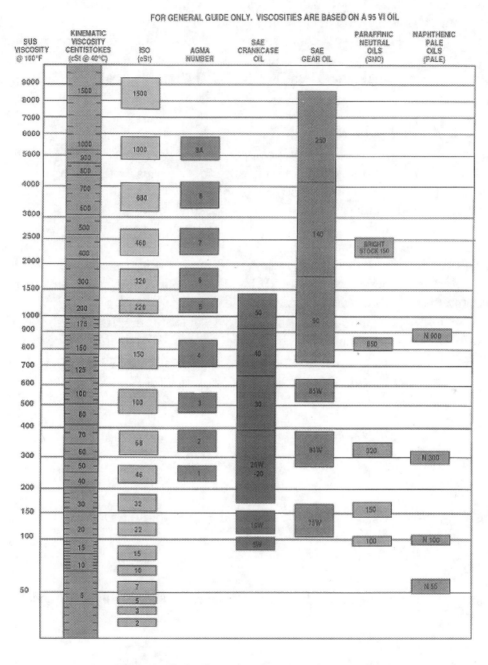

Figure C-1 Viscosity Systems (Texaco)

One should know that there is a viscosity range for every viscosity grade oils. Also, the kinematic viscosity is used for the grade system. However, the dynamic (absolute) viscosity is used in the Reynolds equation and the shear force calculation, therefore, one should pay attention when different oils are used. Typically, the synthetic oils have much higher mass density than the mineral oils; therefore, cautions must be taken when switching the oils from either mineral to synthetic or vice versa.

The viscosity of a lubricant is dependent on the temperature and pressure. In typical journal bearing applications, the hydrodynamic pressure is below 1500 psi. The effect of pressure on viscosity is very small and can be neglected. The most commonly used viscosity-temperature relation is known as the ASTM (American Society for Testing Materials) chart. The ASTM chart shows a straight line drawn through kinematic viscosities of an oil at any two temperatures. Such a straight line relates kinetic viscosity ν in cSt to absolute temperature T in $^{\circ}$K ($^{\circ}$K= $^{\circ}$C+273.15) by Walther equation (Booser, 1983):

$$\log\log(\nu + C) = A - B\log(T) \tag{C-1}$$

where A and B are constants for a given oil and C varies with viscosity level. The constant C is 0.7 for oil viscosities above 2.0 cSt, which covers almost all the oils used in turbomachinery, and increases gradually for viscosities below 2.0 cSt. For a given two viscosities, the constants A and B can be determined and the viscosities at other temperatures can be interpolate or extrapolate by using the Walther equation. In general, the Walther equation (ASTM chart) works very well for mineral oils under normal conditions for the temperature range of 20°F (11°C) to 350°F (177°C). It also can be used to predict the viscosities for many synthetic oils within 5% margin.

The density (ρ) of a lubricant is useful in determining the weight and volume relationship and is very important in bearing performance calculation since the kinematic viscosity (ν) is measured in the laboratory and the dynamic viscosity ($\mu=\rho\nu$) is used in the Reynolds equation. Specific gravity is defined as the ratio of the weight of a given volume of product at 60°F (15.6°C) to the weight of an equal volume of water at the same temperature:

$$SpGr_{60} = \frac{\rho_{60}}{\rho_{W,60}} \tag{C-2}$$

The density of the water at 60°F is given as (Cameron Hydraulic Data, 1988):

$$\rho_{W,60} = 0.9991 \; grams/cc \tag{C-3}$$

The American Petroleum Institute has defined the API gravity by the following equation:

$$^{\circ}API = \frac{141.5}{SpGr_{60}} - 131.5 \quad \text{or} \quad SpGr_{60} = \frac{141.5}{(131.5 + ^{\circ}API)} \tag{C-4}$$

A high API gravity value matches a low specific gravity and vice versa. Density changes with temperature by the coefficient of thermal expansion. The density of an oil at temperature T ($^{\circ}$F) other than 60°F may be calculated by use of the following equation (Wilcock and Booser, 1957):

$$\rho_T = \rho_{60}\left(\frac{1}{1 + \alpha(T - 60)}\right) \tag{C-5}$$

For mineral oil lubricants, the coefficient of expansion can be estimated from ASTM table, as shown in Table C-2.

Coefficient of expansion for mineral oil lubricants estimated from ASTM tables

Specific gravity at 60°F	α (per °F)	α (per °C)
1.076-0.967	0.00035	0.00063
0.966-0.850	0.00040	0.00072
0.850-0.776	0.00050	0.00090
0.775-0.742	0.00060	0.00108

Table C-2 Coefficient of expansion

In general, the specific gravity at 60°F of mineral oil lubricants ranges from 0.82 to 0.92. However, the synthetic lubricants have higher specific gravity. For synthetic lubricants, a similar straight-line relationship also exists between the density and temperature. Density also offers a simple way of identifying the lubricant base stock in petroleum and hydrocarbon-based lubricants.

The density varies linearly with temperature. The relationship can also be written in °C as: (Careron, 1981):

$$\rho_T = \rho_{15.6}\left[1 - \alpha\left(T(^\circ C) - 15.6\right)\right] \tag{C-6}$$

For typical mineral oils the viscosity below 3000 centiPoise (cP), the coefficient of expansion is:

$$\alpha \text{ per } ^\circ C = \left(10 - \frac{9}{5}\log_{10} cP\right) \times 10^{-4} \tag{C-7}$$

For light turbine oils, the typical viscosity is around 32 cP; the coefficient of expansion obtained from the above equation is in agreement with the ASTM table. It is always a good practice to ask the lubricant suppliers for two points of density to correlate the coefficient of expansion. Once the density and kinematic viscosity at a given temperature are known, the dynamic viscosity is given as:

$$\mu = \rho v \tag{C-8}$$

In additional to the density and viscosity of the lubricant, the specific heat is also required in the bearing performance calculation for the temperature rise. The specific heat is also varies linearly with temperature for mineral oils and can be approximated in the following equation (Wilcock and Booser, 1957):

$$C_p = \frac{\left(0.388 + 0.00045\, T(^\circ F)\right)}{\sqrt{\rho_{60}}} \quad (\text{Btu/Lbm-}^\circ F) \tag{C-9}$$

or in SI units (Cameron, 1981):

$$C_p = \frac{\left(1.63 + 0.0034 \, T(^{\circ}C)\right)}{\sqrt{SpGr_{15.6}}} \quad (\text{kJ/kg-}^{\circ}\text{C}) \tag{C-10}$$

Typically, the specific heat of a synthetic oil at three or four different temperatures is provided by the manufacturers. A polynomial can be used to interpolate or extrapolate these values.

An accurate lubricant dynamic viscosity, density, and specific heat are essential to the calculation of bearing performance. The basic properties of a number of commonly used lubricants are collected and stored in the *DyRoBeS-BePerf* lubricant library. The properties for Mobil DTE Light calculated using *DyRoBeS* are listed below for reference:

```
------------------- Lubricant Properties -------------------------------
Mobil DTE Light (VG 32)
Specific Gravity @ 60 deg. F = 0.867
API Gravity @ 60 deg. F......= 31.71
Coefficient of Expansion.(/F)= 0.000433768
Pour Point deg. F............= 19.4
Flash Point deg. F...........= 399.2
Viscosity, cSt (centiStoke)
cSt @ 104 deg. F.............= 30.4
cSt @ 212 deg. F.............= 5.1
Specific Heat Cp (BTU/(Lbm-F)) Coefficients
0.416886,   0.000483502,   0,    0
------------------- Calculated Properties -------------------------------
```

	Absolute		Kinematic			
	======= Viscosity =======		Viscosity		Specific	
	___Reyns___	_centiPoise_	_centiStoke_	Density	Heat	
deg F	(Lbf-s/in^2)	(Pa-s/1000)	(mm^2/s)	(Grams/CC)	(BTU/Lb/F)	deg C
75.00	8.29357E-06	57.182	66.44	0.8606	0.45315	23.89
80.00	7.13137E-06	49.169	57.26	0.8588	0.45557	26.67
85.00	6.17054E-06	42.544	49.65	0.8569	0.45798	29.44
90.00	5.37086E-06	37.031	43.31	0.8551	0.46040	32.22
95.00	4.70104E-06	32.412	37.99	0.8533	0.46282	35.00
100.00	4.13659E-06	28.521	33.50	0.8514	0.46524	37.78
105.00	3.65818E-06	25.222	29.69	0.8496	0.46765	40.56
110.00	3.25048E-06	22.411	26.43	0.8478	0.47007	43.33
115.00	2.90121E-06	20.003	23.64	0.8460	0.47249	46.11
120.00	2.60051E-06	17.930	21.24	0.8442	0.47491	48.89
125.00	2.34039E-06	16.136	19.15	0.8425	0.47732	51.67
130.00	2.11437E-06	14.578	17.34	0.8407	0.47974	54.44
135.00	1.91712E-06	13.218	15.76	0.8389	0.48216	57.22
140.00	1.74427E-06	12.026	14.37	0.8372	0.48458	60.00
145.00	1.59221E-06	10.978	13.14	0.8354	0.48699	62.78
150.00	1.45793E-06	10.052	12.06	0.8337	0.48941	65.56
155.00	1.33893E-06	9.232	11.10	0.8319	0.49183	68.33
160.00	1.23311E-06	8.502	10.24	0.8302	0.49425	71.11
165.00	1.13869E-06	7.851	9.48	0.8285	0.49666	73.89
170.00	1.05420E-06	7.268	8.79	0.8268	0.49908	76.67
175.00	9.78345E-07	6.745	8.18	0.8251	0.50150	79.44
180.00	9.10057E-07	6.275	7.62	0.8234	0.50392	82.22
185.00	8.48409E-07	5.850	7.12	0.8217	0.50633	85.00
190.00	7.92609E-07	5.465	6.66	0.8200	0.50875	87.78
195.00	7.41973E-07	5.116	6.25	0.8183	0.51117	90.56

200.00	6.95911E-07	4.798	5.88	0.8166	0.51359	93.33
205.00	6.53913E-07	4.509	5.53	0.8150	0.51600	96.11
210.00	6.15532E-07	4.244	5.22	0.8133	0.51842	98.89
215.00	5.80383E-07	4.002	4.93	0.8116	0.52084	101.67
220.00	5.48126E-07	3.779	4.67	0.8100	0.52326	104.44
225.00	5.18464E-07	3.575	4.42	0.8084	0.52567	107.22

The results can also be graphically displayed as shown below:

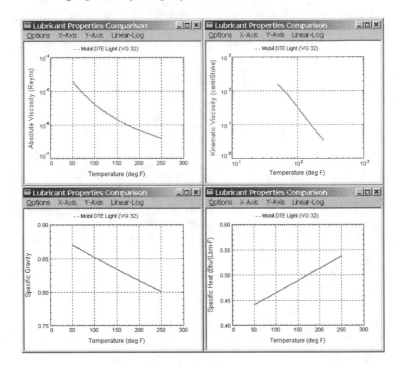

Figure C-2 Lubricant property chart

DyRoBeS also allows you to compare the oil properties. Following figures show the dynamic viscosity and density of two synthetic oils (Exxon ETO 2389 and Exxon ETO 2380; mil-7808J and 23699C) and two mineral oils (Mobil DTE Light and Medium; ISO VG32 and 46). In general, the synthetic oils have higher viscosity index and higher density.

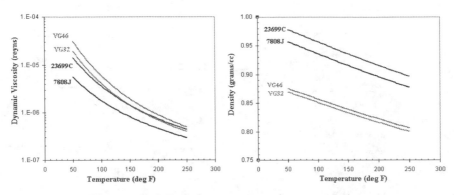

Figure C-3 Lubricant properties comparison

Sample data sheets from various manufacturers are listed below for reference:

Mobil DTE® Oil
Named Series – Circulation Oils

	Light	Medium	Heavy Medium	Mobil DTE Oil Heavy	Extra Heavy	BB	AA	HH
Product Number	60014-5	60015-5	60016-3	60018-9	60020-5	60022-1	60025-4	60027-0
Specific Gravity	.867	.873	.877	.881	.887	.890	.897	.900
Pour Point, °C (°F)	-7 (20)	-7 (20)	-7 (20)	-7 (20)	-4 (25)	-4 (25)	-4 (25)	-4 (25)
Flash Point, °C (°F)	204 (400)	204 (400)	204 (400)	210 (410)	227 (440)	227 (440)	238 (460)	271 (520)
Viscosity								
cSt at 40 °C	30.4	43.7	64.6	95	157.5	230	335	437
cSt at 100 °C	5.1	6.5	8.4	11.4	14.7	19.3	24.7	31.8
SUS at 100 °F	158	226	335	495	840	1245	1815	2338
SUS at 210 °F	44	48.9	55	65.5	80.5	99.1	147	153
Viscosity Index	95	95	95	95	96	95	95	95
ISO VG	32	46	68	100	150	220	320	460
Color, ASTM D 1500	1.5	2.0	2.5	2.5	4.5	5.0	6.0	6.5
Neut. No.	0.20	0.25	0.30	0.30	1.2	1.2	1.2	1.2
Rust Protection, ASTM D 665 A&B	Pass	Pass	Pass	Pass	Pass	Pass	Pass	Pass
Demulsibility, ASTM D 1401, min to 3 ml emulsion								
at 54 °C (130 °F) after 30 min	Pass	Pass	Pass	Pass	—	—	—	—
at 82 °C (180 °F) after 1 hour	—	—	—	—	Pass	Pass	Pass	Pass

PDS I-02

AMOKON Oils meet or exceed all major steam turbine manufacturers specifications including General Electric GEK 28143 (1) and Westinghouse 55125 AC as well as GEK 32568 for gas turbines.

AMOKON Oils provide excellent service in Joy, Sullair, LeRoi, and other rotary air compressors.

AMOKON Oils

Method	Property	Typical [1]			
	ISO-VG	32	46	68	100
D-92	Flash Point, F	430	442	446	500
	Flash Point, C	221	228	230	260
D-97	Pour Point, F	-30	-10	-10	-15
	Pour Point, C	-34	-22	-22	-26
D-287	Gravity, API	32.5	31.7	31.0	30.5
	Specific Gravity, 60/60 F	0.863	0.867	0.871	0.874
D-445	Viscosity,				
	40 C, cSt	32.6	47.9	68.2	99.7
	100 C, cSt	5.61	7.16	8.92	11.8
	100 F, SUS	168	247	353	518
	210 F, SUS	45.0	50.2	56.3	67.1
D-2270	Viscosity Index	110	108	104	108
D-665A	Rust Test, Distilled Water	Pass	Pass	Pass	Pass
D-665B	Synthetic Sea Water	Pass	Pass	Pass	Pass
D-892	Foaming Characteristics, Tend./Stab., Seq.I, ml	10/0	10/0	10/0	20/0
D-943	Oxidation Stability, Hrs. to 2.0 Acid Number	30,000	30,000	35,000	20,000 +
D-974	Acid Number, mg KOH/g	0.10	0.10	0.09	0.10
D-1401	Emulsion, 3 ml or less at 15 minutes	Pass	Pass	Pass	Pass
D-2272	Oxidation Stability, Rotating Bomb, Minutes	2,000	2,000	2,000	2,000
D-524	Carbon Residue, Ramsbottom, %	0.07	0.10	0.08	0.10
D-189	Carbon Residue, Conradson, %	0.02	0.04	0.03	0.04

Appendix A

MIL-L-7808J Specifications vs. ETO 2389 Typical Inspections

The values shown here are representative of current production. Some are controlled by manufacturing specifications, while others are not. All of them may vary within modest ranges

MILITARY SPECIFICATION TESTS	ETO 2389	MIL-L-7808J		ASTM STANDARD	FED. TEST METHOD STD. NO 791
		Min.	Max.		
Particulate contamination, contaminant mg/litre	2.0	——	5.0	Appendix A of 7808 Spec.	
Filtering time, min/qt	14	——	30		
Neutralization no., mg KOH/g	0.23	——	0.30	D664*	
Viscosity at 100°C, cSt	3.19	3.0	——	D445	
Viscosity at 40°C, cSt	12.46	Report	——	D445	
Viscosity at –53.9°C, cSt				D2532	
at 35 minutes	12,700	——	17,000		
at 3 hours..................................	12,750	——	17,000		
at 72 hours.................................	12,850	——	17,000		
Viscosity stability @ –53.9°C, % (2 determinations)..	0.18	——	6	D2532	
Flash point (C.O.C.), °C	220	210	——	D92	
Evaporation loss, 6½ hrs., @ 210°C, wt. %	20.0	——	30	D972	
Foaming characteristics (static)				3213	
Foam volume, ml	30	——	100		
Foam colapse time sec	0	——	60		
Foaming characteristics (dynamic)					3214
Foam volume, ml/collapse time, sec.					
80°C @ 1000 cc/min	15/8	——	100/60		
80°C @ 1500 cc/min	45/8	——	150/60		
80°C @ 2000 cc/min	105/15	——	Report/60		
110°C @ 1000 cc/min	20/8	——	100/60		
110°C @ 1500 cc/min	55/8	——	150/60		
110°C @ 200 cc/min	170/18	——	Report/60		
Deposition test, average deposition rating	0.59	——	1.5		5003
Neut. no., change..............................	11.2	Report	——		
Viscosity @ 40°C, % change	96	Report	——		
Oil consumption, ml............................	100	Report	——		
SOD Pb corrosion, 1 hr. @ 162.8°C, wt change, g/m²	+0.64	——	9.3		5321
Bz corrosion, 50 hrs. @ 323°C, wt. change. g/m² ...	0.0	——	±4.5		5305
Ag corrosion, 50 hrs. @ 232°C, wt. change, g/m⁴ ...	0.0	——	±4.5		5305
Ryder gear load, avg. % relative rating × 508 kN/m					
(avg. of 2 det)	462	420	——	D1947	
Oxidation & corrosion stability @ 200°C for 96 hrs.:					5307
Al, wt. change, mg/cm²	0.00	——	±0.2		
Ag, wt. change, mg/cm²	–0.02	——	±0.2		
Bz, wt. change, mg/cm²	+0.04	——	±0.4		
Fe, wt. change, mg/cm²	+0.02	——	±0.2		
M-50, wt. change, mg/cm²	–0.02	——	±0.2		
Mg, wt. change, mg/cm².........................	–0.02	——	±0.4		
Ti, wt. change, mg/cm²...........................	0.00	——	±0.2		
Kinematic viscosity change, % @ 40°C	+9.5	–5	+25		
Neut. no., change	0.96	——	4.0		

*MODIFIED

MIL-L-7808J Specifications (Cont'd.)

MILITARY SPECIFICATION TESTS	ETO 2389	MIL-L-7808J Min.	MIL-L-7808J Max.	ASTM STANDARD	FED. TEST METHOD STD. NO 791
NBR-H rubber swell, % (168 hrs. @ 70°C)	27.9	12	35		3604
F-A rubber swell, % (72 hrs. @ 175°C)	14.6	2	25		3432
Tensile strength, % change	+6.7	—	50		
Elongation, % change	+13.0	—	50		
Hardness, durometer number change	0	—	20		
FS rubber swell, % (72 hrs. @ 150°C)	5.5	2	25		3432
Tensile strength, % change	-3.9	—	50		
Elongation, % change	+1.1	—	50		
Hardness, durometer number change	0.0	—	20		
QV I rubber swell, % (72 hrs. @ 150°C)	20.4	2	30		3432
Tensile strength, % change	-39	—	50		
Elongation, % change	-3	—	50		
Hardness, durometer number change	-10	—	20		
Bearing deposition				Appendix 1A of 27502 Spec. per Sec. 4.6.4 of 7808 Spec.	
Average deposit rating	49	—	60		
Filter deposit, gm	1.7	—	2.0		
Oil consumption, ml.	864	—	1440		
Viscosity change, %	+7	—	±25		
Neutralization number change	0.2	—	1.0		
Metal specimen weight change, mg/cm²	Pass	—	±25		
Compatibility, turbidity, ml sediment/200 ml of oil	Nil	—	0.005		3403*
Compatibility, intermixing	Passes	Pass		Sec. 3.4.1.4 of 7808 Spec.	
Storage stability, 1 hr. Pb corrosion @163C, after 2 days, g/m²	±0.2	—	40	Sec. 4.6.8.1 of 7808 Spec.	
after 7 days, g/m²	-24.6	—	230		
Trace element content, ppm				Sec. 4.6.2 of 7808 Spec.	
Al	0	—	2		
Fe	1	—	2		
Cr	0	—	2		
Ag	0	—	1		
Cu	0	—	1		
Sn	9	—	11		
Mg	1	—	2		
Ni	0	—	2		
Ti	1	—	2		
Si	1	—	2		
OTHER TESTS					
Coefficient of expansion (avg: -40° to 300°F)	0.00045	—	—	Exxon Test Method	
Specific heat, Btu/lb/°F					
at 100°F	0.444	—	—	D2766	
at 200°F	0.493	—	—		
at 300°F	0.533	—	—		
at 400°F	0.565	—	—		
Thermal conductivity, Btu/(hr)(sq ft)(°F/ft)					
at 100°F	0.087	—	—	D2717	
at 400°F	0.083	—	—		

*MODIFIED

Appendix D: Systems of Units

The data input in **DyRoBeS** is not restricted to a specific system of units. In addition to the consistent units of user's choice, four sets of system of units have been introduced in **DyRoBeS**:

System of Units

Unit = 0 => Consistent Units of User's Choice
Unit = 1 => Consistent English Units (sec, in, Lbf, Lbf-s^2/in)
Unit = 2 => Engineering English Units (sec, in, Lbf, Lbm)
Unit = 3 => Consistent SI Units (sec, m, N, kg)
Unit = 4 => Engineering Metric Units (sec, mm, N, kg)

For Unit =0, consistent units, it is essential that all of the input parameters be expresses in a consistent set of units of the user's choice. For Unit =1 and 3, these are two sets of consistent units. For Unit =2 and 4, these are two sets of inconsistent units that are commonly used in the engineering field. Since they are not consistent units, their input units are explicitly specified on the input screen. Users must pay attention to their units.

Units Systems	Consistent English Unit = 1	Engineering English Unit = 2	Consistent SI Unit = 3	Engineering Metric Unit = 4
Basic Quantities				
Time	Second (s)	Second (s)	Second (s)	Second (s)
Length	in	in	m	**mm**
Force	Lbf	Lbf	Newton (N)	Newton (N)
Mass	Lbf-s^2/in	**Lbm**	kg = N-s^2/m	**kg**
Inputs				
Material Properties				
Density	Lbf-s^2/in^4	**Lbm/in^3**	kg/m^3	**kg/m^3**
Modulus	Lbf/in^2 (psi)	Lbf/in^2 (psi)	N/m^2 (Pa)	N/mm^2 (MPa)
Shaft Elements				
Length, Diameter	in	in	m	mm
Disks				
Mass	Lbf-s^2/in	**Lbm**	kg	**kg**
Inertia	Lbf-s^2-in	**Lbm-in^2**	kg-m^2	**kg-m^2**
Skew Angle	Degree	Degree	Degree	Degree
Unbalance				
Imbalance (*me*)	Lbf-s^2	**oz-in**	kg-m	**kg-mm**
Angle	Degree	Degree	Degree	Degree
Flexible Supports				
Mass	Lbf-s^2/in	**Lbm**	kg	**kg**
Damping	Lbf-s/in	Lbf-s/in	N-s/m	N-s/mm
Stiffness	Lbf/in	Lbf/in	N/m	N/mm
Forces/Moments				
Forces	Lbf	Lbf	N	N
Moments/Torque	Lbf-in	Lbf-in	N-m	**N-mm**

Misalignment/Bow				
Deflection: x, y	in	in	m	mm
Theta: x, y	degree	degree	degree	degree
Time Forcing				
Force or	Lbf	Lbf	N	N
Moment	Lbf-in	Lbf-in	N-m	N-mm
Time Constant	1/s	1/s	1/s	1/s
Excitation Freq.	rpm	rpm	rpm	rpm
ϕ - Phase	degree	degree	degree	degree
Bearings - Lateral				
Stiffness – Kt	Lbf/in	Lbf/in	N/m	N/mm
Damping – Ct	Lbf-s/in	Lbf-s/in	N-s/m	N-s/mm
Stiffness – Kr	Lbf-in/rad	Lbf-in/rad	N-m/rad	N-mm/rad
Damping – Cr	Lbf-in-s/rad	Lbf-in-s/rad	N-m-s/rad	N-mm-s/rad
Length	in	in	m	**mm**
Lubricant Viscosity	Reyn $(Lbf\text{-}s/in^2)$	Reyn $(Lbf\text{-}s/in^2)$	Pascal-second $(N\text{-}s/m^2)$	**CentiPoise** =1.0E03 * Pa-s
Lubricant Density	$Lbf\text{-}s^2/in^4$	**Lbm/in^3**	kg/m^3	**kg/m^3**
Pressure Drop Across seal	Psi	Psi	Pa	Bar =1.0E-05 * Pa
Linear PID + Low Pass Filter Active Magnetic Bearings				
Proportional Gain	Lbf/in	Lbf/in	N/m	N/mm
Integral Gain	Lbf/(in-s)	Lbf/(in-s)	N/(m-s)	N/(mm-s)
Derivative Gain	Lbf-s/in	Lbf-s/in	N-s/m	N-s/mm
Cut-Off Freq.	Hz	Hz	Hz	Hz
Non-Linear Active Magnetic Bearings				
Proportional Gain	A/in	A/in	A/m	A/mm
Integral Gain	A/(in-s)	A/(in-s)	A/(m-s)	A/(mm-s)
Derivative Gain	A-s/in	A-s/in	A-s/m	A-s/mm
Force Constant	$Lbf\text{-}in^2/A^2$	$Lbf\text{-}in^2/A^2$	$N\text{-}m^2/A^2$	$N\text{-}mm^2/A^2$
Air Nominal Gap	in	in	m	mm
Current	A	A	A	A
Torsional				
Stiffness – K	Lbf-in/rad	Lbf-in/rad	N-m/rad	N-mm/rad
Damping – C	Lbf-in-s/rad	Lbf-in-s/rad	N-m-s/rad	N-mm-s/rad
Axial				
Stiffness – K	Lbf/in	Lbf/in	N/m	N/mm
Damping – C	Lbf-s/in	Lbf-s/in	N-s/m	N-s/mm
Gravity - g_o	386.088 in/s^2	386.088 in/s^2	9.8066 m/s^2	9806.6 mm/s^2
Outputs				
Displacements	in	in	m	mm
Velocity	in/s	in/s	m/s	mm/s
Acceleration	in/s^2	in/s^2	m/s^2	mm/s^2
Rotations	rad	rad	rad	rad
Force	Lbf	Lbf	N	N
Moments/Torque	Lbf-in	Lbf-in	N-m	N-mm

The conversions between English and Metric units are list below for quick reference.

Unit	English	Metric	Conversions (* = multiply)
Time	second (s)	second (s)	
Length	in	mm	mm = 25.4 * in
	in	m	m = 0.0254 * in
Force	Lbf	Newton (N)	N = 4.448222 * Lbf
Moment	Lbf-in	N-m	N-m = 0.1129846 * Lbf-in
Density	$Lbf\text{-}s^2/in^4$	kg/m^3	kg/m^3 = 2.767990E+04 * Lbm/in^3
			= 1.068688E+07 * $Lbf\text{-}s^2/in^4$
Mass	$Lbf\text{-}s^2/in$	$kg = N\text{-}s^2/m$	kg = 175.1266 * $Lbf\text{-}s^2/in$
			= 0.4535924 * Lbm
Inertia	$Lbf\text{-}s^2\text{-}in$	$kg\text{-}m^2$	$kg\text{-}m^2$ = 0.1129846 * $Lbf\text{-}s^2\text{-}in$
Moduli	Lbf/in^2 (psi)	kN/m^2 (kPa)	kPa = 6.894757 * psi
		N/m^2 (Pa)	Pa = 6.894757E+03 * psi
Temperature	°F	°C	°C = (°F-32) * 5/9
Viscosity	Reyn ($Lbf\text{-}s/in^2$)	centiPoise	cP = 6.894757E06 * Reyn
Flow Rate	GPM (gal/min)	m^3/hour	m^3/hour = 0.2271 * GPM
Power	hp (horsepower)	kWatt	kWatt = 0.7457 * hp
Stiffness	Lbf/in	N/mm	N/mm = 0.1751266 * Lbf/in
Damping	Lbf-s/in	N-s/mm	N-s/mm = 0.1751266 * Lbf-s/in
Torsional K	Lbf-in/rad	N-m/rad	N-m/rad = 0.1129846 * Lbf-in/rad
Gravity - g	386.088 in/s^2	9.8066 m/s^2	
Misc.			
			Lbm = 386.088 * $Lbf\text{-}s^2/in$
			1 N = 1kg * $1m/s^2$
			cP = grams/CC * cSt (mm^2/s)

The units conversion can be easily implemented by the displacement and force units in the equation of motion:

$$M\ddot{x} + C\dot{x} + Kx = F$$

For Unit = 0, 1, and 3, no data conversion is required. They are consistent units.

For Unit 1 and 2: English Units

The displacement and force units are:

$$x = \begin{Bmatrix} x \\ y \\ \theta_x \\ \theta_y \end{Bmatrix} \quad unit = \begin{Bmatrix} in \\ in \\ 1 \\ 1 \end{Bmatrix}, \qquad F = \begin{Bmatrix} F_x \\ F_y \\ M_x \\ M_y \end{Bmatrix} \quad unit = \begin{Bmatrix} Lb_f \\ Lb_f \\ Lb_f - in \\ Lb_f - in \end{Bmatrix},$$

$$M = \begin{bmatrix} Lb_f - s^2/in & & & \\ & Lb_f - s^2/in & & \\ & & Lb_f - in - s^2 & \\ & & & Lb_f - in - s^2 \end{bmatrix}$$

$$C = \begin{bmatrix} Lb_f - s/in & & & \\ & Lb_f - s/in & & \\ & & Lb_f - in - s & \\ & & & Lb_f - in - s \end{bmatrix}$$

$$K = \begin{bmatrix} Lb_f/in & & & \\ & Lb_f/in & & \\ & & Lb_f - in & \\ & & & Lb_f - in \end{bmatrix}$$

To convert Lbm: $\dfrac{Lb_f - s^2}{in} = \dfrac{Lb_m}{386.088}$

To convert oz-in: $Lb_f - s^2 = \dfrac{oz - in}{386.088 \times 16}$

For Unit 3, Consistent SI Units (sec, m, N, kg), no data conversion is required.

The displacement and force units are:

$$x = \begin{Bmatrix} x \\ y \\ \theta_x \\ \theta_y \end{Bmatrix} \qquad unit = \begin{Bmatrix} m \\ m \\ 1 \\ 1 \end{Bmatrix}, \qquad F = \begin{Bmatrix} F_x \\ F_y \\ M_x \\ M_y \end{Bmatrix} \qquad unit = \begin{Bmatrix} N \\ N \\ N - m \\ N - m \end{Bmatrix},$$

$$M = \begin{bmatrix} kg & & & \\ & kg & & \\ & & kg - m^2 & \\ & & & kg - m^2 \end{bmatrix} \qquad \text{Note:} \quad kg = N - s^2/m$$

$$C = \begin{bmatrix} N-s/_m & & & \\ & N-s/_m & & \\ & & N-m-s & \\ & & & N-m-s \end{bmatrix}$$

$$K = \begin{bmatrix} N/_m & & & \\ & N/_m & & \\ & & N-m & \\ & & & N-m \end{bmatrix}$$

For Unit 4, Engineering Metric Units (sec, mm, N, kg), data conversion is required.

The displacement and force units are:

$$x = \begin{Bmatrix} x \\ y \\ \theta_x \\ \theta_y \end{Bmatrix} \quad \text{unit} = \begin{Bmatrix} mm \\ mm \\ 1 \\ 1 \end{Bmatrix}, \qquad F = \begin{Bmatrix} F_x \\ F_y \\ M_x \\ M_y \end{Bmatrix} \quad \text{unit} = \begin{Bmatrix} N \\ N \\ N-mm \\ N-mm \end{Bmatrix},$$

$$M = \begin{bmatrix} N-s^2/_{mm} & & & \\ & N-s^2/_{mm} & & \\ & & N-mm-s^2 & \\ & & & N-mm-s^2 \end{bmatrix} \quad kg = N-s^2/_m$$

$$C = \begin{bmatrix} N-s/_{mm} & & & \\ & N-s/_{mm} & & \\ & & N-mm-s & \\ & & & N-mm-s \end{bmatrix}$$

$$K = \begin{bmatrix} N/_{mm} & & & \\ & N/_{mm} & & \\ & & N-mm & \\ & & & N-mm \end{bmatrix}$$

To convert mass kg :

$$N - s^2 \Big/ mm = grams = 10^{-3} \times kg$$

To convert Inertia kg-m^2:

$$N - s^2 - mm = 10^3 \times kg - m^2$$

To convert density:

$$N - s^2 \Big/ mm^4 = g \Big/ mm^3 = 10^{-12} \times kg \Big/ m^3$$

To convert unbalance:

$$N - s^2 = 10^{-3} \times kg - mm$$

To convert viscosity:

$$N - s \Big/ mm^2 = 10^{-9} \times centipoise$$

References

Adams, M. L., and Padovan, J., 1981, "Insights into Linearized Rotor Dynamics," *Journal of Sound and Vibration*, Vol. 76, pp. 129-142.

Alford, J., 1965, "Protecting Turbomachinery from Self-Excited Rotor Whirl," *Journal of Engineering for Power*, pp.333-334.

Allaire, P. E., Nicholas, J. C., and Gunter, E. J., 1977, "Systems of Finite Elements for Finite Bearings," *ASME Journal of Lubrication Technology*, pp. 187-197.

Barrett, L. E., 1984, "Turbulent Flow Annular Pump Seals: A Literature Review," *Shock and Vibration Digest*, pp. 3-13.

Barrett, L. E., and Gunter, E. J, 1975, "Steady State and Transient Analysis of a Squeeze Film Damper Bearing for Rotor Stability," *NASA CR-2548*.

Bathe, K. J., 1995, *Finite Element Procedures*, Prentice-Hall, Inc., New Jersey.

Black, H. F., and Jenssen, D. N., 1970, "Dynamic Hybrid Bearing Characteristics of Annular Controlled Leakage Seals," *Proc Instn Mech Engrs*, Vol. 184, pp. 92-100.

Booker, J. F., 1965, "A Table of the Journal-Bearing Integral," *Journal of Basic Engineering*, pp. 533-535.

Booser, E. R., 1983, *Handbook of Lubrication Theory and Practice of Tribology*, Vol. II, CRC Press, Inc., Boca Raton, FL.

Bureau of Standards, *Miscellaneous Publication of the Bureau of Standards* No. 97, "Specific Heats of Lubricating Oils,"

Cameron, A., 1966, *The Principles of Lubrication*, Wiley, New York,

Chen, W. J., and Chen, H. M., 1999, "Rotor Transient Response with Fault Tolerant Magnetic Bearings," *Proceedings of Asia-Pacific Vibration Conference,* Nanyang Technological University, pp. 53-58.

Chen, W. J., 1998, "A Note on Computational Rotor Dynamics," *ASME Journal of Vibration and Acoustics*, Vol. 120, pp. 228-233.

Chen, W. J., 1997, "Energy Analysis to the Design of Rotor-Bearing Systems," *ASME Journal of Engineering for Gas Turbines and Power*, Vol. 119, pp.411-417.

Chen, W. J., 1996, "Instability Threshold and Stability Boundaries of Rotor-Bearing Systems," *ASME Journal of Engineering for Gas Turbines and Power*, Vol. 118, pp.115-121.

Chen, W. J., 1995, "Torsional Vibrations of Synchronous Motor Driven Trains Using p-Method," *ASME Journal of Vibration and Acoustics*, Vol. 117, pp.152-160.

Chen, W. J., 1995, "Bearing Dynamic Coefficients of Flexible-Pad Journal Bearings," *STLE Tribology Transaction*, Vol. 38, pp.253-260.

Chen, W. J, Zeidan, F. Y., and Jain, D., 1994, "Design, Analysis, and Testing of High Performance Bearings in a High Speed Integrally Geared Compressor", *Proceedings of 23rd Turbomachinery Symposium*, Texas A&M University, pp. 31-42.

Chen, W. J., Rajan, M., Rajan, S. D., and Nelson, H. D., 1991, "Application of Nonlinear Programming for Balancing Rotor Systems," *Engineering Optimization*, Vol. 17, pp. 79-90.

Chen, W. J., Rajan, M., Rajan, S. D., and Nelson, H. D., 1988, "The Optimal Design of Squeeze Film Dampers for Flexible Rotor Systems," ASME Journal of Mechanism, Transmission, and Automation in Design, Vol. 110, pp. 166-174.

Childs, D. W., 1993, *Turbomachinery Rotordynamics: Phenomena, Modeling, and Analysis*, John Wiley & Sons, Inc.

Childs, D. W., 1983, "Dynamic Analysis of Turbulent Annular Seals Based On Hirs' Lubrication Equation", *ASME Journal of Lubrication Technology*, Vol. 105, pp.429-436.

Childs, D. W., 1972, "A Simulation Model for Flexible Rotating Equipment," *ASME Journal of Engineering for Industry*, pp. 201-209.

Constantinescu, V. N., 1973, "Basic Relationships in Turbulent Lubrication and Their Extension to Include Thermal Effects," *ASME Journal of Lubrication Technology*, Vol. 95.

Corbo, M. A., and Malanoski, S. B., 1996, "Practical Design Against Torsional Vibration," *Proceedings of the 25th Turbomachinery Symposium*, Texas A&M University, College Station, TX.

Corbo, M. A., and Malanoski, S. B., 2003, "Pump Rotordynamics Made Simple", *Proceeding of The 15th International Pump Users Symposium*, pp. 167-204.

Cowper, G. R., 1966, "The Shear Coefficient in Timoshenko's Beam Theory," *ASME Journal of Applied Mechanics*, pp. 335-340.

Craig, R. R., Jr., 1981, *Structural Dynamics: An Introduction to Computer Methods*, John Wiley & Sons, Inc.

Cunningham, R. E., Gunter, E. J., and Fleming D. P., 1975, "Design of an Oil Squeeze Film Damper Bearing for a Multimass Flexible-Rotor Bearing System," *NASA TN D-7892*.

Diewald, W., and Nordmann, R., 1989, "Dynamic Analysis of Centrifugal Pump Rotors with Fluid-Mechanical Interactions", *ASME Journal of Vibration, Acoustics, Stress, and Reliability in Design*, Vol. 111, pp.370-378.

Dimentberg, F. M., 1961, *Flexural Vibrations of Rotating Shafts*, Butterworth, London.

Florjancic, S., and McCloskey, T., 1991, "Measurement and Prediction of Full Scale Annular Seal Coefficient," *Proceeding of The 8th International Pump Users Symposium*, Texas A&M University, pp. 71-83.

Gardner, W. W., *Hydrodynamic Oil Film Bearings – Fundamentals, Limits, and Application*, Waukesha Bearings Corporation.

Gargiulo, E. P. Jr., "A Simple Way to Estimate Bearing Stiffness", *Machine Design*, 1980, pp.107-110.

Garner, D. R., 1981, *The Use of Design Procedures for Plain Bearings*, The Glacier Metal Company Limited, England.

Glacier Metal Company Limited, 1980, *Tilting Pad Journal Bearings – Designers' Handbook* No. 10, England.

Greenhill, L. M., Bickford, W. B., and Nelson, H. D., 1985, "A Conical Beam Finite Element for Rotor Dynamics Analysis," *ASME Journal of Vibration, Acoustics, Stress and Reliability in Design*, Vol. 107, pp.421-430.

Gross, W. A., Matsch, L. A., Castelli, V., Eshel, A., and Vohr, J., 1980, *Fluid Film Lubrication*, John Wiley & Sons, Inc.

Gunter, E. J., and Chen, W. J., 2005, "Dynamic Analysis of a Turbocharger in Floating Bushing Bearings," *ISCORMA-3*, Cleveland, OH.

Gunter, E. J., and Chen, W. J., 2005, "Dynamic Analysis of An 1150 MW Turbine Generator," *ASME Power Conference*, PWR2005-50142.

Gunter, E. J, and Jackson, C., 1992, "Balancing of Rigid and Flexible Rotors," *Handbook of Rotordynamics*, Editor: Ehrich, F. F., McGraw-Hill, Inc., New York.

Gunter, E. J., and Gaston, C. G., 1987, "CRITSPD-PC User's Manual", RODYN Vibration Analysis, Inc.

Gunter, E. J., and Gaston, C. G., 1987, "ROTORBAL-PC User's Manual", RODYN Vibration Analysis, Inc.

Gunter, E. J., Barrett, L. E., and Allaire, P. E., 1977, "Design of Nonlinear Squeeze Film Dampers for Aircraft Engines," *ASME Journal of Lubrication Technology*, Vol. 92, No. 1, pp. 57-64.

Gunter, E. J., 1966, *Dynamic Stability of Rotor-Bearing Systems*, SP-113, NASA.

Guyan, R. J., 1965, "Reduction of Stiffness and Mass Matrices, " *AIAA Journal*, pp. 380.

Hamrock, B. J., 1991, *Fundamentals of Fluid Film Lubrication*, NASA RP-1255.

Ingersoll-Rand Company, 1988, *Cameron Hydraulic Data*, Ingersoll-Rand Company, Woodcliff Lake, New Jersey.

Jones, G. J. and Martin, F. A., 1979, "Geometry Effects in Tilting-Pad Journal Bearings," *ASLE Tribology Transactions*.

Ker Wilson, W., 1956, *Practical Solution of Torsional Vibration Problems*, Vol. 1, 2, and 3, John Wiley and Sons, New York, New York

Kirk, R. G., Gunter, E. J., Chen, W. J., 2004, "Transient Rotor Drop Evaluation of AMB Machinery," *Proceedings of Conference on Standardization for AMB Rotors*, NEDO-ISO Joint workshop, pp. 89-113.

Kirk, R. G., and Donald, G.H., 1983, "Design Criteria for Improved Stability of Centrifugal Compressors", *ASME Rotor Dynamical Instability*, AMD-Vol. 55.

Kirk, R. G., and Gunter, E. J., 1970, *Transient Journal Bearing Analysis*, NASA CR-1549.

Klaus, E. E., and Tewksbury, E. J., 1983, "Liquid Lubricants," *CRC Handbook of Lubrication*, Vol. II, pp.229-240.

Kramer, E., 1993, *Dynamics of Rotors and Foundations*, Springer-Verlag Berlin Heidelberg.

Lalanne, M., and Ferraris, G., 1990, *Rotordynamics Prediction in Engineering*, John Wiley & Sons, Ltd.

Lund, J. W., 1987, "Review of the Concept of Dynamic Coefficients for Fluid Film Journal Bearings," *ASME Journal of Tribology*, Vol. 109, pp. 37-41.

Lund, J. W., and Thomsen, K. K., 1978, "A Calculation Method and Data for the Dynamic Coefficients of Oil-Lubricated Journal Bearings," *ASME Topics in Fluid Film Bearing and Rotor Bearing System Design and Optimization*, pp.1-28.

Lund, J. W., 1974, "Stability and Damped Critical Speeds of a Flexible Rotor in Fluid-Film Bearings", *ASME Journal of Engineering for Industry*.

Lund, J. W., and Tonnesen, J., 1972, "Analysis and Experiments on Multi-Plane Balancing of a Flexible Rotor," *ASME Journal of Engineering for Industry*, pp 233-242.

Lund, J. W., 1964, "Spring and Damping Coefficients for the Tilting-Pad Journal Bearing," *ASLE Trans.*, Vol.7, No.4.

Meirovitch, L., 1980, *Computational Methods in Structural Dynamics*, Sijthoff & Noordhoff International Publishers, The Netherlands.

Muszynska, A., 1986, "Whirl and Whip – Rotor/Bearing Stability Problems," *Journal of Sound and Vibration*, Vol. 110, pp. 443-462.

Nelson, H. D., and Chen, W. J., 1993, "Undamped Critical Speeds of Rotor Systems Using Assumed Modes," *ASME Journal of Vibration and Acoustics*, Vol. 115, pp. 367-369.

Nelson, H. D., Chen, W. J., and Nataraj, C., 1989, *Periodic Motion of Mechanical Systems with Nonlinear Components*, NASA Report NAG 3-580.

Nelson, H. D., 1980, "A Finite Rotating Shaft Element Using Timoshenko Beam Theory," *ASME Journal of Mechanical Design*, Vol. 102, pp.793-803.

Nelson, H. D., and McVaugh, J. M., 1976, "The Dynamics of Rotor-Bearing Systems Using Finite Elements," *ASME Journal of Engineering for Industry*.

Nicholas, J. C., Gunter, E. J., and Allaire, P. E., 1976, "Effect of Residual Shaft Bow on Unbalance Response and Balancing of a Single Mass Flexible Rotor: Part I– Unbalance Response; Part II-Balancing," *Journal of Engineering for Power*.

Nicholas, J. C., Gunter, E. J., and Allaire, P. E., 1979, "Stiffness and Damping Coefficients for the Five-Pad Tilting-Pad Bearing," *ASLE Trans.*, Vol.22, No. 2

Nicholas, J. C., 1977, "A Finite Element Dynamic Analysis of Pressure Dam and Tilting-Pad Bearings," Ph.D. Dissertation, University of Virginia, Charlottesville, VA.

Orcutt, F. K., 1967, "The Steady-State and Dynamic Characteristics of Tilting-Pad Journal Bearing in Laminar and Turbulent Flow Regimes," *Journal of Lubrication Technology*.

Peng, J. P., and Carpino, M., 1997, "Finite Element Approach to the Prediction of Foil Bearing Rotor Dynamic Coefficients," *ASME Journal of Tribology*, Vol. 119, pp. 85-90.

Peng, J. P., 1993, "Theoretical Predication of Rotor Dynamic Coefficients for Foil Bearings," Ph.D. Dissertation, The Pennsylvania State University.

Proctor, M. P., and, Gunter, E. J., 2005, "Nonlinear Whirl Response of a High-Speed Seal Test Rotor with Marginal and Extended Squeeze-Film Dampers," *ISCORMA*-3, Cleveland, OH, 19-23.

Raimondi, A. A., 1961, "A Numerical Solution for the Gas Lubricated Full Journal Bearing of Finite Length," *ASLE Transactions*, Vol. 4, pp. 131-155.

Rao, J. S., 1983, *Rotor Dynamics*, Wiley Eastern Limited.

Reddy, J. N., 1993, *An Introduction to the Finite Element Method*, McGraw-Hill, New York.

Salamone, D. J., and Gunter, E. J., 1979, "Synchronous Unbalance Response of an Overhung Rotor with Disk Skew," *ASME* 79-GT-135

Schenck Trebel Corporation, 1983, *Fundamentals of Balancing*, Second Edition, New York.

Shang, L., and Dien, I. K., 1989, "A Matrix Method for Computing the Stiffness and Damping Coefficients of Multi-Arc Journal Bearings," *Tribology Transactions*, Vol. 32, pp. 396-404.

Shapiro, W., and Colsher, R., 1977, "Dynamic Characteristics of Fluid Film Bearings, "*Proceedings of the 6th Turbomachinery Symposium*, Texas A&M University, pp. 39-53.

Shigley, J. E., and Mischke, C. R., 1989, *Mechanical Engineering Design*, McGraw-Hill, Inc.

Someya, T., 1989, *Journal-Bearing Databook*, Springer-Verlag Berlin, Heidelberg.

Szeri, A. Z., 1980, *Tribology: Friction, Lubrication, and Wear*, McGraw-Hill Book Company.

Taylor, D. L., and Kumar, B. R. K., 1980, "Nonlinear Response of Short Squeeze Film Dampers," *ASME Journal of Lubrication Technology*, Vol. 102, pp.51-58.

Tessarzik, J. M., Badgley, R. H., and Anderson, W. J., 1972, "Flexible Rotor Balancing by the Exact Point-Speed Influence Coefficient Method," *ASME Journal of Engineering for Industry*, 1972, pp 148-158.

Thomson, W. T., 1981, *Theory of Vibration with Applications*, Prentice-Hall Inc.

Timoshenko, S., Young, D. H., and Weaver W., Jr., 1974, *Vibration Problems in Engineering*, 4th Edition, John Wiley & Sons.

Tondl, A., 1965, *Some Problems of Rotor Dynamics*, Chapman & Hall Limited.

Vance, J. M., 1988, *Rotordynamics of Turbomachinery*, John Wiley & Sons, Inc.

Wachel, J. C., 1983, "Compressor Case Histories," presented at *Rotating Machinery and Controls (ROMAC)* Short Course, University of Virginia, June 8-10.

Wachel, J. C., and Szenasi, F. R., 1993, "Analysis of Torsional Vibrations in Rotating Machinery" *Proceedings of the 22nd Turbomachinery Symposium*, Texas A&M University, College Station, TX.

Waukesha Bearings Corporation, *Tilting Pad Journal Bearing Selection Guide*, Waukesha, WI.

Wilcock, D, F., and Booser, E. R., 1957, *Bearing Design and Application*, McGraw-Hill, New York.

Yamamoto, T., and Ishida, Y., 2001, *Linear and Nonlinear Rotordynamics: A Modern Treatment with Applications*, John Wiley & Sons, Inc.

Young, W. C, 1989, *Roark's Formulas for Stress & Strain*, Sixth Edition, McGraw-Hill Book Company, pp.650-652.

Zienkiewicz. O. C., and R. L. Taylor, 2000, *Finite Element Method: Vol. 1 – The Basis, Vol. 2 – Solid Mechanics*, Butterworth-Heinemann.

Index

About *DyRoBeS©*

DyRoBeS© is a pair of powerful, yet easy to learn and use, engineering design/analysis software tools authored by Wen Jeng Chen, Ph.D., P.E., for complete rotor dynamics analysis and comprehensive bearing performance calculations. The Finite Element based codes combine a user interface intuitive enough for the occasional user, with a powerful set of modeling and analysis capabilities for the most demanding requirements. Extensive built-in modeling, analysis, post-processing and visualization tools allow the user to work within a single integrated environment. To aid the user, the programs include context sensitive help, a standard *Windows* help system, and an extensive set of examples.

DyRoBeS-Rotor is the rotor-dynamics component. Developed with the basic philosophy of providing powerful, but easy to use tools for the working engineer, *DyRoBeS-Rotor,* is capable of linear and non-linear analyses of free and forced vibrations (lateral, torsional, and axial) as well as static deflections of multi-shaft and multi-branch flexible rotor-bearing-support systems. More than ten different types of bearings, dampers, and seals are built in. In additional, a number of analysis and design tools, including aerodynamic cross-coupling calculations, rolling element deflection and stiffness estimation, liquid annular seal analysis, calculation of inertia properties of a solid, equivalent impedance of a bearing-support system, rotor orbit analysis, and balancing calculations are provided.

DyRoBeS-BePerf complements *DyRoBeS-Rotor* by calculating <u>Be</u>aring steady state and dynamic <u>Perf</u>ormance. The wide range of bearing configurations which can be analyzed include plain cylindrical, partial arc, elliptical, lobed, multi-pocket, pressure dam, general fixed profile lobed, flexural/tilting pad with or without flexible pivots, floating ring bushings, gas lubricated bearings, and thrust bearings. The solution includes automatic calculation of effective film temperatures at operating conditions. In addition to bearing analysis, the program also performs lubricant properties analysis and oil flow calculations. The bearing dynamic coefficients calculated by using *DyRoBeS-BePerf* can be easily integrated into a *DyRoBeS-Rotor* model.

Under continuous development since 1991, *DyRoBeS* has been rigorously validated and extensively tested by many academic researchers and industrial engineers. It is widely used by government agencies, universities, and various industries around the world.

For additional information on *DyRoBeS*, contact
RODYN Vibration Analysis, Inc.
Charlottesville, VA 22903
www.RODYN.com
Tel: 434-296-3175

About the Authors

Wen Jeng Chen, Ph.D., P.E., Fellow ASME, is the developer of *DyRoBeS©* and the founder of Eigen Technologies, Inc. Dr. Chen has been working in industry after receiving his Ph.D. in 1987. He is a registered Professional Engineer in the State of Illinois. Dr. Chen was awarded the H. H. Jeffcott Award for outstanding paper on rotating machinery. He has made significant contributions in the area of rotating machinery as a practical industrial engineer and outstanding researcher. His achievements have been both analytical and design-oriented and have spanned a broad spectrum of topics within the general area of rotating machinery. Dr. Chen is clearly a versatile and prolific researcher and is well respected in the rotordynamics communities, both academic and industrial.

Edgar J. Gunter, Jr., Ph.D., Fellow ASME, is the founder of RODYN Vibration Analysis, Inc. He has worked in the field of vibrations of rotor machinery and fluid bearings for over 40 years and written over 150 technical papers and reports on various aspects of the dynamics of rotating machinery, fluid film bearings and balancing. Dr. Gunter is now a professor emeritus, after teaching for 34 years in the Department of Mechanical, Aerospace and Nuclear Engineering at the University of Virginia. He has worked with industry and government on various rotor dynamic and vibration problems. He also consulted with NASA and participated in hands-on work on the space shuttle program in its early days. NASA engineers utilize *DyRoBeS* software extensively, and Dr. Gunter has presented numerous on-site training courses at NASA on the use of both the rotor dynamics and bearing components of the software.

Selected Color Figures

CHAPTER 1

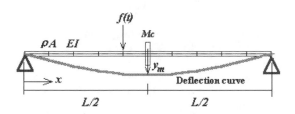

Figure 1.2-2 Simply supported beam with a concentrated mass

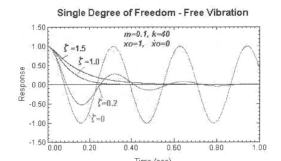

Figure 1.4-3 Damping effect on free vibration

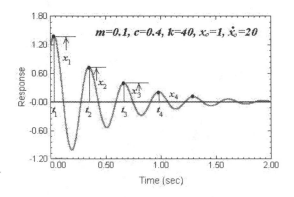

Figure 1.4-1 Logarithmic decrement for underdamped case ($\xi < 1$)

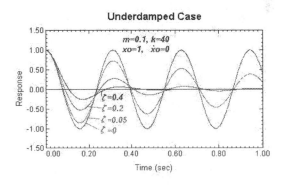

Figure 1.4-4 Damping effect – underdamped case

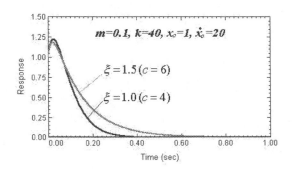

Figure 1.4-2 Critically damped and overdamped cases

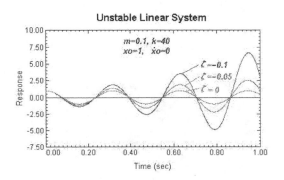

Figure 1.4-5 Unstable linear systems

449

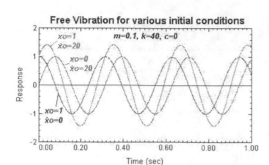

Figure 1.4-6 Free Vibration with various
initial conditions

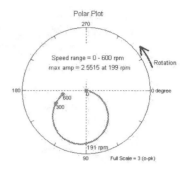

Figure 1.5-5 Polar plot for steady state
unbalance response

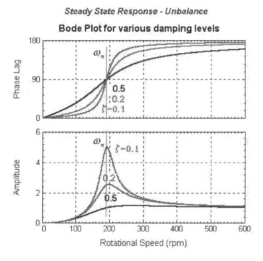

Figure 1.5-2 Steady state response due to
mass unbalance

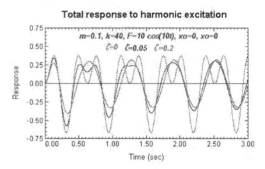

Figure 1.6-1 Total response to harmonic
excitation for various damping factors

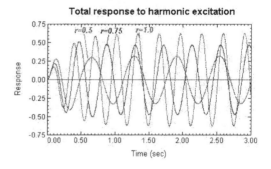

Figure 1.6-2 Total response to harmonic
excitation for various excitation frequencies

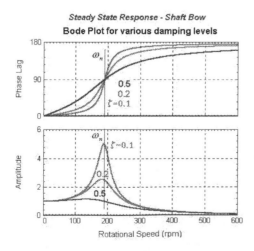

Figure 1.5-3 Steady state response due to
shaft initial bow

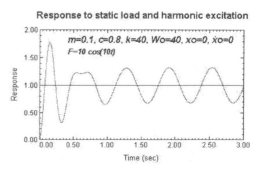

Figure 1.6-4 Total response to static load and
harmonic excitation

CHAPTER 2

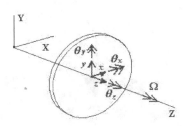

Figure 2.1-1 Coordinate system and degrees of freedom

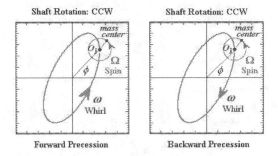

Figure 2.3-2 Forward and backward precessions

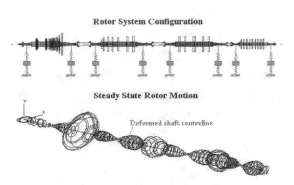

Figure 2.2-1 Rotor steady state Motion (courtesy of Malcolm E. Leader)

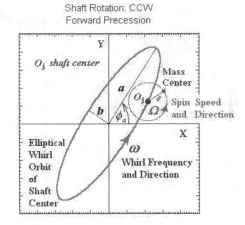

Figure 2.3-3 Elliptical orbit properties

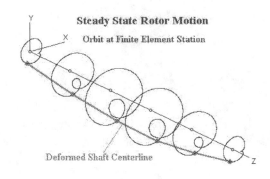

Figure 2.2-2 Rotor limit cycle motion

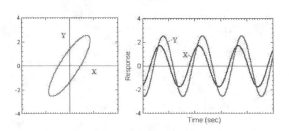

Figure 2.3-1 Rotor elliptical orbit

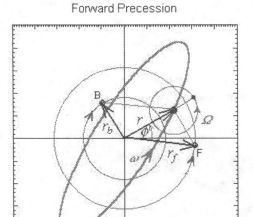

Figure 2.3-4 Elliptical orbit consists of two circular motions

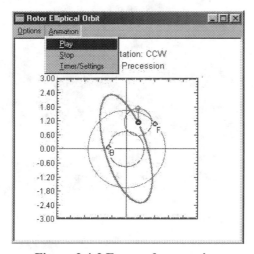

Figure 2.4-1 Orbit data input and output

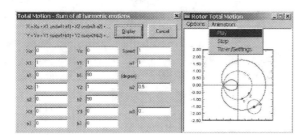

Figure 2.4-2 Forward precession

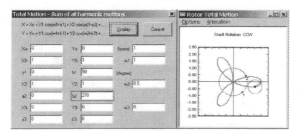

Figure 2.4-7 Sub-harmonic component is forward

Figure 2.4-8 Sub-harmonic component is backward

CHAPTER 3

The Laval-Jeffcott Rotor

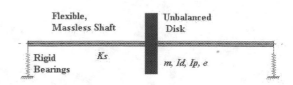

Figure 3.1-1 A simple Laval-Jeffcott rotor

Figure 3.1-2 Two fundamental motions for a symmetric flexible rotor

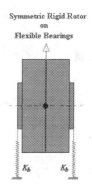

Figure 3.1-3 A symmetric rigid rotor with flexible bearings

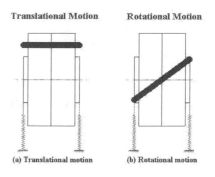

Figure 3.1-4 Two fundamental motions for a symmetric rigid rotor

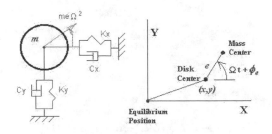

Figure 3.2-2 A two-degrees-of-freedom model

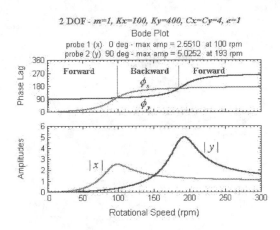

Figure 3.4-3 Bode plot for 2DOF system with unbalance

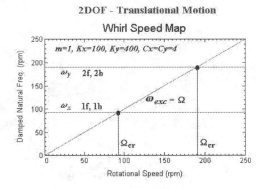

Figure 3.3-1 Whirl speed map for a 2DOF system

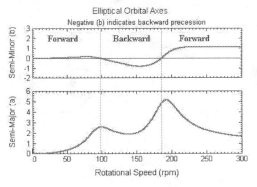

Figure 3.4-4 Elliptical orbital axes

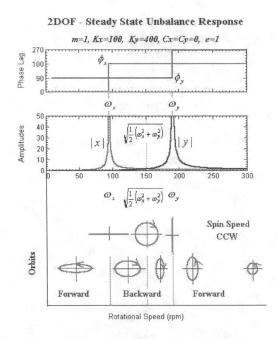

Figure 3.4-1 The steady state unbalance response and orbit analysis

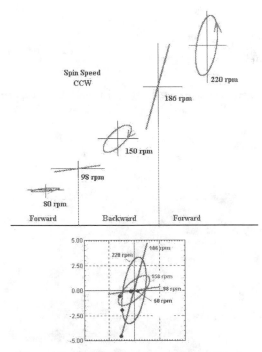

Figure 3.4-5 Steady state rotor orbits at different rotor speeds

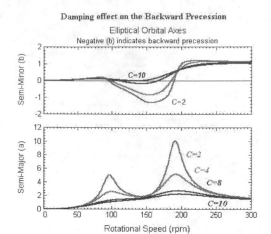

Figure 3.4-9 Elliptical orbital axes for various damping levels

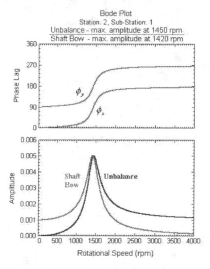

Figure 3.5-3 Comparison of unbalance response and shaft bow response

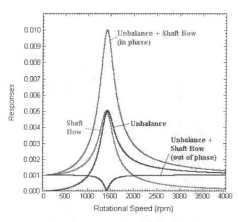

Figure 3.5-4 Combined effect of unbalance and shaft bow

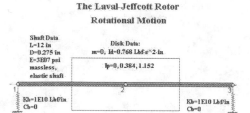

Figure 3.7-3 Rotation motion for the Laval-Jeffcott rotor

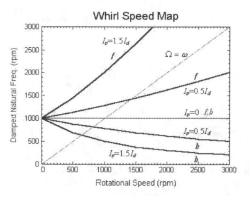

Figure 3.7-4 Whirl speed map with three different polar moments of inertia

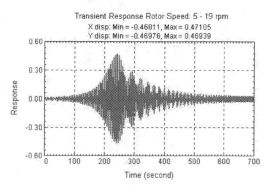

Figure 3.9-4 Time transient response vs. time

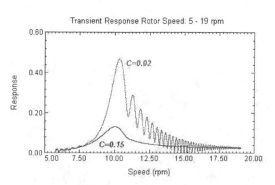

Figure 3.9-6 Transient responses for various damping levels

CHAPTER 4

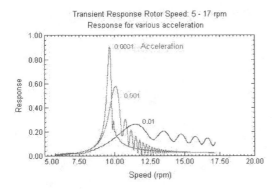

Figure 3.9-8 Time transient responses for various acceleration rates

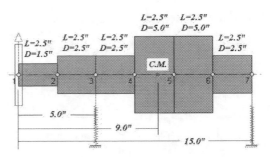

Figure 4.2-2 A rigid rotor system

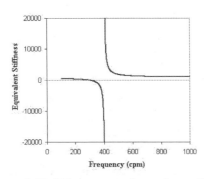

Figure 3.10-7 Steady state journal relative displacement to the support

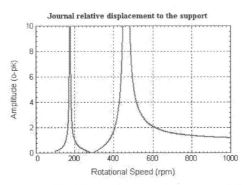

Figure 3.10-8 Frequency dependent effective stiffness

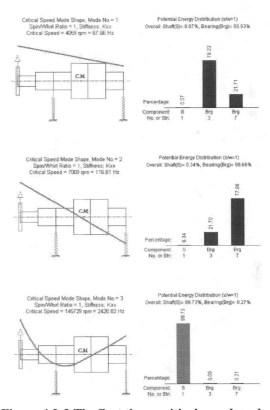

Figure 4.2-3 The first three critical speed mode shapes and associated potential energy

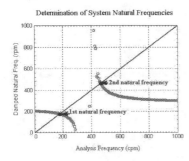

Figure 3.10-10 Determination of system natural frequencies

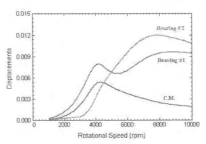

Figure 4.2-5 The steady state unbalance response

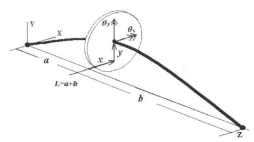

Figure 4.2-6 (c) The steady state unbalance response at 10000 rpm

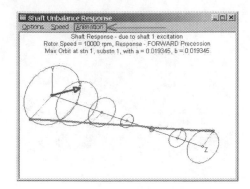

Figure 4.3-1 An off-centered rigid disk mounted on a flexible rotor

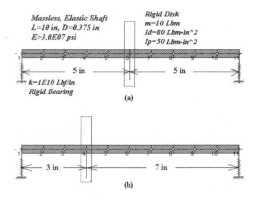

Figure 4.3-4 A rigid disk on a flexible rotor

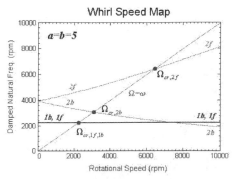

Figure 4.3-5 Whirl speed map for the centrally mounted disk

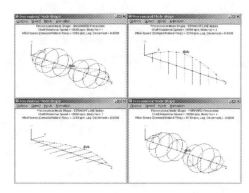

Figure 4.3-7 The first four mode shapes at 10,000 rpm

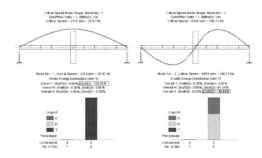

Figure 4.3-8 The first two forward synchronous critical speed mode shapes and kinetic energies.

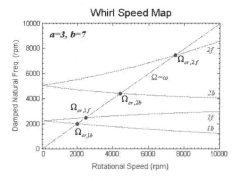

Figure 4.3-10 Whirl speed map for off-centered mounted disk

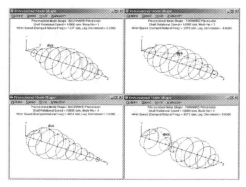

Figure 4.3-12 The first four mode shapes at 10,000 rpm

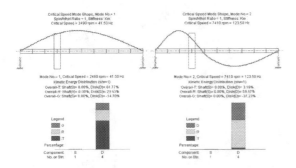

Figure 4.3-14 The first two forward synchronous critical speed mode shapes and kinetic energies.

CHAPTER 5

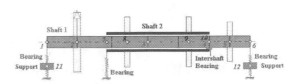

Figure 5.1-1 A dual-shaft system

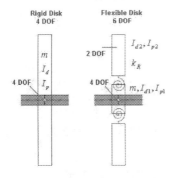

Figure 4.4-1 Models for rigid and flexible disks

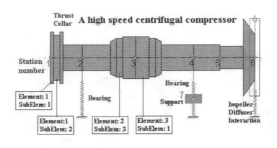

Figure 5.1-2 A typical single shaft system

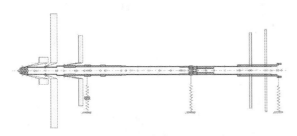

Figure 4.5-2 Offset disks in an aircraft engine model

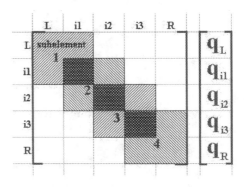

Figure 5.5-2 Elemental matrix

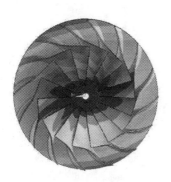

First Diametral Disk Mode

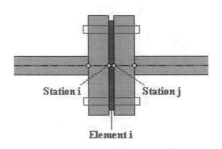

Figure 5.6-2 Coupling model

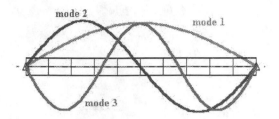

Figure 5.7-5 The first three mode shapes for a simply supported beam

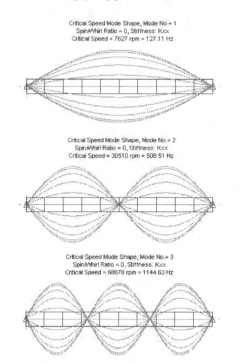

Figure 5.7-6 Mode animation

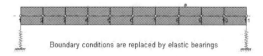

Boundary conditions are replaced by elastic bearings

Figure 5.7-14 A uniform beam supported by two elastic springs

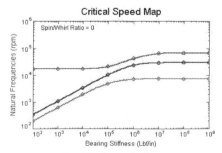

Figure 5.7-15 The first three natural frequencies vs. bearing stiffness

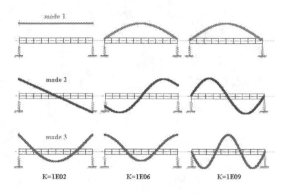

Figure 5.7-16 Mode shapes for various bearing stiffness

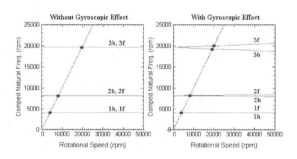

Figure 5.7-18 Whirl speed map without and with gyroscopic effect

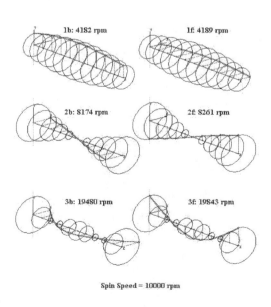

Figure 5.7-20 Mode shapes with gyroscopic effect

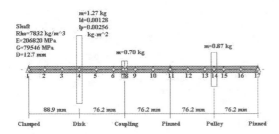

Shaft
Rho=7832 kg/m^3
E=206820 MPa
G=79546 MPa
D=12.7 mm

m=1.27 kg
Id=0.00128
Ip=0.00256
kg-m^2

m=0.70 kg

m=0.87 kg

88.9 mm   76.2 mm   76.2 mm   76.2 mm   76.2 mm

Clamped   Disk   Coupling   Pinned   Pulley   Pinned

Figure 5.7-21 Two shafts connected by a coupling

CHAPTER 6

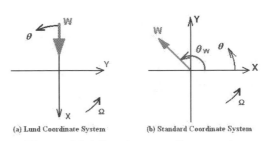

(a) Lund Coordinate System

(b) Standard Coordinate System

Figure 6.2-1 Bearing coordinate systems

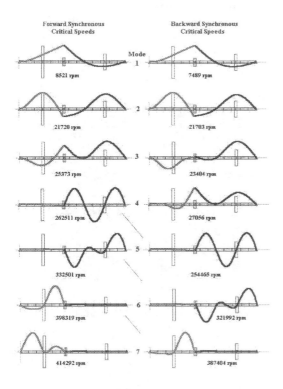

Forward Synchronous
Critical Speeds

Backward Synchronous
Critical Speeds

Mode
1

8521 rpm   7489 rpm

2

21720 rpm   21703 rpm

3

25373 rpm   23404 rpm

4

262511 rpm   27056 rpm

5

332501 rpm   254465 rpm

6

398319 rpm   321992 rpm

7

414292 rpm   387404 rpm

Figure 5.7-22 Critical speed mode shapes

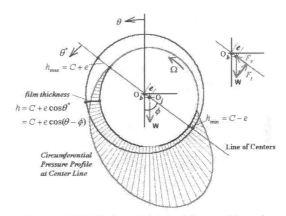

θ

θ^*

$h_{max} = C + e$

Ω

film thickness
$h = C + e\cos\theta^*$
$= C + e\cos(\theta - \phi)$

$h_{min} = C - e$

Circumferential
Pressure Profile
at Center Line

Line of Centers

Figure 6.3-1 Plain cylindrical journal bearing

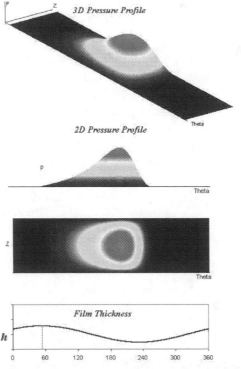

3D Pressure Profile

2D Pressure Profile

Film Thickness

Figure 6.3-2 Pressure profiles and film thickness

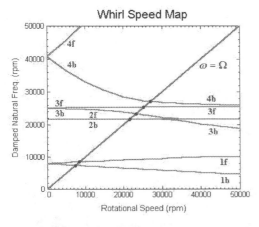

Whirl Speed Map

4f

4b

$\omega = \Omega$

4b

3f

3f
3b
2f

2b

3h

1f

1b

Rotational Speed (rpm)

Damped Natural Freq. (rpm)

Figure 5.7-23 Whirl speed map

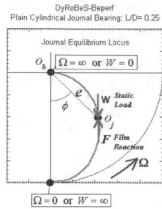

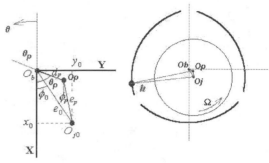

Figure 6.6-2 The relationship between the journal center, bearing center, and lobe center

Figure 6.3-3 Journal equilibrium locus

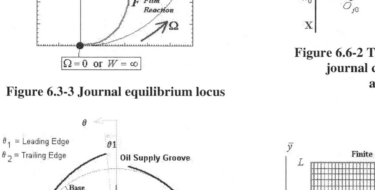

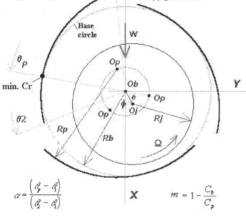

Figure 6.4-2 A three-lobe bearing

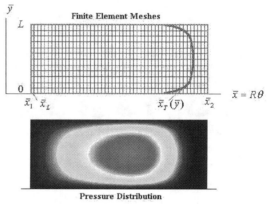

Figure 6.7-4 Boundary conditions

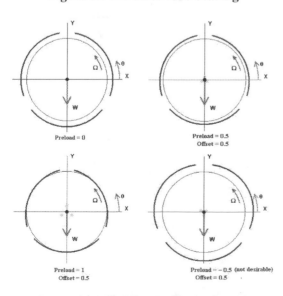

Figure 6.4-3 Bearing configuration for different preloads

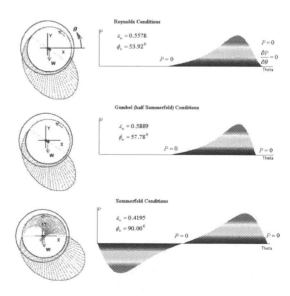

Figure 6.7-5 Three circumferential boundary conditions

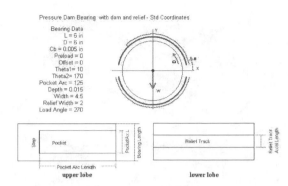

Pressure Dam Bearing with dam and relief - Std Coordinates

Bearing Data
L = 6 in
D = 6 in
Cb = 0.005 in
Preload = 0
Offset = 0
Theta1 = 10
Theta2 = 170
Pocket Arc = 125
Depth = 0.015
Width = 4.5
Relief Width = 2
Load Angle = 270

upper lobe   lower lobe

Figure 6.7-7 Pressure dam bearing example

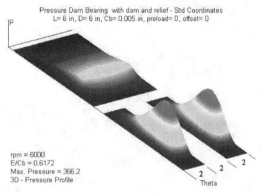

Pressure Dam Bearing with dam and relief - Std Coordinates
L= 6 in, D= 6 in, Cb= 0.005 in, preload= 0, offset= 0
Speed = 6000 rpm
Load = 1000 Lbf
W/LD = 27.7778 psi
Vis. = 2E-06 Reyns
Sb = 2.592
E/Cb = 0.6172
Att. = 56.03 deg
hmin = 1.914 mils
Pmax = 366.165 psi
Hp = 18.0709 hp
Stiffness (Lbf/in)
 2.176E+06   2.918E+06
 3.289E+04   1.670E+06
Damping (Lbf-s/in)
 6.505E+03   2.713E+03
 2.715E+03   2.979E+03
Critical Journal Mass
 16.39

Pressure Dam Bearing with dam and relief - Std Coordinates
L= 6 in, D= 6 in, Cb= 0.005 in, preload= 0, offset= 0

rpm = 6000
E/Cb = 0.6172
Max. Pressure = 366.2
3D - Pressure Profile

Figure 6.7-10 Results at 6000 rpm for centered relief

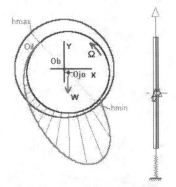

Figure 6.8-1 A single journal bearing

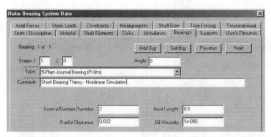

Figure 6.8-2 Bearing data for nonlinear analysis

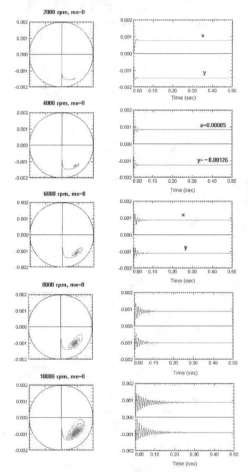

Figure 6.8-4 Journal equilibrium positions

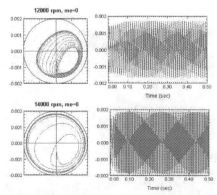

Figure 6.8-5 Journal equilibrium motion

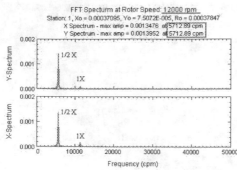

Figure 6.8-6 FFT at 12,000 rpm

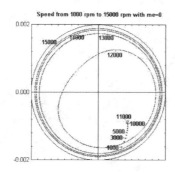

Figure 6.8-8 Journal equilibrium motion

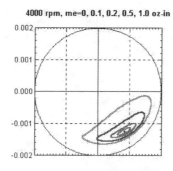

Figure 6.8-9 Journal steady state orbits

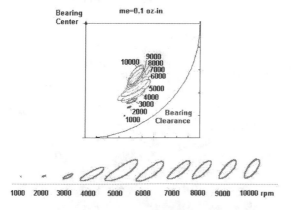

Figure 6.8-10 journal orbits versus speed

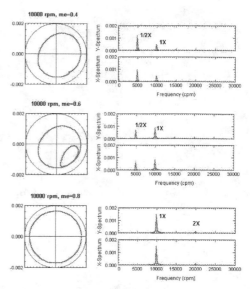

Figure 6.8-12 Journal responses at 10,000 rpm with different unbalances

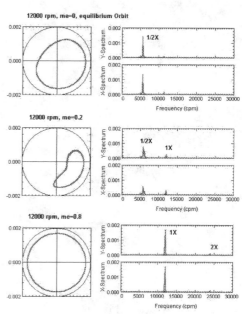

Figure 6.8-13 Journal responses at 12,000 rpm with different unbalances

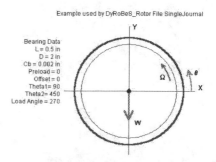

Figure 6.8-14 Plain cylindrical bearing

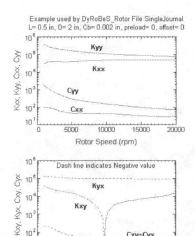

Example used by DyRoBeS_Rotor File SingleJournal
L= 0.5 in, D= 2 in, Cb= 0.002 in, preload= 0, offset= 0

Dash line indicates Negative value

Figure 6.8-17 Bearing linear dynamic coefficients

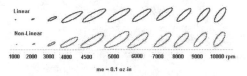

Linear

Non-Linear

1000 2000 3000 4000 4500   5000   6000   7000   8000   9000   10000 rpm

me = 0.1 oz-in

Figure 6.8-21 Comparison of liner and nonlinear orbits

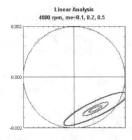

Linear Analysis
4000 rpm, me=0.1, 0.2, 0.5

Figure 6.8-22 Steady state unbalance response orbits

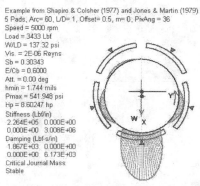

Example from Shapiro & Colsher (1977) and Jones & Martin (1979)
5 Pads, Arc= 60, L/D= 1, Offset= 0.5, m= 0, PivAng = 36
Speed = 5000 rpm
Load = 3433 Lbf
W/LD = 137.32 psi
Vis. = 2E-06 Reyns
Sb = 0.30343
E/Cb = 0.6000
Att. = 0.00 deg
hmin = 1.744 mils
Pmax = 541.948 psi
Hp = 8.60247 hp
Stiffness (Lbf/in)
  2.264E+05 0.000E+00
  0.000E+00 3.008E+06
Damping (Lbf-s/in)
  1.867E+03 0.000E+00
  0.000E+00 6.173E+03
Critical Journal Mass
Stable

Figure 6.9-9 Bearing performance at 5,000 rpm – load on pivot

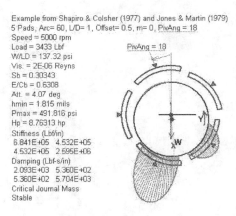

Example from Shapiro & Colsher (1977) and Jones & Martin (1979)
5 Pads, Arc= 60, L/D= 1, Offset= 0.5, m= 0, PivAng = 18
Speed = 5000 rpm
Load = 3433 Lbf
W/LD = 137.32 psi
Vis. = 2E-06 Reyns
Sb = 0.30343
E/Cb = 0.6308
Att. = 4.07 deg
hmin = 1.815 mils
Pmax = 491.816 psi
Hp = 8.76313 hp
Stiffness (Lbf/in)
  6.841E+05 4.532E+05
  4.532E+05 2.595E+06
Damping (Lbf-s/in)
  2.093E+03 5.360E+02
  5.360E+02 5.704E+03
Critical Journal Mass
Stable

PivAng = 18

Figure 6.9-13 Bearing performance at 5,000 rpm – load on an arbitrary angle

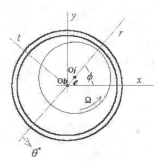

Figure 6.11-3 Coordinate systems

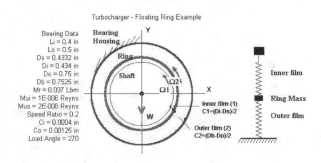

Turbocharger - Floating Ring Example

Bearing Data
Li = 0.4 in
Lo = 0.5 in
Ds = 0.4332 in
Di = 0.434 in
Do = 0.75 in
Db = 0.7525 in
Mr = 0.037 Lbm
Mui = 1E-006 Reyns
Muo = 2E-006 Reyns
Speed Ratio = 0.2
Ci = 0.0004 in
Co = 0.00125 in
Load Angle = 270

Bearing Housing
Ring
Shaft
Inner film (1)
C1=(Di-Ds)/2
Outer film (2)
C2=(Db-Do)/2

Inner film
Ring Mass
Outer film

Figure 6.12-1 Floating ring bearing

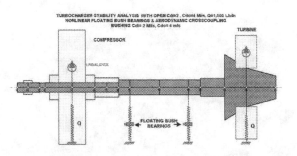

TURBOCHARGER STABILITY ANALYSIS WITH OPEN Cdi=2 , Cdo=4 Mils, Q=1,000 Lb/in
NONLINEAR FLOATING BUSH BEARINGS & AERODYNAMIC CROSSCOUPLING
BUSHING Cdi= 2 Mils, Cdo= 4 mils

COMPRESSOR   TURBINE

FLOATING BUSH
BEARINGS

Figure 6.12-3 Turbocharger with floating ring bearings

CHAPTER 7

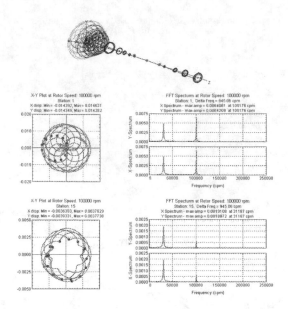

Figure 6.12-4 Limit cycle motions

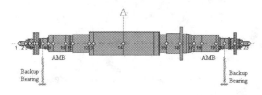

Figure 6.14-4 The NEDO ISO test rotor

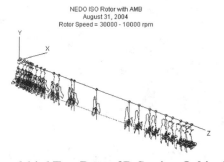

Figure 6.14-6 Test Rotor 3D Station Orbits for Drop from 30k to 10k rpm

Figure 6.14-9 Test Rotor 3D Station Orbits for Drop from 50k to 40k rpm

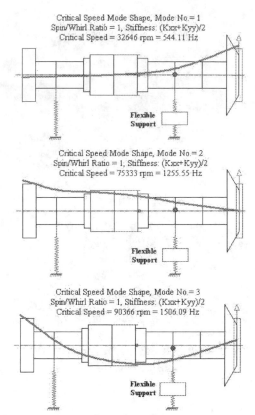

Figure 7.3-1 The first three critical speed mode shapes

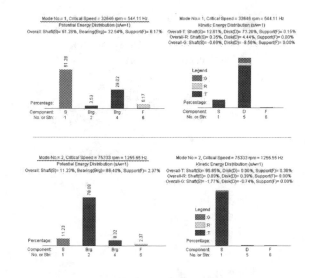

Figure 7.3-2 Energy distribution

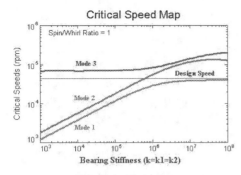

Figure 7.3-3 Critical speed map with change in both bearing stiffnesses

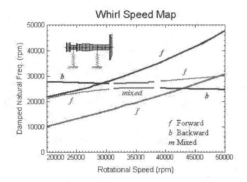

Figure 7.4-3 Whirl speed map

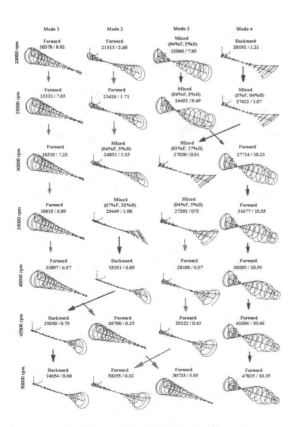

Figure 7.4-4 Mode shapes

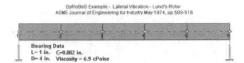

Figure 7.7-1 Example 7.1

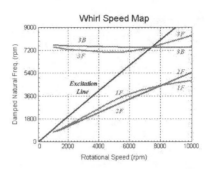

Figure 7.7-11 Whirl speed map

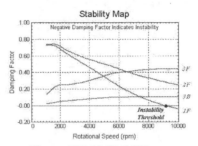

Figure 7.7-12 Stability map

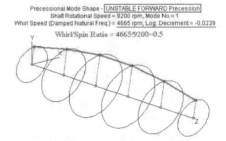

Figure 7.7-14 Unstable forward precessional mode

Figure 7.7-27 Shaft steady state response at 9200 rpm – 0.3 oz-in unbalance

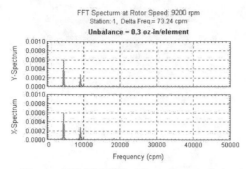

Figure 7.7-28 FFT at bearing station at 9200 rpm – 0.3 oz-in unbalance

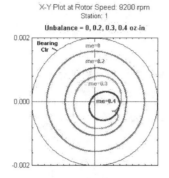

Figure 7.7-31 Rotor limit cycle motions at bearing station for various unbalance forces

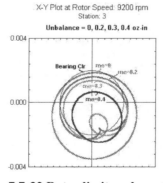

Figure 7.7-32 Rotor limit cycle motions at station 3 for various unbalance forces

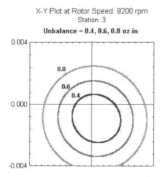

Figure 7.7-34 Rotor limit cycle motions at station 3 for unbalance greater than 0.4 oz-in

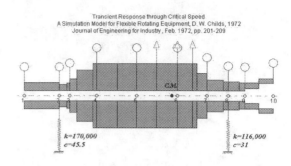

Figure 7.7-35 Rotor model for example 7.2

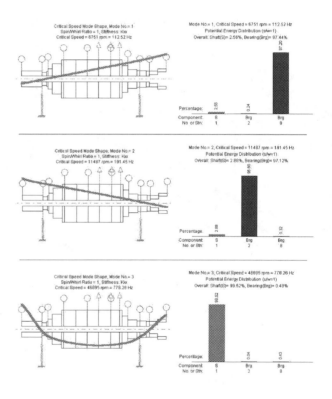

Figure 7.7-36 Critical speed mode shapes and potential energies

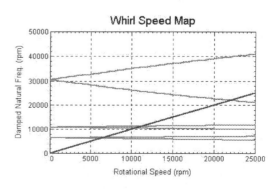

Figure 7.7-37 Whirl speed map for example 7.2

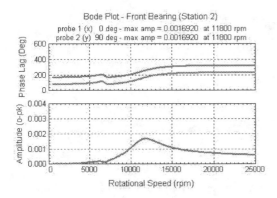

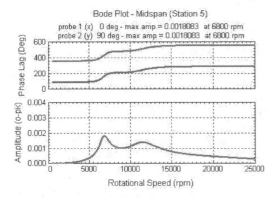

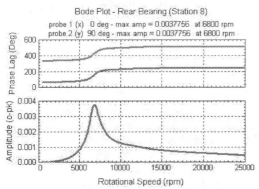

Figure 7.7-38 Bode plots

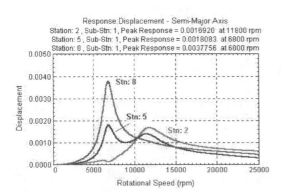

Figure 7.7-39 Steady state unbalance responses
at stations 2, 5, and 8

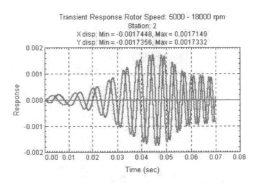

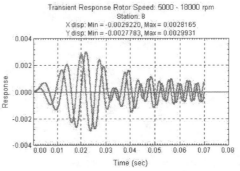

Figure 7.7-40 Time transient responses at
front and rear bearings

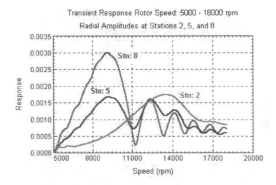

Figure 7.7-42 Amplitudes of the transient
response

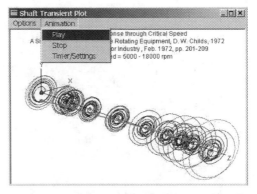

Figure 7.7-43 Animation of shaft transient
response during startup

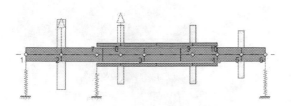

Figure 7.7-44 Dual rotor system

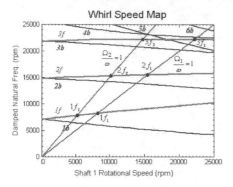

Figure 7.7-46 Whirl speed map

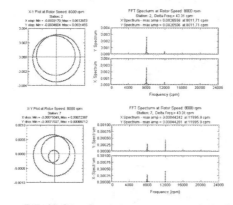

Figure 7.7-49 Steady state responses at stations 2 and 7 due to shafts 1 and 2 unbalances

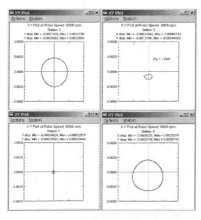

Figure 7.7-52 Steady state response with Fy = — 100 Lbf side load

CHAPTER 8

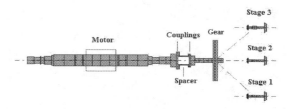

Figure 8.1-1 A geared train for torsional vibration

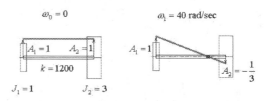

Figure 8.3-2 Mode shapes for the two-inertia system

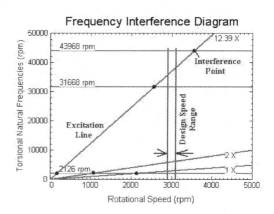

Figure 8.7-1 Torsional frequency interference diagram

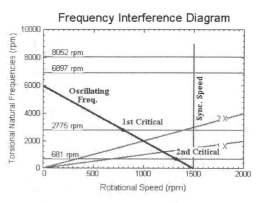

Figure 8.7-2 Frequency interference diagram for a synchronous motor

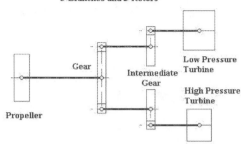

Figure 8.10-1 Marine steam turbine propulsion system

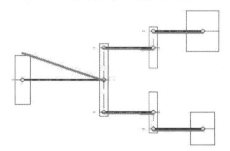

Figure 8.10-3 The first mode shape

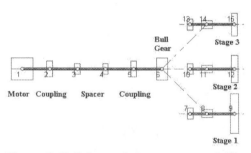

Figure 8.10-9 A synchronous motor driven 3-stages compressor

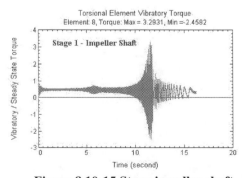

Figure 8.10-15 Stage impeller shaft vibratory torque

CHAPTER 9

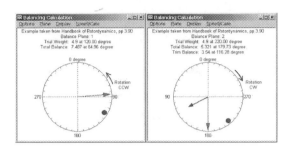

Figure 9.3-6 Balance vectors

APPENDIX

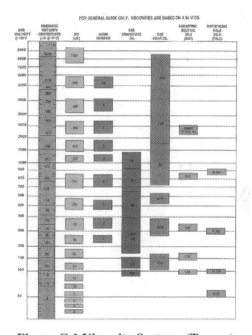

Figure C-1 Viscosity Systems (Texaco)

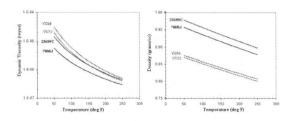

Figure C-3 Lubricant properties comparison

Printed in the United States
By Bookmasters